Mine Mapping and Layout

Mine Mapping and Layout

W. RANDOLPH WILLIAMS, P.E.

Associate Professor, Engineering Technology
Fairmont State College
and
Adjunct Associate Professor of Education
West Virginia University

Prentice-Hall, Inc., Englewood Cliffs, New Jersey 07632

Library of Congress Cataloging in Publication Data

Williams, W. Randolph.
Mine mapping and layout.

Includes bibliographical references and index.
1. Mine surveying. 2. Mine Maps. I. Title.
TN273.W48 622 81-23545
ISBN 0-13-583617-4 AACR2

Editorial/production supervision by Anne Simpson
Manufacturing buyer: Joyce Levatino
Page layout: Bruce Kenselaar

Printed in the United States of America

10 9 8 7 6 5 4 3 2 1

ISBN 0-13-583617-4

Prentice-Hall International, Inc., *London*
Prentice-Hall of Australia Pty. Limited, *Sydney*
Prentice-Hall Canada Inc., *Toronto*
Prentice-Hall of India Private Limited, *New Delhi*
Prentice-Hall of Japan, Inc., *Tokyo*
Prentice-Hall of Southeast Asia Pte. Ltd., *Singapore*
Whitehall Books Limited, *Wellington, New Zealand*

Dedicated to three outstanding educators

Helen L. Plants — the teacher
Harold Fairbank — the researcher
Thomas R. Long — the advisor
and
Patricia

Contents

CHAPTER 3 OVERVIEW OF DRAFTING AND CARTOGRAPHY 68

CHAPTER 4 PROCEDURES OF MINE MAP CONSTRUCTION 100

CHAPTER 5 FUNCTIONAL PERFORMANCE 144

CHAPTER 6 PREMINING DATA AND LAYOUT TECHNIQUES 174

Preface

The most scholarly underrepresented activity within the mine engineering profession today is that of mine mapping and design layout. Several contributing causes can be suggested for this situation. First, mining technology is a highly eclectic field requiring some degree of proficiency in structural design, solid and fluid mechanics, electrical system distribution, surveying, geology, environmental systems, safety design, economics, coal preparation, labor management, transportation systems, operations research, and graphic analysis. These topics all compete with one another for the precious little time allotted educators for the instruction of students. As a result of this competition, some topics must suffer in their curricular treatment. Often those that do suffer are the less exotic topics which have been "around a while" and offer little or no research potential. Mine mapping and graphical methods of layout are two such topics that have suffered.

Another contributing cause to the lack of suitable mine graphics literature has been the withdrawal of mining and engineering schools alike from drawing board solutions in favor of the "higher-technology" computers or mathematical solutions. The unfortunate result of this trend is the increasing graduation of students who are generally unaware of the potential of graphical solutions and, indeed, the need for these methods in the mining industry.

This book was written in an attempt to provide a study aid combining graphical techniques of mine mapping, mine design, and mine layout. My colleagues and I realized the need for such a textbook when we could not locate an adequate source to recommend to our students. First and foremost, this book is a treatment of mine mapping.

The first chapter presents an overview of cartographic techniques used in mine map production. Chapter 2 reviews the fundamentals of plane surveying and presents the fundamentals of mine surveying. Chapter 3 reviews standard drafting room practice and develops guidelines for mine map symbolization and production. Chapter 4 reviews mine entry nomenclature and presents the details of topographic and mine map production.

The second goal of this book is to present the supplemental techniques needed for adequate mine design and layout. Chapter 5 develops the methods of achieving functional performance, safety, reliability, maintainability, and value in the layout of the mine. Chapter 6 details the nature of raw explanatory data, statistical mapping of this data, and techniques of descriptive geometry needed for mine design and layout. Chapters 7 through 10 present design and operational layouts of the major functional operations of ventilation, transportation, drainage, and electrical distribution, respectively.

Applicable problems have been compiled at the end of each chapter. However, because of the nature of field notes, exploratory field data, and maps, the problem format can meet the needs of students only partially.

The motivation for the writing of this text comes from those former students who were made to suffer through classes with no suitable reference materials or text addressing the subject. I wish to thank Thomas M. Nunnally and Dotson Cather, whose notes and constant encouragement have served me since I first undertook the instruction of mine mapping and layout classes.

The content of this text is founded upon the efforts of those who came before, and it is to them that much of the credit belongs. Particularly, I would like to thank personnel within the Mine Extension Service of West Virginia University, the U.S. Geological Survey, the U.S. Bureau of Mines, the Mine Safety and Health Administration, Joy Machinery Company, Consolidation Coal Company, and Eastern Associated Coal Company.

The initial editing credit belongs entirely to my wife, Patricia, without whose aid this manuscript would be undecipherable. I wish to thank Robert Hefner of Fairmont State College's Media Section for his photographic enhancement of certain maps and illustrations.

I would also like to thank Dr. James Hales, Director of the Division of the Technology at Fairmont State College, for his facilitation of the preparation of this manuscript.

Finally, to Mrs. Mary Lewis, for her repeated typing of this manuscript under quite adverse conditions, I wish to extend my thanks and appreciation.

W. R. Williams

Mine Mapping and Layout

1

Introduction: Maps, Diagrams, and Layouts

1.0 INTRODUCTORY COMMENTS

Human beings have an intellectual need to represent the world around them. Although the format of expression changes—the need persists. Art is the aesthetic indication of this need; technology is its pragmatic manifestation. It is the latter endeavor with which this book is concerned, specifically as it regards the operations and practices of mining.

Drafting is an aspect of technology that employs graphic symbols and techniques as methods for representing and communicating information about the physical world. *Mine drafting* is a more specialized subdivision of drafting which is directed principally toward the representation of surface and subsurface features occurring within the boundaries of a mine. Mine drafting entails, but is not limited to, the construction of surface contour maps, mine contour maps, plats of property surveys, plots of mine surveys, construction of exploration and geologic maps, construction of functional layouts (examples being ventilation, transportation, electrical power, and drainage), and the construction of support drawings (such as thematic maps, flow diagrams, isometric diagrams, and so on). The use of each of these drawings is essential to the safety, productivity, and longevity of any mining operation.

Mine cartography, the preparation and construction of mine maps and projections, constitutes the greatest majority of mine drafting needs. For the purposes of this text, a *map* is defined as a two-dimensional representation (usually drawn on paper) of some portion of the earth depicting surface or subsurface features existing in certain spatial relationships. The specific purposes for such representations vary from situation to situation; however, certain common uses or purposes that all maps share are as follows:

1. Maps are used to compile, arrange, and store information about a region.
2. Maps are used to simplify the "visualization" or "conception" of a region by reducing the total number of elements portrayed.
3. Maps are used to clarify relationships existing in a region by making the more important aspects stand out.
4. Maps are used by the cartographer to represent physically nonexistent concepts or future plans for a given area.
5. Maps, particularly three-dimensional maps or cartographic models, provide the mapper with the ability to manipulate or experiment on a region physically, in a controlled manner.

The last use describes the specialized fields of probative modeling and engineering similitude which are beyond the introductory goals of this text. The first four purposes for maps are contained within the field of paradigmatic or descriptive modeling, the principles of which are presented in the next section.[1]

1.1 CONSTRUCTION PRINCIPLES

Maps are in fact "models," displayed in graphical two-dimensional form. A *model* is simply a representation of an original. A designer's sketch of an automobile is to a greater or lesser extent a model of a car. Similarly, a child's plastic airplane is merely a model of a real airplane. In this same vein, a map is a model of some portion of the earth, the original. The map portrays this portion of the earth by depicting symbols that represent real-world features and their spatial relationships. Maps obviously differ from the original since the real world is three dimensional (or four dimensional, when you consider changes over time) whereas the map is probably two dimensional. However, the differences between the map and the original extend far beyond these simple dimensional variations.

These other differences result from the cartographer's treatment of the "raw" data acquired about the area to be mapped. Raw data are facts and information that have been acquired from field surveys, laboratory experiments, or other data-gathering operations. The purpose of data gathering is to describe, in as much detail as possible, the current status of the original. After the data have been gathered, the first role of the cartographer is to manipulate or treat the data in such a way that he or she can then use them to construct a map.

Treated data are any raw or other treated data that have been refined, organized, restructured, or in any way manipulated to facilitate their subsequent storage, interpretation, publication, or mapping. Most of the data we see in our lifetime are treated data.

The specific nature and types of treatment processes are discussed in more detail in later chapters, but five general treatment classes are used: (1) simplification, (2) enhancement, (3) distortion, (4) manipulation, and (5) convergence.

1.1.1 Data Simplification

Any region or portion of the earth is exceedingly more complex than any map could possibly portray. That is, the number of elements, features, and relationships at the original site far exceed our ability to represent them. The role of the cartographer is to compare the type and extent of available data about a region with the purpose(s) of the map and to then eliminate all but the most essential information. If a map of the vertical relief (rise and fall of the surface) of an area is all that is needed or desired, the placement of subsurface geologic features would unjustifiably complicate the map. Thus the cartographer must eliminate these subsurface data from consideration. This elimination process is a data-treatment procedure called *simplification.*

No hard and fast rules can be proposed for the degree of simplification. Topher and Pillwyer developed a simplification equation based upon what they termed "the principle of selection," which purportedly determined the number of items on a newly compiled map that could be transferred from an original source map based upon the ratio of map scale.[2] Although theoretically interesting, such attempts at quantifying the nature of data simplification of an original are usually impractical. Pragmatically, you should remember that a map is a simplification of an original and has a distinct purpose. When possible, avoid overcrowding by eliminating those data that are not directly related to that purpose.

1.1.2 Data Enhancement

Regardless of how complete or extensive a field survey is, some needed data points will invariably be missing. These points may be too costly to justify additional field effort or may be altogether impossible to obtain. *Enhancement* is the treatment by the cartographer that makes additions, usually based on trends or educated estimates, to the data. *Contour lines* are examples of enhancement. A contour line is a line drawn on a map which represents *all* adjacent points on the earth's surface at an equal elevation above a reference. Since most surveys are "point related," the cartographer must estimate the nature of elevation between any two known points. The line he or she draws is an enhancement process; that is, the cartographer adds information to the drawing which was not contained in the survey data. The purpose of enhancement is to make the data appear more continuous and more complete. The important point to remember is that the effect of enhancement is to alter the appearance of the raw data, regardless of the intentions of the mapmaker.

1.1.3 Data Distortion

Occasionally, raw data are obtained that conflict with accepted theories, practices, or other raw data. *Distortion* is the purposeful action of the cartographer toward resolving this apparent conflict in the data by selectively eliminating or altering a portion of the raw data. This action differs from simplification in that the data that are eliminated or added by distortion are usually essential to the purpose of the map. If a contour map were desired, all elevation data are relevant to the purpose of the map. If all elevation data were within 10 ft of one another except for one datum point which was 500 ft higher, the cartographer may justifiably eliminate this one point. Even though the elimination was valid, the raw data have been distorted. (Distortions resulting from the modeling procedures are discussed in Section 1.2).

1.1.4 Data Manipulation

The three treatment techniques described above are often termed "passive" data treatments since the nature and dimensional character of the original raw data are unchanged. Data manipulation is an "active" treatment of the raw data, in that the cartographer applies a manipulative process that converts the raw data to a different form or to entirely different dimensions. Various manipulative processes exist, most of which have statistical or other mathematical roots. Except in special cases these treatment processes are beyond the scope of this text and are not discussed further.

1.1.5 Convergence of Data

The last treatment procedure initiated by the cartographer entails the positioning and coding of the treated data onto the map. This procedure of projection is the actual process of mapping and is called *convergence in modeling*. The principal convergence problem faced by the cartographer is that of how to project points from the spherical three-dimensional surface of the earth onto a flat two-dimensional map without erroneously altering the usefulness of the data. Many types of graphical projections and mathematical data transformations have been and are used to reduce the convergence error. Some of the fundamentals of map convergence are described below. A more rigorous analysis is left to more specific texts.[2-4]

1.2 MAP CONVERGENCE

Two aspects of map convergence are (1) projection of the spatial relationships between physical elements from the original to the model (map), and (2) specification of intended distortions made during projection: (a) scales and (b) feature symbolization.

1.2.1 Element Projection: Points of View

Several projection techniques exist for transferring points from one surface, the original, to another, the model. You are perhaps familiar with *orthographic projection*, in which a model surface is positioned perpendicular to the line of sight of an observer. The observer standing at an assumed infinite distance from the model and original transfers the points from the original to the modeling surface along parallel extension lines. Such a projection is shown in Fig. 1-1. When a flat surface is orthogonally projected, a one-to-one correspondence between the original and the model occurs. That is, for every point and every relationship on the original, one and only one point and similar relationship exists on the model surface. The convergence correspondence is unity. However, when a curved original such as the earth is orthogonally projected, the distances between points on the more oblique (tangents) surfaces to the lines of sight appear closer together on the map surface. Each element in the original appears on the model; however, the element relationships, such as distances, areas, and angles, are no longer the same. Thus a convergence error or deformation exists. The amount of this deformation is proportional to the obliqueness of the tangent of the surface relative to the projection lines and thus increases as we approach the outer edges of the model.

Another method of projection is called a *central perspective projection*, in which the center of the earth becomes the projection point of view and the extension lines are radiated perpendicular to the surface of the earth. This type of projection is illustrated in Fig. 1-2. Note that when this type of

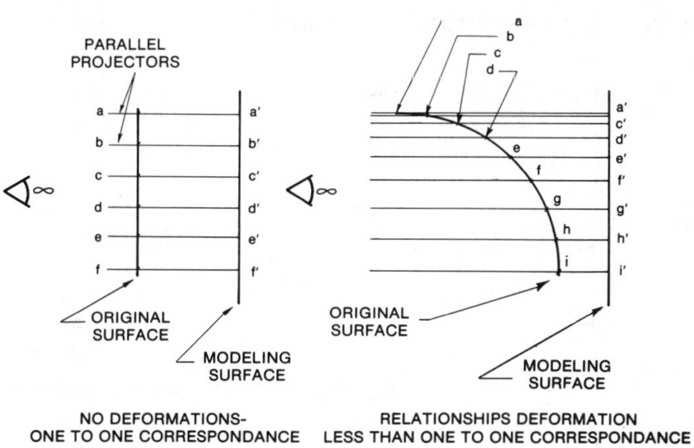

Figure 1-1 Orthogonal projection: effects of surface curvature.

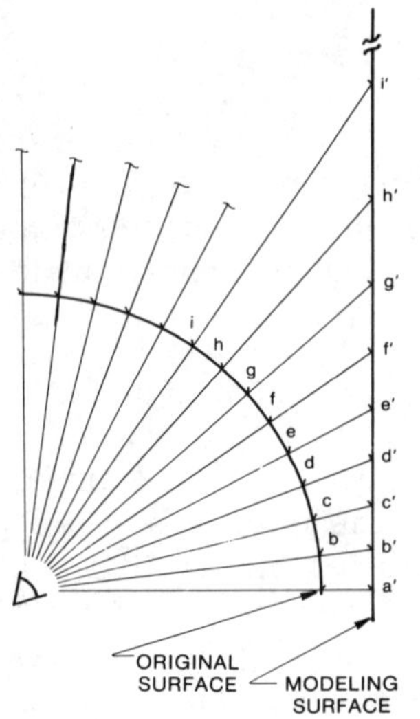

Figure 1-2 Central projection effect.

projection is used, the distance between points lying along a north-south line becomes greater on the map as you move farther away from the center. This type of projection point of view also produces distorted distances, areas, and angles.

1.2.2 Element Projection: Developable Surfaces

The method of projection is dependent upon where we locate the point of origin and on the way in which we deploy the modeling surface. The modeling surface (map) is called a *developable surface* by mathematicians because we are able to flatten the surface after the projection process without distorting the relationships projected. These surfaces include the plane, the cylinder, and the cone. (Note that a spherical surface cannot be flattened without distortion.) Figure 1-3 shows the three major arrangements of developable model surfaces relative to the earth. Each has been used to effectively map portions of the earth. The practical use of each is discussed in Section 1.2.4.

In addition to the three surface configurations, the type and amount of

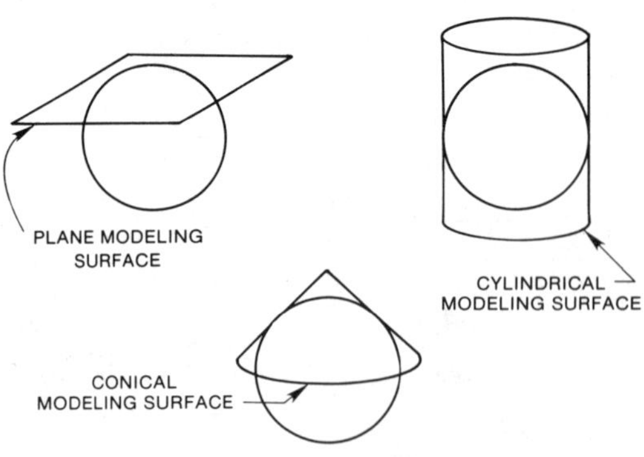

Figure 1-3 Development modeling surfaces.

relationship deformation depends on the relative position between the model surface and the original. This is illustrated for a cylindrical surface in Fig. 1-4. When the position of the cylinder is tangent to the earth, a one-to-one correspondence occurs at the equator, but distortion increases proportionately the farther one moves toward the poles. When the cylinder is positioned within the sphere, the deformation is least at the intersection of the model surface and the surface of the original and increases to either side of these lines.

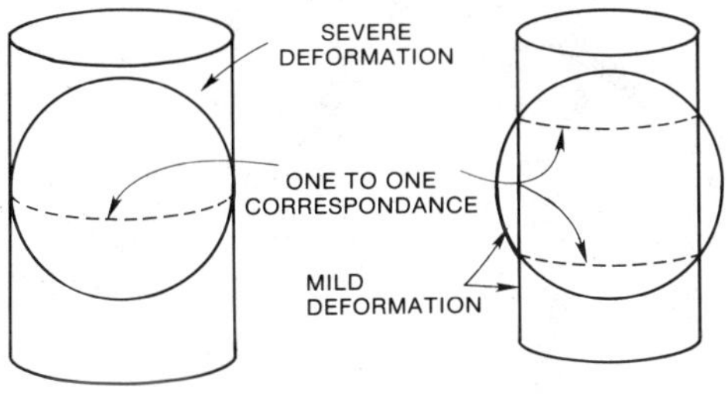

Figure 1-4 Original modeling surfaces arrangement and effect upon extent of deformation.

Various combinations of these two projection origins have been used to gain greater control over the correspondence between the model and the original. Usually, the cartographer is most interested in depicting a true portrayal of areas, distances, and angles (directions) shape or scale. Once one trait of the original is projected at a one-to-one correspondence, the others become deformed.

1.2.3 Positioning: Reference Systems

Together with the possible types of projection systems a referencing method for positioning the original points on the model surface is needed. Various schemes for doing this are possible; however, each depends on assuming one point to be fixed and all other positions to be referenced to that point. Two common reference systems are the spherical coordinate and the plane rectangular coordinate systems.

The *spherical coordinate system* is based on two sets of lines that are formed by the intersection of the earth and two sets of perpendicular planes as illustrated in Fig. 1-5. The imaginary axis of the earth from the north to the south pole forms the initial reference for the orientation of the planes. One set, *A*, is arranged perpendicular to the axis and separated by amounts providing equal surface (arc) distances on the earth. The other set, *B*, is arranged parallel to and extending radially outward from the axis. The set of lines formed by the intersection of the earth's surface and the first set of planes are called *parallels of latitude*. The identification of the relative position of the parallels of latitude is based on the angular relationship between the axis of the earth and the tangent to the earth's surface. The position of the parallel at the north pole is described at 90°, at the equator

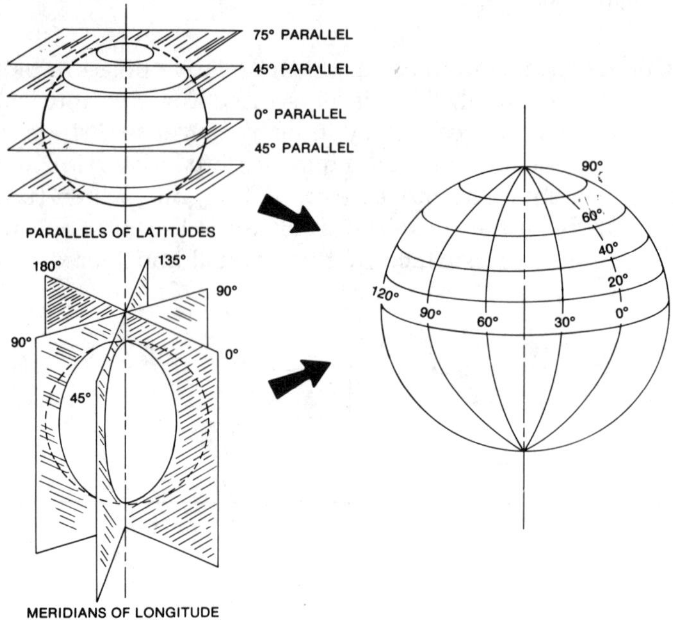

Figure 1-5 Development of spherical coordinates.

the position is described at 0°. Describing the degrees, minutes, and seconds of any parallel along any given north-south line a point position is specified.

The set of lines resulting from the intersection of the earth's surface and the second set of planes (*B*) are called *meridians of longitude.* The identification of meridians of longitude is based on an assumed reference or prime meridian and the angular departure of the other meridians measured relative to it. The location of the prime meridian, numerically the 0° longitude, is Greenwich, England. Other meridian locations are measured by degree, minutes, and seconds from 0° to 180° either east or west of this prime meridian.

The second and more useful reference system for mapping is the *rectangular coordinate system.* A *coordinate* can be defined as a set of two numbers, (x,y). The first number, x, represents a distance measured relative to an origin (often a known landmark) in the east-west direction. The second number, y, represents a distance measured relative to the origin in the north-south direction. The x values are called *departures;* the y values are called *latitudes.*

Coordinates are used in conjunction with a rectangular grid system. A *rectangular grid* is two sets of two parallel, uniformly spaced lines with one set of lines perpendicular to the other. Together the two constitute the *rectangular coordinate system,* which is used to specify any point within the extent of the map relative to the origin. The distance between any two points can be simply found by calculating the absolute departure (the difference in x values) and absolute latitude (the difference in y values) and computing the horizontal distance using the Pythagorean theorem (Eq. 1-1) as follows (see Fig. 1-6):

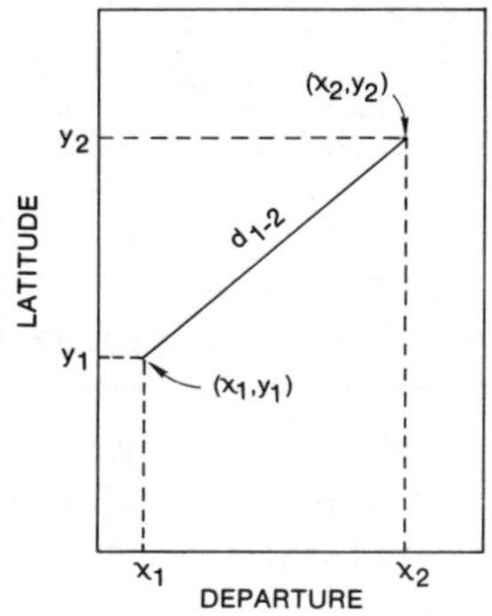

Figure 1-6 Rectangular coordinate plot of line 1–2.

$$d_{1\text{-}2} = \sqrt{(x_2 - x_1)^2 + (y_2 - y_1)^2} \tag{1-1}$$

where the coordinates of

$$\text{point } 1 = (x_1, y_1)$$
$$\text{point } 2 = (x_2, y_2)$$

and

$x_2 - x_1$ = absolute value of departure

$y_2 - y_1$ = absolute value of latitude

$d_{1\text{-}2}$ = horizontal distance between points 1 and 2

1.2.4 Major Projections

Of the many permutations that can occur between the projection view point, the deployment of the modeling surface, and the types of reference systems, five major projections are of interest in mine mapping: (1) the simple conic projection, (2) Lambert's conformal conic projection, (3) the Mercator projection, (4) the traverse Mercator projection, and (5) the universal traverse Mercator projection.

Simple conic projection. The simple conic projection, depicted in Fig. 1-7, uses a tangent conical surface that intersects the surface of the earth at one standard parallel when the apex of the cone is oriented above a pole. The line of intersection is the one line that is free of distortion as points are projected. Projection scales above and below this line are distorted proportionate to the distance away from the line. Because of the accuracy of this type of projection in the middle latitude (e.g., 30 to 60°), it is used most

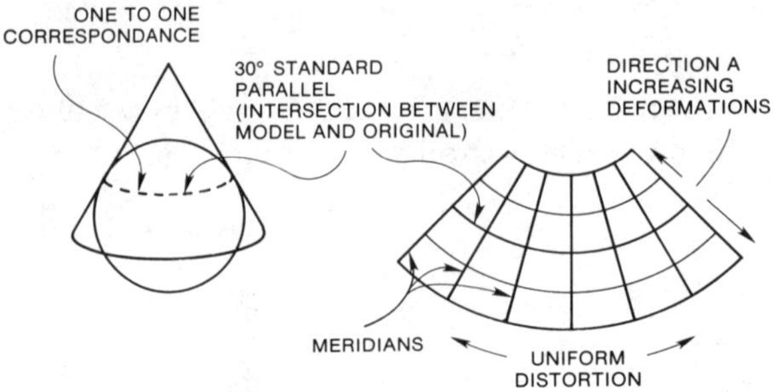

Figure 1-7 Simple conical projection with intersection tangent at the 30° standard parallel.

often for regions within those latitudes which are predominantly east-west in extent. Other characteristics of this projection are that the meridians are shown as straight lines that converge toward the apex of the cone or as one moves toward the poles, and that the parallels are equally spaced concentric circular arcs, and for small regions (in the middle latitudes) the amount of area, distance, and scale are minimal.

Lambert's conformal conic projection. Lambert's projection, also described as a secant projection, uses a conical surface to intersect the sphere at two locations or two parallels (the 33rd and 45th), as shown in Fig. 1-8. As shown, the scale is exact at the two intersected parallels. At areas between the points of intersection, the scale is too small; at areas outside the points of intersection, the scale is too large. The scale factor (SF), which is the ratio of map distances to actual differences, is used to depict the amount of distortion. Thus the SF for the two parallels is exactly 1, outside the parallels the SF is greater than 1, and between them the SF is less than 1.

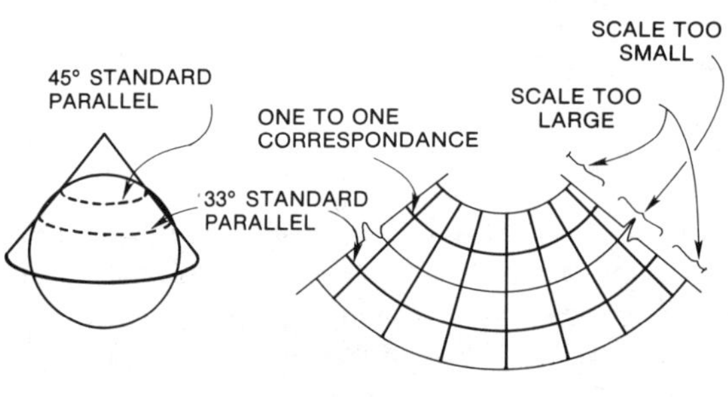

Figure 1-8 Lambert's conic conformal projection strategy.

Lambert's plot is also conformal; that is, it models shape on a one-to-one correspondence and directions are true. The meridians and parallels are assumed to cross at right angles; thus all areas defined by parallel of longitude and meridian of latitude appear as rectangular areas.

Lambert's conformal conic projection provides the same characteristics as the simple conic projections except that the SF across the country is more normalized. The characteristics make Lambert's projection ideal for mapping the United States and is one of the two projections used by the U.S. Geological Survey (USGS) in their National Mapping Program, for the 15- and 7.5-minute map series describing the state coordinate system.[5]

Mercator projection. The Mercator projection uses a cylindrical surface tangent to the earth at the equator (therefore no distortion), with the axis of the earth also acting as the axis of the cylinder, as shown in Fig. 1-9. Using a cylinder, the parallels of latitude, which are parallel circles on the globe, are depicted as parallel on the modeling surface. Meridians of longitude, which converge on the globe, are also depicted as parallel on the model. Thus the meridians and parallels are perpendicular, which is one of the characteristics of conformal projections.

If an orthographic projection is used, the result is severe compression of parallels, distorting shape and direction as the poles are approached. If a central projection view is assumed, a severe elongation between the parallels as one approaches the poles produces a deformed shape and direction distortion. Mercator blended these two projection extremes mathematically by increasing the distance between the parallels at any given latitude proportional to the increase that occurs between the meridians at that same latitude. The result (Fig. 1-9) is a logarithmic elongation of the distance between parallels, producing a one-to-one shape correspondence between the earth and the map. Note that this one-to-one correspondence in shape is between the model and the original and does not indicate any relative scale correspondence on the map. Thus, although Greenland appears on the map to be larger than South America, in reality South America is over nine times larger.[6] The principal advantage of this map is that true courses or bearings appear as straight lines.

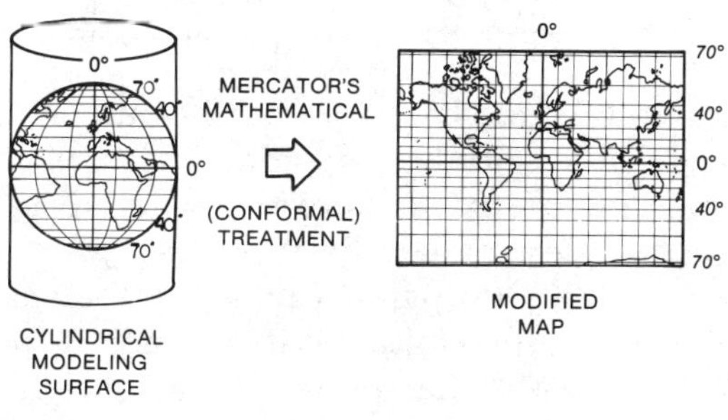

Figure 1-9 Mercator's conformal treatment strategy.

Transverse Mercator projection. The transverse Mercator projection is produced by rotating the axis of the cylindrical modeling surface 90° so that the intersection of the surface and the earth is now along a meridian, as shown in Fig. 1-10. Of course, the points along this meridian are projected without distortion. The conformal advantages of the Mercator projection still hold, but now the true scale characteristics that occurred at the equator on the Mercator projection can be afforded any meridian. Therefore, as long as the region to be mapped does not extend too far in an east-west direction, the map will provide a minimum of distortion. This trait makes the transverse Mercator projection ideal for mapping states that have a predominantly north-south extent (e.g., Illinois).

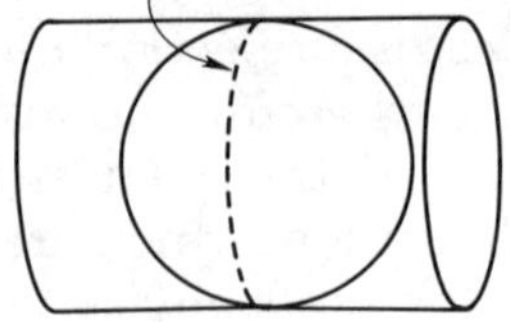

Figure 1-10 Transverse (modeling surface) projection strategy.

The transverse Mercator, combined with Lambert's projection systems, form the projection bases for the state plane coordinate system, discussed in Section 1.3.

Universal transverse Mercator projection. As the diameter of the cylindrical modeling surface used in the transverse Mercator projection decreases, the result is a globe intersected at two parallel positions on the earth's surface, as illustrated in Fig. 1-11. This secant modification to the transverse Mercator projection is known as the universal transverse Mercator projection (UTM). It is used by most military and disaster relief agencies (for latitudes between 84° N and 80°S[2]) and can locate to the nearest meter any point on the earth's surface by using a 15-digit sequence of numbers.

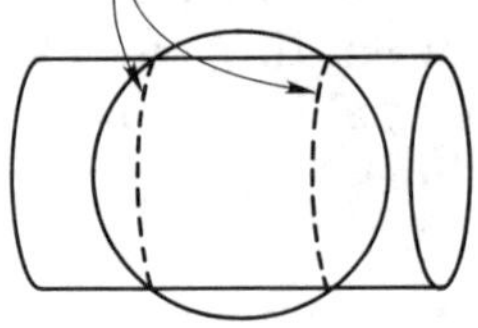

Figure 1-11 Universal transverse Mercator projection strategy.

The globe is divided into 60 equally spaced north-south segments called *zones.* Each zone is 6° wide in longitude. The zone lines, which are meridians, constitute the boundaries of the east-west extents. The north-south control begins at the equator (designated the 0 Northing for the northern hemisphere) and progresses in 100,000 parallel meter increments northward and southward. Figure 1-12 shows the United States, which is spanned by 10 zones (zone numbers 10 through 19). Each zone has a centerline or central meridian, called the *500,000 meterline*, which bisects each zone. Other north-south lines, called *easting lines*, are drawn parallel to the central meridian at 100,000 parallel meter distances, thereby forming a second set of uniformly parallel lines. The *northing lines*, which are parallel to the equator, are, of course, perpendicular to the easting lines. Together the two sets constitute a rectangular coordinate system.

Any point can be located to the nearest meter using a 15-digit number. The first two integers denote the zone number, the next seven indicate the number of meters from the equator, and the last six represent the number of meters from the left boundary of the zone. For example, 13,4760000,550000 designates a local point in zone 13, 4,760,000 meters

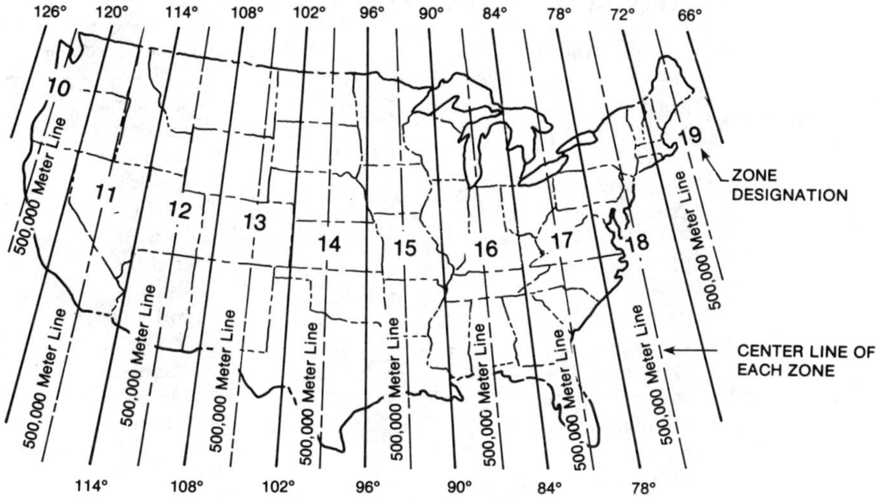

Figure 1-12 UTM zones and control medians spanning the United States.

north of the equator and 50,000 meters to the right of the zone's central meridian.

1.3 MAPPING SPATIAL CONTROL

The spatial orientation of a map is the method employed by the cartographer for controlling the horizontal and vertical positions of features on a map. Several methods exist and are commonly employed by cartographers for showing the exact position and orientation of points of interest on a map. The UTM grid described above is one method of horizontal control. Another is the state plane coordinate system.

1.3.1 State Plane Coordinate System: Horizontal Control

The state plane coordinate system uses a network of known triangulation stations, established in 1935 by the U.S. Coast and Geodetic Survey, now the U.S. Geodetic Survey, to provide geodetic control. Each state has its own separate set of coordinates. The coordinates represent the mathematical projection of that state on a plane surface based upon the feet north-south and east-west of a known but arbitrarily specified origin. Both the transverse Mercator projection and Lambert's conformal conic projection are used to project the geodetic control points. The transverse Mercator projection is used for those states having a north-south extent, whereas Lambert's projection is used for those states that have a predominate east-west extent. Table 1-1 is a tabulation of the projection(s) used in each state.

The state plane coordinate system has distinct advantages over plane projections for use in mine surveying and subsequent mapping:

TABLE 1-1 STATE PLANE COORDINATE SYSTEM

Lambert system	Transverse Mercator system
Alaska*	Alabama
Arkansas	Alaska*
California	Delaware
Colorado	Florida*
Connecticut	Georgia
Florida*	Hawaii
Iowa	Idaho
Kansas	Illinois
Kentucky	Indiana
Louisiana	Maine
Maryland	Mississippi
Massachusetts	Missouri
Michigan	Nevada
Minnesota	New Hampshire
Montana	New Jersey
Nebraska	New Mexico
New York*	New York*
North Carolina	Rhode Island
North Dakota	Vermont
Ohio	Wyoming
Oklahoma	
Oregon	
Pennsylvania	
South Carolina	
South Dakota	
Tennessee	
Texas	
Utah	
Virginia	
Washington	
West Virginia	
Wisconsin	

*Utilize both systems.

1. The use of this system solves the mapper's problem of how to represent horizontal control with the minimum amount of deformation. Each point on the surface has one and only one corresponding point on the map; thus the data from a plane survey can be tied into the plot using state plane coordinates, thereby increasing the generality of these data.
2. This system provides a positive check for the surveyor, thereby preventing the accumulation of angular and horizontal error which can occur when large areas are traversed using only plane surveying techniques and references.
3. Mining operations adjacently located, which utilize state plane coordinates, can correlate and control their surveys and maps with each other. Thus property boundaries can be better established and maintained.
4. Once a survey station, called a *spad* when positioned underground, has been located in the state system, the coordinates of that point can easily

be recovered, even if all evidence of the point is destroyed (e.g., as a result of a roof fall).

5. The use of this system offers the cartographer the inherent accuracy of the state plane projection method.
6. The system is easily translated into the UTM system, and the universal geodetic coordinates of latitude and longitude.

Other horizontal callouts. One simple alternative method of horizontal control used most often in plane mapping involves the identification of both a true meridian (a true north-south line) and a commonly known landmark which is contained within the boundaries of the map. Using this system, the mapper or reader can locate the position of any point relative to the commonly known landmark by denoting the bearing angle (an angular deflection measured relative to either the north or south line which passes through the known point) and the true horizontal distance (which is scaled on the map) between the landmark and the unknown point.

Another method of horizontal control is to use a rectangular coordinate system referenced to an assumed origin. The position of any point (x_1,y_1) then depends only on the relative horizontal departure x_1 and vertical latitude y_1 from the origin (0,0).

The disadvantages of both alternative methods are (1) that the point of origin must be known by the surveyor and any subsequent reader, and (2) that large maps thus established incur significant position deformation between the original surface and that represented on the map.

1.3.2 Vertical Control

The vertical control of a map defines the topographical relief or changes in elevations of the surface. Several methods for representing topographic relief are in common use. These methods include (1) contour lines, (2) trachographic relief, (3) shadient relief, (4) perspective relief, and (5) photo-image map.

Contour lines. By far the most prevalent method of depicting vertical relief is by using contour lines. Contour lines are lines drawn on a plane view which represent all adjacent points of equal evaluation. Figure 1-13 shows a contour map. The rules, methods, and purposes of contour and topographic mapping are discussed in greater detail in Chapter 4.

Trachographic relief. Trachographic relief is a method of depicting the relative elevation or contrasts in elevation in a region by using small, hill-shaped lines to indicate relative relief. The height of the curved lines represents the approximate elevation. Unlike contour lines, this method is used only for showing qualitative, not quantitative values. Figure 1-14 shows a trachographic relief map.

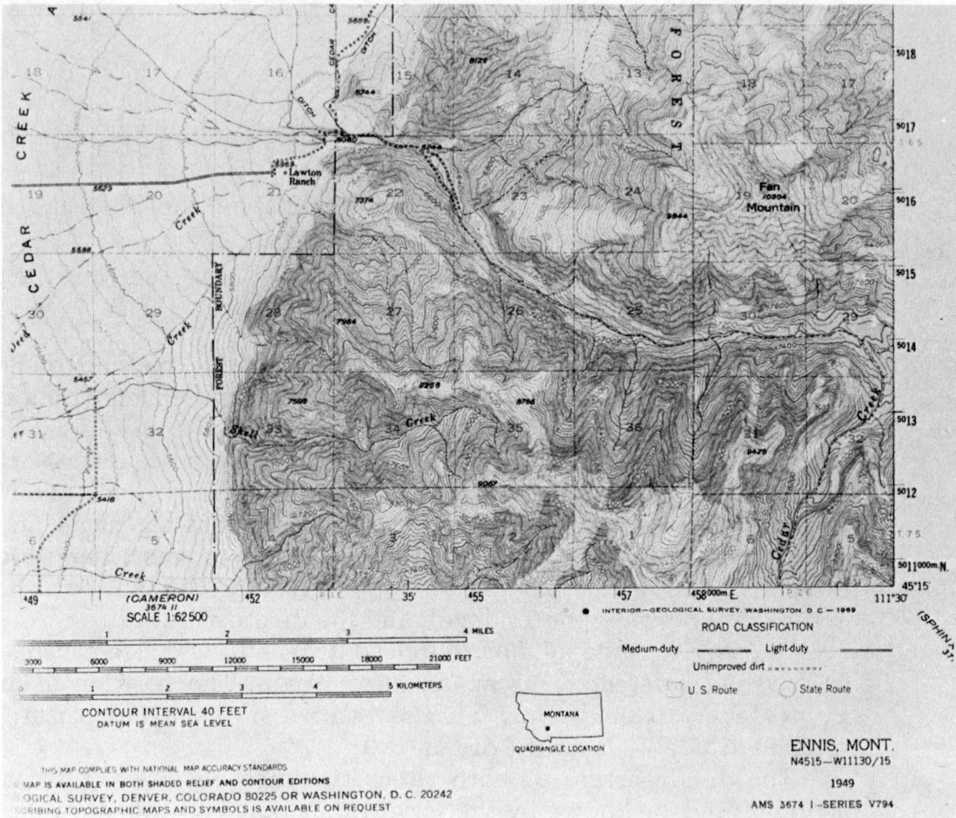

Figure 1-13 Contour map illustrating margin data. (Courtesy of USGS.)

Figure 1-14 Method of trachographic relief.

Shadient or chorochromatic relief. This type of relief is another qualitative method of depicting relative relief in an area. Various tints, usually varying tones of gray, are used to represent the relative height, as depicted in Fig. 1-15. Normally, these methods of showing relief are used only for maps of large extent, maps with little quantitative detail, or to enhance contour mapping.

Perspective relief. Perspective relief is an isometric illustration of a three-dimensional perspective view of a given region. Often these methods are combined with sectional views to show the stratigraphic positioning of the subsurface, as shown in Fig. 1-16.

Photoimages or orthophoto maps. Produced from line enhancement of

Figure 1-15 Black-and-white "chromatic relief." (Courtesy of USGS.)

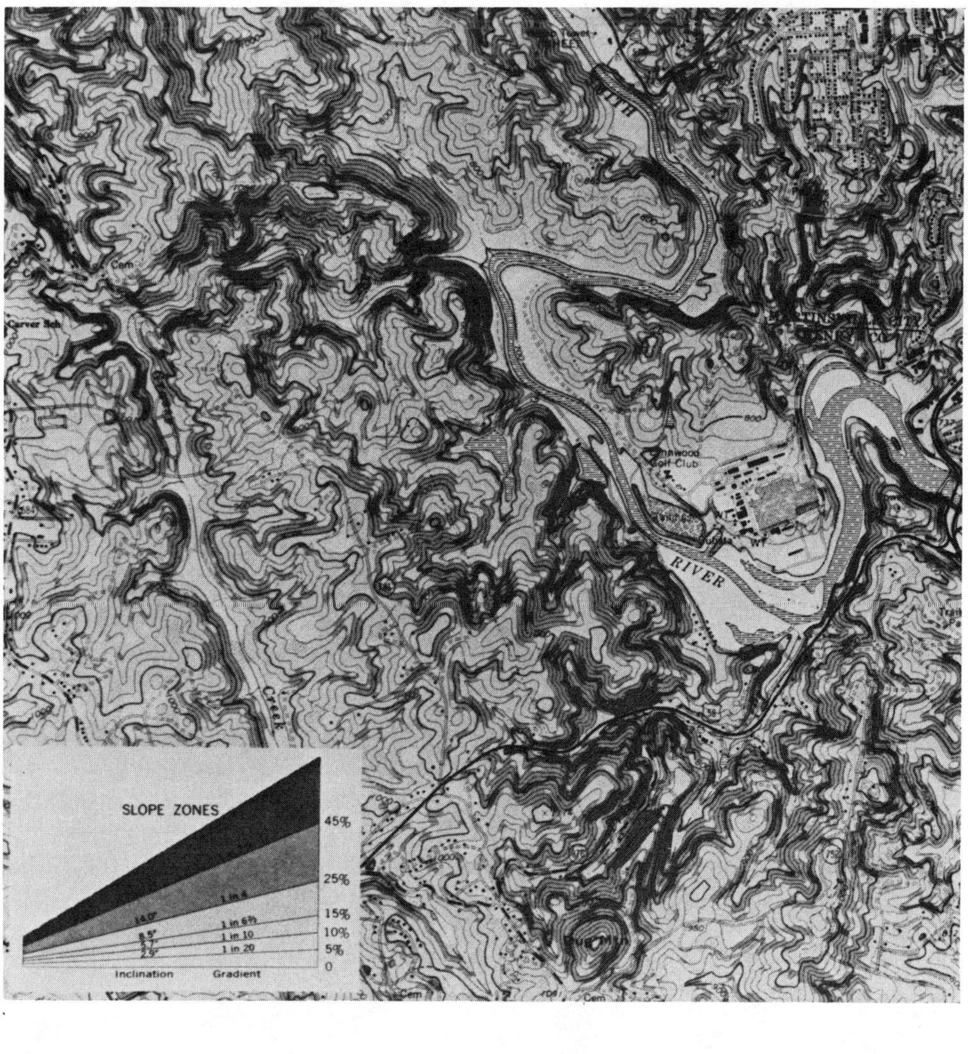

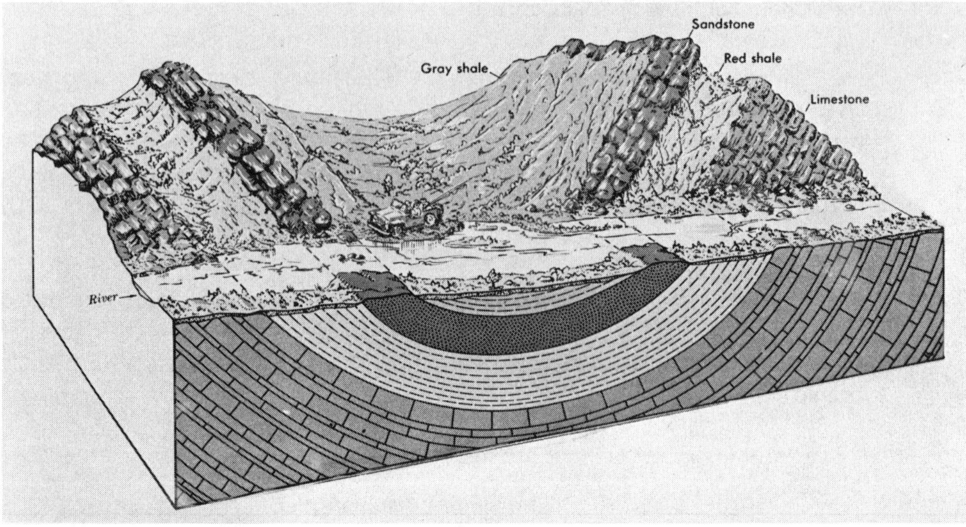

Figure 1-16 Perspective relief depicting a geological profile. (Courtesy of USGS.)

Figure 1-17 Photomap depicting relief. (Courtesy of USGS.)

available photographs, photoimages or orthophoto maps effectively replace trachographic methods of depicting relief-representation. Figure 1-17 shows a portion of a typical photoimage map produced by the USGS. Photoimage maps are most applicable where few contours and a low density of cultural feature make conventional mapping of little value. Photoimage maps can also be used to define vegetation, surface water, and even geological anomalies such as faults (see Chapter 6).

1.3.3 Map Scale

Another convergence commonality shared by all maps is the use of a *scale*. Scale is a somewhat arbitrarily specified ratio between the linear distances (measured in units, which may be inches, feet, miles, meters, etc.) on the map and the corresponding distances (measured in the same units) at the original site. A scale of 1:6000, therefore, means that 1 ft on the map is equal to 6000 ft at the site or 1 in. on the map equals 6000 in. at the site. The greater the ratio, the "smaller" the scale. A map with a scale between 1:125,000 and 1:1,000,000 is classified by the USGS as a *small*-scale map, one with a scale between 1:50,000 and 100,000 as a *medium-scale* map, and one with a scale less than 1:30,000 as a *large*-scale map. The topographic series produced by the USGS has a scale range from 1:24,000 to 1:1,000,000.[7]

The selection of the most appropriate scale for any map depends on (1) the size of the area to be represented, (2) the scale of available base maps from which new maps are to be produced, and (3) the intended uses for the map. The larger the region to be mapped, the smaller will be the scale. If a 30-square-mile subsurface layout of a geologic condition is desired, a convenient base map would be a 7.5-minute topographic map available from the USGS. Thus the scale would be fixed at 1:24,000. The more detail and accuracy that are required for the information represented, the larger should be the scale.

Construction, drainage, and excavation maps are examples of maps that require large scales. Mining maps are also large-scale maps because of the detail usually required. By law[8] the scale available to the mining cartographer ranges between 1 in. = 100 ft and 1 in. = 500 ft or between 1:1200 and 1:6000. To show an overview of mining operations, the 1:6000 scale is commonly used, whereas for detailed layout the 1:1200 scale is more appropriate. It should be noted that this scale range applies to those maps submitted to government regulatory agencies but does not restrict the use of smaller or larger scales for company maps and layouts.

The most common methods of expressing map scale are *mathematical representation*, described above, graphic representation, and equivalent areas. *Graphic representation* uses a bar scale similar to those illustrated in Fig. 1-18 to signify the relationship of map dimensions to the corresponding distances of the original. The *equivalent area* is depicted as a ratio, similar

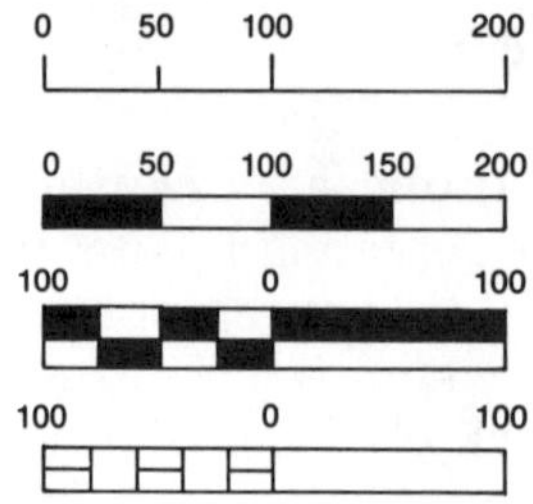

Figure 1-18 Common graphical methods for depicting a scale of 1 in. = 100 ft.

to the linear mathematical representation. For example, $1{:}2000^2$ or 1:2000 sq means that 1 square inch on the map equals 2000 square inches on the earth's surface.

1.3.4 Feature Symbolization

Maps are representations of an original; therefore, all marks, points, and figures on a map only symbolize the original features. In general, the *symbols* used should be (1) *simple* in composition (size, outline, clarity), (2) *obvious* in what they are intended to represent, and (3) *easily drawn.* Based on these guidelines, symbols are not details of the specific feature but are generalizations of the nature of the feature. As the scale increases, the degree of generalization increases. This change in generalization results in a corresponding change in the symbol, thus complicating any efforts to compile an "absolute" set of standard symbols.

Symbols are subdivided into the following categories:

1. Hydrographic (water-related) features
2. Human-made features
3. Surface relief features
4. Surface vegetation
5. Special symbols (e.g., geologic, mining)

Each of these can be further divided into various categories; for example, human-made features can be divided into building, structures (e.g., bridges), transportation symbols, boundaries, and so forth. The meaning of various symbols, like all drafting symbols, has changed considerably throughout history. This, in addition to the preferences of the cartographer, has contributed to the general lack of consistency and standardization of symbols within the industry. Some standard map symbols are discussed and tabulated in Chapter 3. Mining symbols have yet to be standardized. The Appendix represents an initial effort to apply the three composition guidelines above to the development of a uniform set of mine mapping symbols.

All maps have, or should have, a legend that serves to define the map symbols that appear frequently. The legend is, in essence, the language of the map, and normally only those symbols that may not be commonly understood are included as part of the legend.

1.4 MAJOR TYPES OF MAPS

Many different types of maps exist (Fig. 1-19), each serving a specific purpose. For mining activities, five basic types of maps are commonly used: (1) the topographic or contour map, (2) the cadastral or property development map, (3) the mine development map, (4) functional layouts (including premining layouts), and (5) thematic maps and diagrams. Each type of map is constructed by appropriately scaling and plotting the data obtained from a field or other surveying method.

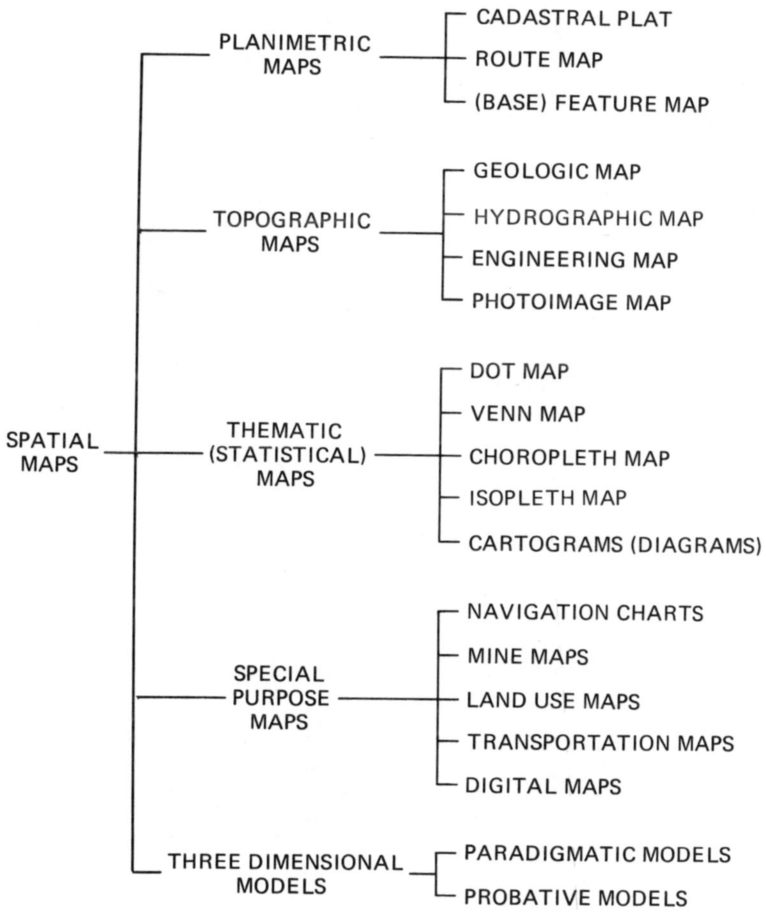

Figure 1-19 Family of spatial maps.

1.4.1 Topographic Maps

The principal purpose of a topographic or contour map is to show the variability of the vertical relief within an area. The vertical relief of an area is described and quantified by elevation. The elevation of any unknown point is determined by measuring the vertical distance from the unknown point to a point of known elevation (a benchmark) or to a point where the elevation is assumed to be zero. If the elevation of the unknown point is determined relative to a known elevation, this elevation is called *absolute elevation* and is the height of that point above sea level (USGS 00.000). If the height of the point is determined relative to an *assumed* zero plane, the elevation is called *relative elevation*. To avoid confusion, most mine maps use absolute elevation when plotting vertical relief.

Figure 1-20 shows a typical contour map and pictorial view of what it represents. The lines shown in the bottom portion are called *contour lines* and represent all adjacent points of equal elevation on the map. The configuration that results from plotting the contour lines represents the *topographic relief* (the relative change in elevation) within the extent of the map.

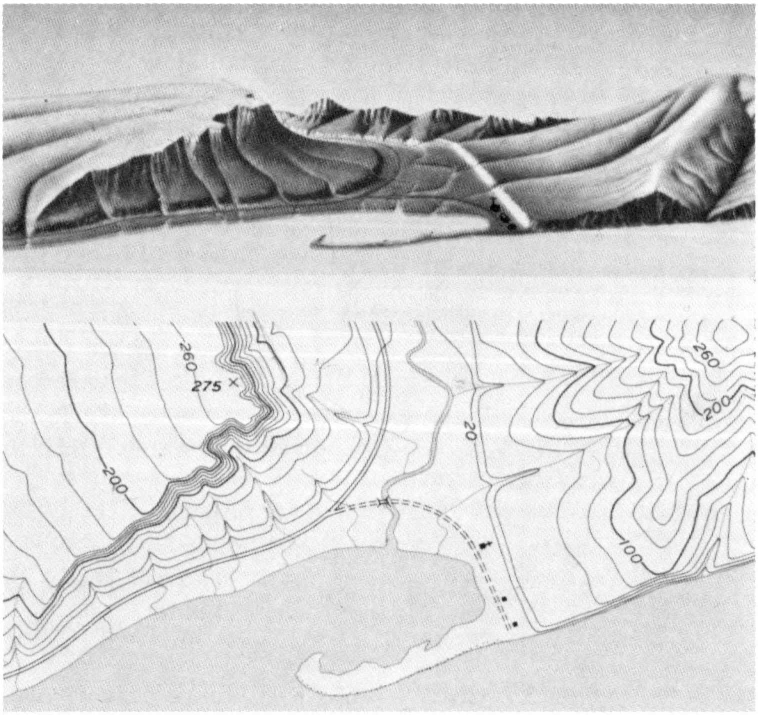

Figure 1-20 Pictorial view of a bay area and the resulting contour map. (Courtesy of USGS.)

1.4.2 Cadastral and Property Maps

The cadastral map is a subgrouping of the planimetric map. The basic purpose of a cadastral map is to show boundaries of ownership. Cadastral maps show real property ownership, leases, rights-of-way, mineral ownership, property descriptions, and human-made developments (such as roads, railroads, utilities, significant buildings, etc.). Because of the detail required of these maps, they usually range in scale factor from 1:12 (1 in. = 1 ft) to 1:2400 (1 in. = 200 ft). Large areas are shown by subdividing the area into quadrants or several sections and mapping each to the scale needed. Normally, contours are not shown on cadastral maps, but elevations of significant points such as summits or depressions are indicated.

1.4.3 Mine Maps

The mine map combines the features of the topographical and cadastral plat maps with specific information regarding the status of mining excavations, including locations and descriptions of entries, bore holes, seam outcrops, elevations or contours of the workings, air courses and ventilation devices, escapeways, hydrographic detail, locations of active, idle, and abandoned sections, other adjacent surface and subsurface excavations, location of gas and oil wells, overburden data, and so forth.

Because of the quantity of information needed to describe fully the status of mining activity and the clarity that must accompany those data, most mine maps rarely include all the content on a single map. More often a general mine map is prepared that includes property information, some basic contouring (or elevation designations) and the outline of excavations to date, that is, the position and designation of headings and pillar outlines. As additional data are needed, overlays are produced. When complete maps are required, the basic map and the overlays can be reproduced to yield a single print depicting all relevant data. Figure 1-21 shows a portion of a gen-

Figure 1-21(a) General mine map depicting extent of detail needed for surface and underground features at 1″ = 500′ scale. (Courtesy of Eastern Associated Coal Corp.)

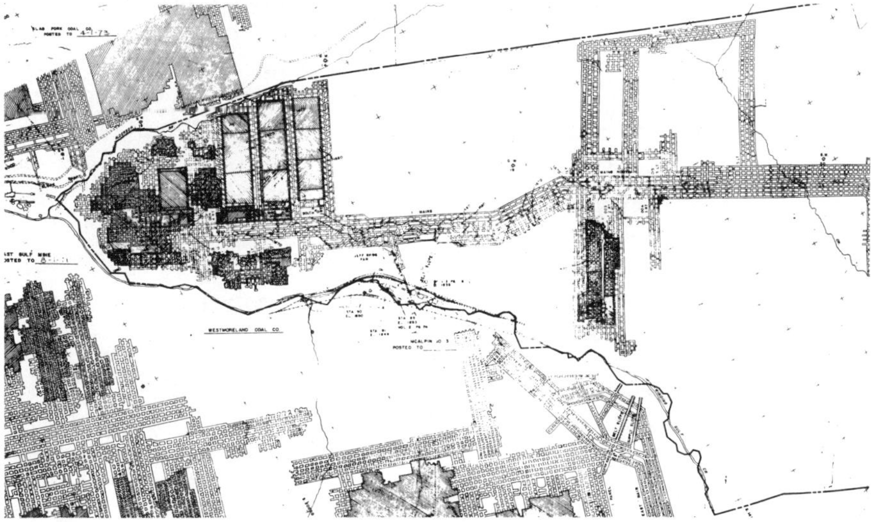

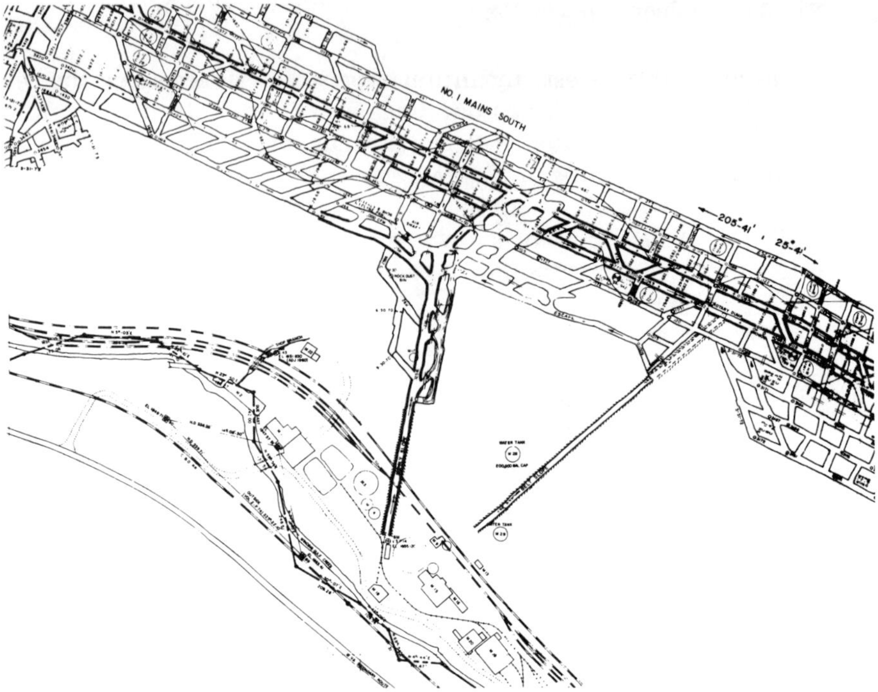

Figure 1-21(b) General mine map depicting extent of detail needed for surface and underground features at 1″ = 100′ scale. (Courtesy of Eastern Associated Coal Corp.)

eral mine map. After the general mine map has been accurately produced, additional excavation information is added as may be required for the updating and functional use of the map.

1.4.4 Functional Layouts

Functional layouts depict critical mine services, or functions, as they relate to and operate within the boundaries of the mine. Functional layouts serve two principal purposes. First, initial layouts, called *design layouts*, provide the details needed to open and develop a mine in a planned fashion. Second, *current layouts* store current information about the operations for optimization analysis, safety, and other operational studies. The major functions discussed in this text are ventilation (Chapter 7), transportation (Chapter 8), drainage (Chapter 9), and electrical distribution (Chapter 10).

1.4.5 Thematic Maps

Thematic maps, also termed single-factor maps, special-purpose maps, or statistical maps, depict both the qualitative and quantitative characteristics of a topic in a region. Qualitative characteristics show general or relative distributions of one or more factors (e.g., rock type, sulfur, ash, etc.) relative to the area of interest. Quantitative characteristics show the relative variations in amount or quantity (e.g., percent sulfur, percent ash, etc.) of the topic in the region. Quantitative maps are presented in three principal forms:

1. *Dot maps.* Dots or small symbols are positioned on a map representing a given number of elements. The density or frequency of dots per unit area represents a given density, and the relative size of the dot represents the strength or intensity of the topic.
2 *Choropleth maps.* Consist of various contrasting colors or tints on a map, representing a given value or percentage of the element of interest.
3. *Isopleth maps.* Isopleths are lines analogous to contour lines, connecting places or positions of equal value. For example, points of equal temperature are represented by isotherms and points of equal pressure are represented by isobars. Thematic maps provide qualitative as well as quantitative information, and are particularly useful to mining exploration. Thematic maps are discussed in greater detail in Chapter 6.

REFERENCES

1. William Randolph Williams, *The Philosophical Foundations of Educational Systems Analysis and Modeling*, Doctoral dissertation, West Virginia University, Morgantown, W.Va., March 1979.
2. Arthur H. Robinson and Randall D. Sale, *Elements of Cartography*, 3rd ed., John Wiley & Sons, Inc., New York, 1969.
3. Erwin Raisz, *Principles of Cartography*, McGraw-Hill Book Company, New York, 1962.
4. David Greenhood, *Mapping*, University of Chicago Press, Chicago, 1964.
5. Morris M. Thompson, *Maps for America*, U.S. Dept. of the Interior, U.S. Geological Survey, U.S. Government Printing Office, Washington, D.C., 1979.
6. Topographic Maps, U.S. Dept. of the Interior, U.S. Geological Survey, U.S. Government Printing Office, Washington, D.C., 1976.
7. Federal Mine Safety and Health Act of 1977, Public Law 91-173 as amended by Public Law 95-164, U.S. Government Printing Office, Washington, D.C., 1977.

PROBLEMS

1. Plot the following data points: (1,1); (2,1); (2,2); (3,3); (4,3); (5,5); (6,6); (7,6); (7,7). A straight line drawn to approximate these data would represent what type(s) of data treatment?
2. Take any road map of your area and list the specific data simplification, enhancement, and distortions that occur.

3. Plot the following departures and latitudes: station 1 (25.2, 32.1); station 2 (10.5, 40.0); station 3 (4.9, 15.2); station 4 (18.7, 5.6); station 5 (20.5, 21.0). Determine the distance between successive stations.
4. What does the following UTM designation mean: 12,422000,5200000?
5. Convert the original site measurements to the representative scaled dimensions for the map using the appropriate scale factor.

Original site measurement	Scale factor	Map dimension
200 ft	1 in. = 200 ft	______ in
630 ft	1 in.= 200 ft	______ in.
250 m	1:1000	______ cm
750 ft	1:1200	______ in.
2 mi	1:6000	______ in.
6 mi	1:125,000	______ in.
10 mi	1: 62,500	______ ft
116 ft	1:2400	______ in.
20 ft	1:1200	______ in.
4 ft	1: 600	______ in.
36 in.	$^1/_2$ scale	______ in.

6. Draw a map of your home from the nearest town, city, or interstate highway. Include major symbolization (see Chapter 3), cultural features, and relief, using one of the techniques described.

2

Overview of Mine Surveying

2.0 INTRODUCTION

Before a map can be produced, the data that are to be represented must first be obtained and then be manipulated into graphic and symbolic language. The operation of obtaining the necessary data is called *field surveying*. The care and accuracy of the field survey will in large part determine the validity and usefulness of any map made from the data collected.

A field survey is a methodical procedure of obtaining data which relates the relative positions of points or features above, on, or beneath the surface of the earth. Various types of field surveys are needed to properly describe and map a given area.

Property (Cadastral) surveys. These are used to establish and fix property boundaries, including ownership, leases, rights-of-way, and data needed for property assessment. These surveys are generally used in the construction of official plats of land ownership but are also used for general map construction. A plat differs from a map in that a plat shows dimensional information of boundaries of a parcel of land on the drawing, whereas a map relies solely upon the scaled representation of dimensions.

Topographic surveys. These are used to obtain the horizontal and vertical control data, together with the data needed to construct a topographic map.

Geologic surveys. These are executed to obtain data about the geologic nature and condition of the surface and subsurface region. Such data include the location and severity of faults, stratigraphic or lithological positions, overburden depths, tectonic forces, and so forth.

Hydrographic surveys. These are initiated to obtain data defining the surface and subsurface hydrology in the area. These data include the occurrence, circulation, distribution, locations, chemical and physical properties, and the relationship to the environment of the water resources in an area.

Engineering control surveys. These operations are directed toward obtaining data that are necessary for successful construction planning or to maintain dimensional control of site activities.

Mine control surveys. These operations are used to locate the position and status of underground workings, or aid in maintaining the dimensional control of mining activities, including the direction and extent of entries, crosscuts, and so forth.

Mine surveying employs the same principles and instruments as surface surveying; however, certain requirements make mine surveying uniquely different from the above field surveys. Underground surveys:

1. Require the running of traverses totally isolated from independent meridian reference points such as the sun or Polaris. As a result, the meridian, a true north-south line, must be carried underground with precise accuracy and the positions of the individual stations must be maintained with an accuracy greater than that of a corresponding surface survey.
2. Require continual updating as the excavations underground proceed.
3. Require many stations and therefore many individual setups, which are usually closely spaced. Both conditions increase the probability of error, thus necessitating greater surveying accuracy.
4. Require multilevel referencing and control to maintain planning coordination with other underground workings and with the surface.
5. Require establishing control on the roof above the surveyors to avoid the loss of station markers resulting from haulage traffic.
6. Require specially designed and protected equipment since the environment is dark, wet, confined (in some cases the height between the top and bottom of the entry closes to 22 in.), dusty, and unstable.
7. Require a referencing system that will enable the surveying team to reestablish stations after they are lost, as frequently occurs from roof falls or contact with haulage equipment.

In addition to these informal but necessary requirements, the federal government and each state government mandate certain surveying requisites.

For example, the federal government in the Federal Coal Mine Health and Safety Act[1] requires that "mine traverses shall be advanced by closed loop methods of traversing or other equally accurate methods of traversing." State agencies independently affect the methods of surveying; for example, the West Virginia Coal Mine Safety Law of 1971, Part 22-2-1,[2] requires the use of "the rectangular coordinate traversing method and meridians carried through and tied between at least two parallel entries. . . ." The survey regulations that apply in your state should be studied in detail to avoid possible conflict or confusion.

Although each of the types of surveys described above is needed to properly plan, lay out, and operate a mine, only simple surface property and mine surveying is presented in this chapter. Subsequent chapters develop the needs and methods for geologic, hydrographic, and engineering surveying.

2.1 VERTICAL AND HORIZONTAL CONTROL

Regardless of the type of survey being conducted, most surveys share the need to establish, maintain, and record vertical and horizontal control in the region of the survey. Several methods for establishing horizontal control were described in Chapter 1, including the state plane coordinate systems (Lambert's conic projection and the transverse Mercator systems), the UTM system, and a rectangular coordinate system using some independently established origin. Although any system properly established and accurately surveyed will probably meet the legal requirements, the best system will be one that provides the greatest survey control and adequate mapping accuracy at the least cost in time and money. The state plane coordinate system is such a system.

Within the United States, elevation is uniformly based on the National Geodetic Vertical Datum of 1929 (the mean-sea-level datum). Elevation monuments, or *bench marks*, are established by both the National Geodetic Survey (NGS) and the U.S. Geological Survey (USGS). The monuments established by the NGS are usually of first- and second-order levels of accuracy, whereas the older and more numerous bench marks established by the USGS are often of the less stringent third-order class of leveling accuracy. The use of NGS and USGS references are usually dictated by the convenience of the station to the mine surveyor rather than the accuracy or order of the station, since the relative elevations on the mine site are more important than the accuracy of absolute elevations.

The initiation of both vertical and horizontal control surveys must be preceded by locating the bench marks and horizontal control monuments within the general region of the survey. The information needed to locate these monuments is obtained by requesting a copy of the "Geodetic Control Diagram" covering that region, from the National Geodetic Survey, C-18, Rockville, Maryland 20852. The control diagram is a system of maps spanning an area 1° in latitude by 2° in longitude (roughly 69 miles by 106 miles). The 1° by 2° region is divided into two 1° by 1° sections, and each 1° section is divided into four 30-minute quadrangles, as illustrated in Fig. 2-1. Each of the 30-minute quadrangle areas is numbered clockwise from 1 to 4 beginning with the northeast region. The individual 30-minute quadrangle that carries the location of the control stations within an area is identified by a six-digit number. The first two digits represent the latitude, the next three the longitude, and the last digit designates the 30-minute quadrangle. The six-digit callout for the shaded portion of Fig. 2-1 would be 380902.

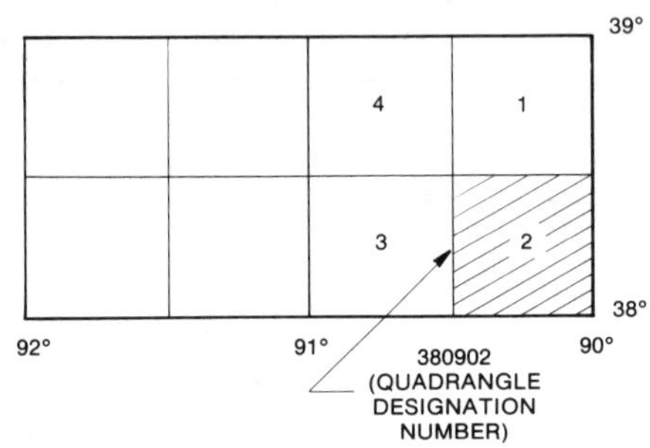

Figure 2-1 Division of the St. Louis, Missouri–Illinois quadrangle: 1:250,000 geodetic control diagram.

2.2 VERTICAL CONTROL

From a review of the 30-minute quadrangle map, the surveyor requests detailed descriptions of the (vertical) control stations that may be required. The bench marks are described[3] according to (1) their distance and direction from a well-known place, (2) their distance and direction from nearby objects, (3) the nature of the object in or on which the mark is placed (for most NGS marks this will be a concrete monument post), and (4) the nature of the mark and how it is stamped. For NGS marks these will usually be a standard bronze disk 4 in. in diameter (Fig. 2-2). It is important to gain as much information as possible about these stations since many are quite old and as a result may not be readily accessible. Although every effort has been made to maintain an accurate accounting of these stations, many of the stations are missing, have been moved, or have been vandalized to the point that they are no longer useful. In these cases and in cases where useful stations still exist, each surveyor or engineer should file a "Recovery Report" with the NGS to facilitate NGS maintenance and subsequent use of the marker by others.

2.2.1 Leveling Procedures

The set of procedural activities required to determine various elevations relative to a known or assumed bench mark is called *leveling*. *Spirit* or *direct*

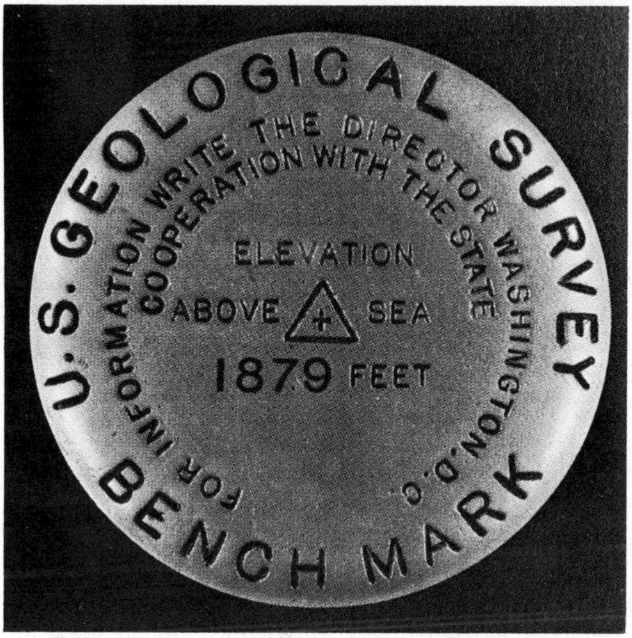

Figure 2-2 Bench mark monument disk. (Courtesy of USGS.)

leveling is a method of determining different elevations by directly reading measurements from leveling rods located at different positions. *Differential leveling*, the most common method of direct leveling, establishes the elevation of any unknown point by first measuring the difference in height between that point and some point of known elevation and then adding (or subtracting) that difference from the known elevation.

Before describing the procedures of differential leveling, some useful terminology should be defined. A *bench mark*, abbreviated in diagrams and notes as B.M., is a point of known or assumed elevation. The use or selection of a point as a bench mark is based on the positional stability of that point over time and the accessibility of the point during the survey. Bench marks, which are referenced to the N.G.V.D. 1929 (sea level), are used to establish the absolute elevation of subsequent points. It is customary practice to carry only absolute elevations on mine surveys. Some states (e.g., West Virginia) make the practice a law.

A *foresight*, abbreviated F.S., represents the information recorded about a previously unknown station or point measured relative to the surveying instrument. In leveling, the information that is recorded is the height difference between the point and the line of sight of the instrument. In horizontal control the F.S. information is used to determine the direction of the line of sight of the instrument relative to a meridian.

A *backsight*, abbreviated B.S., is information recorded about a known station in order that the exact orientation and elevation of the instrument can be determined. For leveling, the information recorded is the difference in height between the bench mark, known point of elevation, and the line of sight of the survey instrument. In horizontal control the B.S. information consists of angular data used to orient the instrument relative to the meridian.

A *turning point*, abbreviated T.P., is the location of the instrument. For leveling the location is used to establish a perfectly level line of sight to which all elevations can be temporarily referenced. The selection of the turning point is dependent on the temporary stability of the location and the ability to maximize the number of foresights taken from it.

The *height of instrument*, abbreviated H.I., is the elevation of the line of sight determined by adding (or in the special case of underground leveling, subtracting) the B.S. measurement and the elevation of the bench mark.

The procedures of differential leveling illustrated in Fig. 2-3 are as follows:

1. Select and establish the location reference bench mark, *A*, for the region to be surveyed. In mine surveying this should be linked to at least one and preferably more NGS bench marks.
2. Select the most appropriate point for setting up the instrument, the turning point. Adjust and firmly locate the level to provide an unvarying plane for the line of sight. Note that for all measurements a slight difference exists between the line of sight, which is a straight line, and the actual level line, which is a curved line corresponding to the line equidistant above the earth's curved surface. However, by positioning the T.P midway between the bench mark and the unknown point, the difference does not introduce an error in the calculations. For a detailed presentation of the operations needed to adjust the level and the error that can result, consult any typical surveying text.[3,4]
3. Leveling rods, graduated in easily read units, are positioned directly above both the B.M. and the forward station.
4. The transitman first turns the leveling telescope to align with the B.M. rod, then reads and records the height indicated. This is, of course, the B.S. Next, the transitman aligns the telescope with the forward rod, then reads and records the height above this unknown point. This reading is the F.S.
5. The B.S. measurement is added to the B.M. to obtain the H.I. The F.S. is then subtracted from the H.I. to obtain the elevation at the forestation. Equations 2-1 and 2-2 summarize these computations.

$$\text{B.M.} + \text{B.S.} = \text{H.I.} \tag{2-1}$$

$$\text{H.I.} - \text{F.S.} = \text{Elev.} \tag{2-2}$$

where B.M. = elevation at the bench mark
B.S. = height of the line of sight of the instrument measured above the B.M.
H.I. = elevation of the line of sight of the instrument
F.S. = height of the unknown station measured below the line of sight at the instrument

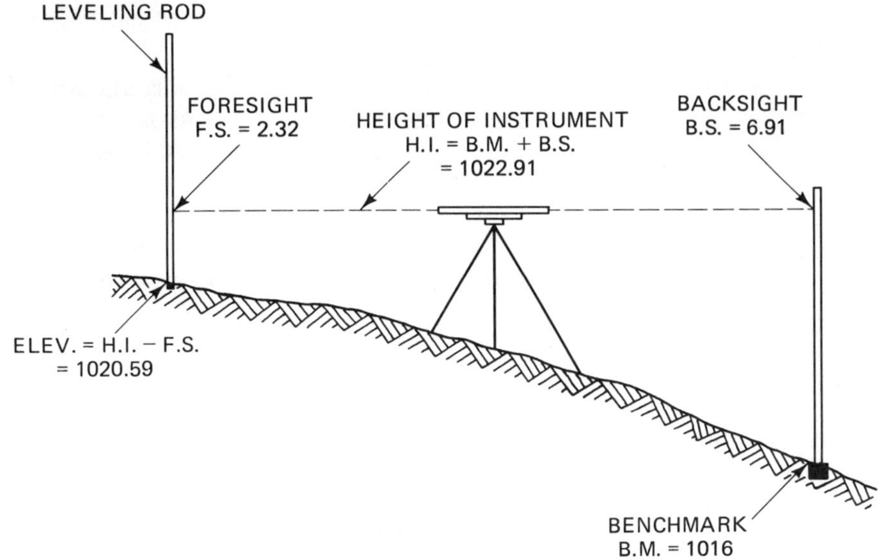

Figure 2-3 Example of simple differential leveling.

As often occurs in hilly terrain, the elevation of the unknown point is substantially higher than can be obtained using a single differential reading. In these cases, illustrated in Fig. 2-4, one or more intermediate T.P.'s need to be established. The individual turning points act to "transfer" elevations from the bench mark to the point of unknown elevation. With more than one turning point, procedures similar to those described above are used. The level is first positioned midway between the bench mark and the first turning point. The backsight and foresight are taken and recorded. The instrument is

Figure 2-4 Example of differential leveling requiring an intermediate turning point.

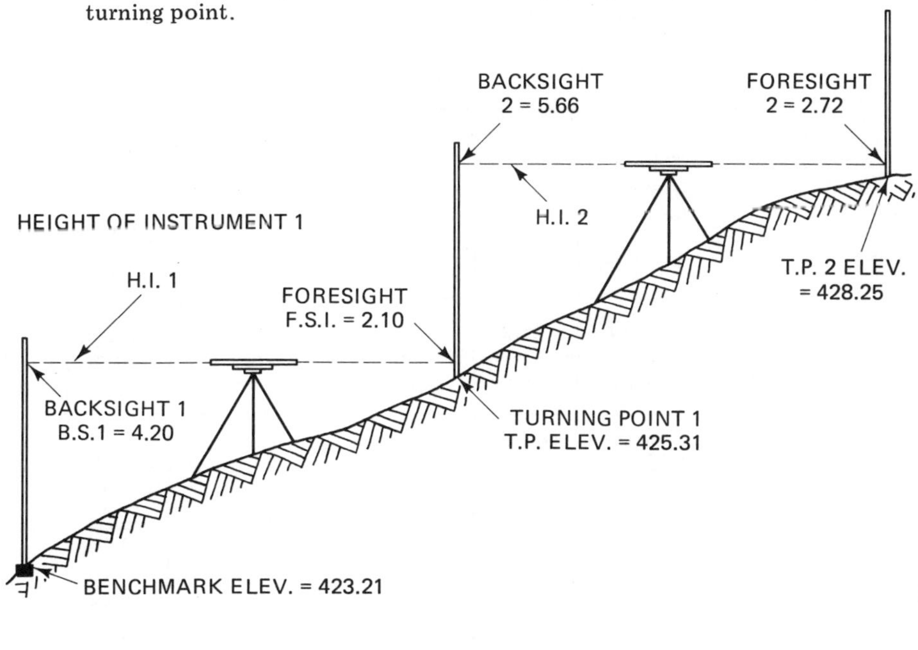

then moved midway between the first and second turning points. Backsight and foresight are again recorded. Finally, the level is positioned between the second T.P. and the unknown elevation. The backsight (on the second turning point) and the foresight (on the point of unknown elevation) are taken and recorded. The elevation of the forestation is the sum of the B.M. and the backsights minus foresights. The procedures are repeated for additional points and the elevations thus determined throughout the extent of the survey region.

2.2.2 Surface Field Notes

Other than correctly setting up the instrument and executing the proper procedures, the most critical task of the survey leader is to prepare accurate, clear, legible, and concise field notes which will be used for making engineering designs, diagrams, and maps. Differential leveling field notes consist of compiling a tabulation of those measurements taken *and* of field computations made regarding the elevations. An example of the format commonly used is illustrated in Fig. 2-5. As a minimum these notes always contain the station number or turning point identification, the backsight (B.S.) measurement, the height of instrument (H.I.) calculation (the height of instrument is simply the addition of the bench mark or known elevation with the rod reading at that point), the foresight (F.S.) measurement, and the elevation (Elev.) calculation of the point of interest (the elevation of the point is, of course, the height of instrument elevation minus the rod reading measured at the unknown elevation). The advent of pocket and now credit-card-size calculators has greatly enhanced the accuracy and speed of field calculations. However, field notes should always be subjected to detailed office tabulation once the field survey has been completed.

Figure 2-5 Sample differential leveling notes.

6

BUFF - ⊼ 9-26-78 JENKINS MINE
ROGERS - φ RUNNING LEVELS

STA.	B.S.	H.I.	F.S.	ELEV.	REMARKS
B.M.	4.75	859.22		854.47	USGS MARKER #K48
T.P.-1	2.11	855.33	6.00	853.22	
T.P.-2	3.75	855.99	3.09	852.24	
T.P.-3	4.02	854.83	5.18	850.81	
T.P.-4	4.03	852.75	6.11	848.72	

2.2.3 Mine Leveling

Vertical control in a mining operation is needed for the engineering of proper drainage and adequate haulage, not to mention general mine layout. Since entries and tunnels are used for frequent travel and haulage, the vertical control stations must be located where they will not be easily damaged or tampered with, namely the mine roof. The location of the stations on the roof, above the surveyor, creates one essential difference between surface and underground leveling operations. This difference is illustrated in Fig. 2-6. Note that the leveling rod is inverted and placed against the top; thus the backsight measurement (on the bench mark) is now subtracted from (rather than added to) the bench mark to determine the height of instrument (H.I.),

Figure 2-6 (a) Example of underground differential leveling. (b) Sample of underground leveling notes.

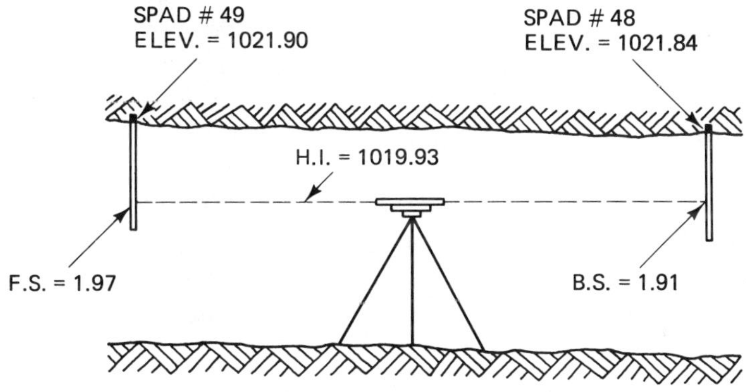

(a)

31

BUFF - ⊼ 3-10-79 JENKINS MINE
JOHNSON - φ UNDERGROUND LEV.

STA	B.S.	H.I.	F.S.	ELEV.	REMARKS
B.M.	1.02	817.48		818.50	PORTAL CHECK
SPAD 48	1.91	817.54	1.97	819.45	
49	2.20	817.31	1.97	819.51	
50	2.06	815.89	0.64	817.95	
51	2.00	814.91	1.02	816.91	

(b)

while the elevation of the unknown point is now determined by adding (rather than subtracting) the rod reading of the foresight. These calculations are summarized in Eqs. 2-3 and 2-4.

$$\text{B.M.} - \text{B.S.} = \text{H.I.} \tag{2-3}$$

$$\text{H.I.} + \text{F.S.} = \text{Elev.} \tag{2-4}$$

Whether above or below ground, the first bench mark is essential for proper vertical control. Above ground the first bench mark is established on the mine property using differential leveling to "carry" the elevation from the NGS monument to a new B.M. monument positioned on mine site. Once the surface bench marks are established, the next task is to "carry" the elevations below ground.

For slope and drift entry portals (see Chapter 4) the procedures of carrying elevations underground are relatively simple. A precise surface bench mark on the mine property is first established, which is visible from the portal opening. This surface bench mark should be tied to several regional NGS bench marks and have the following characteristics:

1. It should be permanent and of substantial construction (e.g., affixed to a foundation).
2. It should be readily visible from the first portal bench mark and in a location that will facilitate its use on the mine site.
3. It should be well described so it can be easily recovered and used. (See the description requirements for NGS bench marks in Section 2.1.1.)

Similarly the portal B.M. monument should be established using the concrete portal structure to anchor a disk or reference pin. This monument must be visible from points within the entry. Entry elevations are then determined according to the leveling procedures described above.

For shaft and steeply inclined slope portals, the problems of carrying elevations underground are somewhat more complex than for drift entries. Two bench mark monuments must be established for these cases, one embedded in the top collar of the shaft and the second embedded in the shaft foundation at the level of the main entries. The collar bench mark is located with the same precision as the portal bench mark described earlier. The second, lower bench mark must be clearly visible from both the collar bench mark and from locations along the main entries.

Several methods can be used to establish the elevation of the second bench mark. One method employs directly taping the vertical distance from the collar bench mark to the first underground bench mark. A plumb wire is used to ensure true vertical measurement. The elevation of the underground benchmark is directly calculated by subtracting the measured height from the elevation of the collar bench mark. Another measurement technique that results in the same type of calculation is the use of vertically mounted *electronic distance measuring* (EDM) equipment to measure the vertical distance. With EDM equipment, described in later sections, errors that can result from taping are avoided.

Another method, popular for steeply inclined slopes, is called *trigonometric leveling*. Trigonometric leveling entails the calculation of elevations using the measured vertical angles and the horizontal or sloped distances between the point of known elevation and that of unknown elevation. Figure 2-7 illustrates the relationship of angles and distances. The vertical angle θ is measured down the slope using a theodolite (see Section 2.4). The sloped distance (S.D.) is measured using a steel tape or EDM equipment. If the EDM equipment is used, the horizontal distance and vertical distance can be computed internally and read directly from the display. If a steel tape is used, the horizontal distance must be calculated according to Eq. 2-5.

$$\text{S.D.}(\cos\theta) = \text{H.D.} \tag{2-5}$$

where S.D. = sloped distance
θ = vertical angle measured between the horizontal and the sloped distance
H.D. = horizontal distance

The elevation of the bottom bench mark is calculated according to Eq. 2-6.

$$\text{Elev. } 2 = (\text{Elev. } 1 - \text{B.S.}) - [\text{H.D.}(\tan\theta) + \text{F.S.}] \tag{2-6}$$

where Elev. 2 = Elev. of the bottom bench mark
Elev. 1 = Elev. of the collar bench mark
B.S. = backsight
H.D. = horizontal distance
θ = angle measured between the horizontal and the sloped distance
F.S. = foresight

Another method that can be used to determine the bottom bench mark elevation is *barometric leveling*. Barometric leveling depends upon the differ-

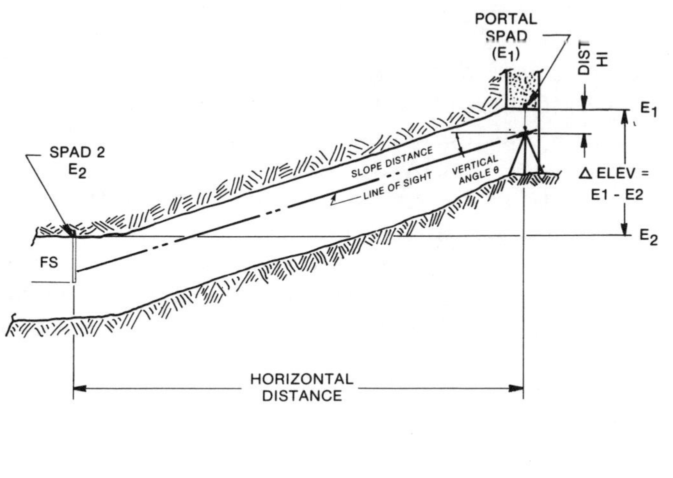

Figure 2-7 Trigonometric leveling down a slope entry.

ence in barometric pressures that occur as a result of the differences in local elevation. Since the weight of a column of air and the barometric pressure are dependent on the height of the column of air, the higher the elevation of a point, the less will be the weight and corresponding barometric pressure. This method is not recommended as the principal method for establishing the bench mark because it is highly dependent on environmental variations and less accurate than other methods.

2.2.4 Bottom Profile Control

Profile leveling is the systematic method of recording elevations of stations along a fixed path at regular intervals. According to the federal mine law,[1] the elevation of the mine floor at intervals of 200 ft or less must be shown on the mine map. To meet this requirement the survey team uses the technique of profile leveling. The elevations of the top are calculated according to the procedures described earlier, then the thickness of the seam is measured. The thickness is subtracted from the corresponding elevations to yield the elevation along the floor. To simplify the mapping process, the survey team will often locate and record whole-number elevations, since only whole-number elevations should be shown on a mine map.

2.3 HORIZONTAL CONTROL

Horizontal control implies the ability to accurately measure the position of all physical points relative to the orientation of the earth's surface. Horizontal control depends on the accuracy of angular and linear measurements. Angular measurements are departures in direction made relative to some known or assumed line on the earth. Linear measurements are straight-line distances between some unknown point and some known or assumed point of origin.

2.3.1 Reference Systems

Although any assumed direction and fixed origin could be used as a reference system for a survey, the usefulness of such a reference system would be quite limited. The most accurate and useful reference system is the network of horizontal control stations established by the National Geodetic Survey. Like the bench marks established for the vertical control network, the horizontal control monuments are first identified on the NGS Geodetic Control Diagram and a description is obtained permitting the surveyor to locate the station in the field. Three markers are used to identify the NGS monuments, two reference markers and a station marker. Reference markers function to ease the location of the station and facilitate the reestablishment of the station in the event of damage. The station marker provides the initial point of origin for distance measurements. The angular control reference line is established from the station to an azimuth marker, normally located at least 1000 ft away from the control marker.

Like the vertical control, several NGS control stations are needed to extend properly the horizontal control from the NGS network to the mine property. Several methods are used to extend surveys over long distances, including triangulation and trilateration. *Triangulation* is a technique of chaining together an arrangement of triangles. The position of the vertices of these triangles are known by accurately measuring the exterior angles and one side, the baseline. Once the baseline and angles have been measured, the lengths of the remaining sides can be calculated and the coordinates of the vertices determined.

Trilateration is another method of extending horizontal control. Instead of measuring angles, trilateration depends on accurate measurement of the distances of simple geometric configurations. Once measured, these distances are used to compute the interior angles and subsequently the coordinates of the control points. The finer points of both triangulation and trilateration are included in most surveying references.[3,4]

2.3.2 Angular Measurement

Angular measurement for horizontal control is predicated on the ability to divide a circle into discrete units. Several systems have been advanced for doing this, including the centesimal or grad system, the radian system, the mil system, and the sexagesimal or degree system.[3] For our purposes and for most surveying, the *sexagesimal system* is the most important. With this system the circle is divided into 360 equal parts called *degrees.* Each degree segment is further divided into 60 parts called *minutes*, and each minute is divided into 60 parts called *seconds.* Thus, to define an angular departure one must specify the angular difference between two lines measured in degrees, xxx°; minutes, xx′; and seconds, xx″.

Procedures. Several different methods or arrangements of measuring angular departure are possible, including (1) angle right, (2) multiple angles right, (3) deflection angle, (4) interior angles, (5) azimuth, and (6) bearing angle. The center of circle or the point from which the angular measurement will be made is the instrument station, I.S. In the first four cases the known direction of the *reference line* is established by sighting from the I.S. to a backsight, B.S. In the latter two cases the *reference line* is a true north-south line called a *meridian.* The various arrangements are described below.

Angle Right. Figure 2-8 depicts the angle-right arrangement, in which the instrument is first aligned on the B.S. and set on zero. Then the angle is turned to the right of this reference line or in a clockwise direction to the foresight, F.S.

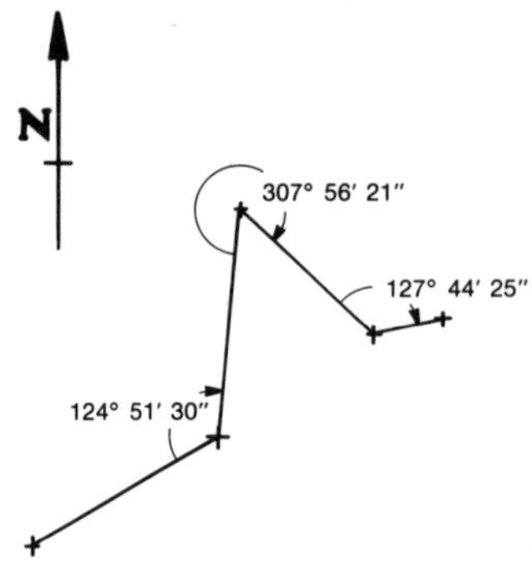

Figure 2-8 Angle-right method of angular measure.

Multiple Angles Right. This method involves the same turning direction as the angle-right method except that accumulative measurements using the same B.S., I.S., and F.S. are made. The actual angle is determined by dividing the accumulative total by the number of times the angle is turned. Only even multiples (two, four, six, etc.) of turns should be used, to avoid unwieldy decimal remainders. The advantage of multiple turning of angles right is that the reading error is averaged on each measurement and regression toward the norm or actual value occurs.

Deflection Angle. Figure 2-9 shows the deflection-angle arrangement, in which the angle from an extension of the backline to the foreline is measured either to the right or left but always at less than 180°. The right or left direction is noted as R or L following the angular reading.

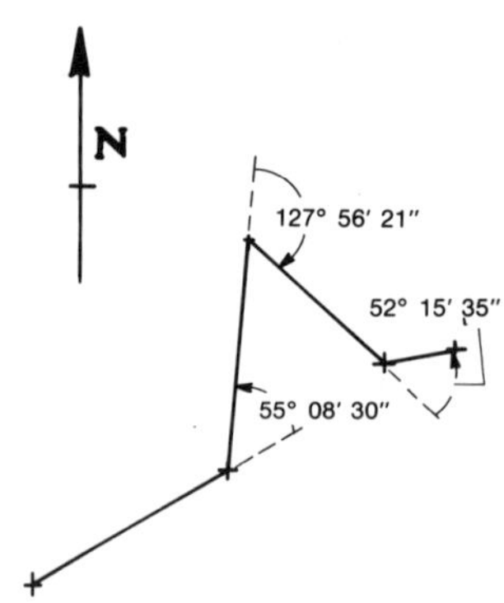

Figure 2-9 Deflection angle method of angular measure.

Interior Angles. Figure 2-10 illustrates a closed traverse in which the interior angles were used to record angular departures. The interior angle method can be used only for closed surveys.

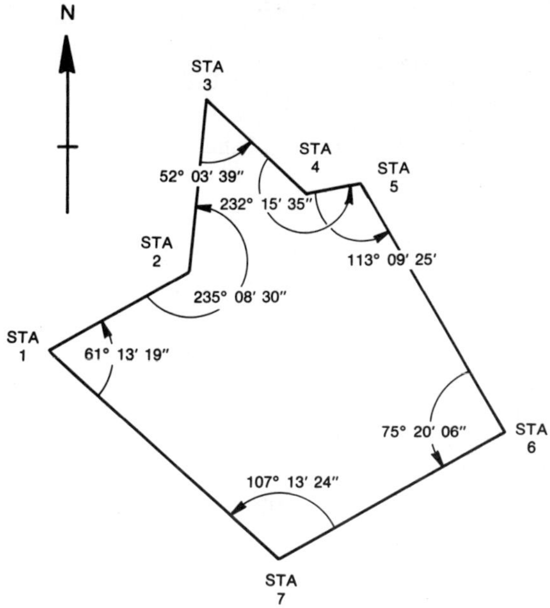

Figure 2-10 Interior angle method of angular measure.

Azimuth. Figure 2-11 shows the measurement of angles using the angular departure of the line from true north. Angles called azimuths are measured only from the meridian, beginning from the north and rotating clockwise to the foreline. The important difference between this method and those previously described is that it is referenced to the physical structure of the earth. (Note that the NGS measures azimuths starting from the south rather than the north.)

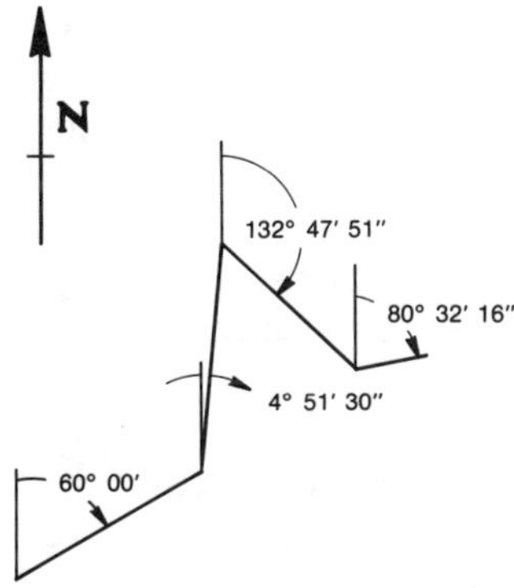

Figure 2-11 Azimuth reference method of angular measure.

Bearings. Figure 2-12 depicts the angular departure of various lines relative to a meridian measured from either the north or the south. The bearing of a line is equal to or less than 90°. Bearings are designated by N (north) or S (south), depending on where the measurement starts; followed by the angle in degrees, minutes, and seconds; followed by an E (east) or W (west) designation which shows the direction the angle is measured from the north or south.

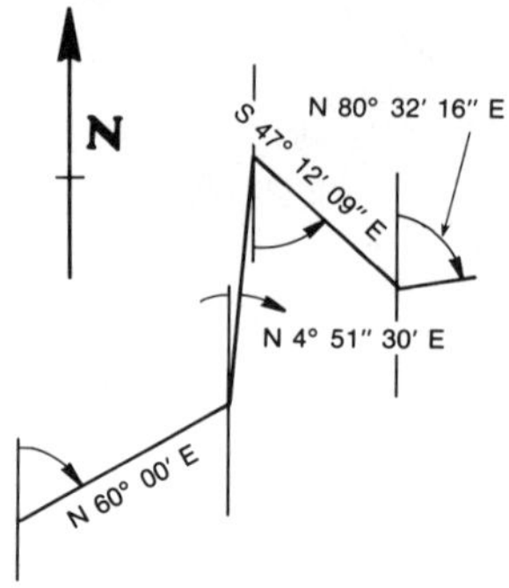

Figure 2-12 Bearing angle method of angular measure.

Mine surveying typically uses angle-right methods; however, double-right-angle measurements are often made to enhance surveying accuracy of underground traverses.[5] The reporting of angular data is usually recorded as either bearing or azimuth formats, which facilitate later calculations. Deflection angles are used for the layout of haulageways (track) and are discussed in detail in Chapter 8.

2.3.3 Linear Measurement

Like angular units of measure, distance can be measured using differing systems of units. The two major systems of units you will encounter are the *English* (E) and the *International* (SI) systems. In the English system, length is measured in feet; in the SI system, length is measured in meters. Historically measurements recorded in feet have dominated the mine surveying field; however, because of the simplicity of SI units, many leaders in the field are promoting the conversion from English to metric measures. The Metric Conversion Act of 1975, which makes the voluntary conversion to the SI system of units a national policy, has done much to assist these efforts. The mathematical conversion from feet to meters is accomplished by multiplying the number of feet by 0.3048. Field conversion necessitates the replacement of instrument scales, and since this is expensive it has met with a great deal of resistance. Since you must be proficient in the use of both English and SI systems, this text will employ both in problems and examples.

Distance is measured using either a steel tape or an electronic distance measuring instrument. The steel tape (described later) is the more traditional instrument. The advantages of the steel tape compared to EDM equipment are its low cost, ease of short-measure setup, and observable accuracy errors (such as slack tape). The major advantages of EDM equipment are its speed in longer measurements and high accuracy. Although the errors affecting EDM instruments are well known, they are not observable. Thus the instrument man must be very careful to program into the instrument the proper environmental and other adjustments prior to making distance readings.

2.3.4 Mine Property Traverses

Three methods of surveying a property are triangulation, trilateration, and traversing. The first two methods were discussed as the best methods for extending the NGS network or horizontal control to a mine site. The best method for surveying property, a route, or mine excavation is traversing. A *traverse* is a consecutive series of lines whose directions (relative to a meridian) and lengths between stations are measured and recorded. A *station* is any point of known orientation and position (established by previous surveying, such as triangulation) which is used to determine the orientation and position of another point.

Open traverses produce a series of lines and stations that do not have common points of closure. *Closed traverses* eventually close on the point of origination. An example of the first is a mine entry or highway survey; an example of the latter is a property perimeter survey.

The traverse of the mine property begins from the triangulation baseline and stations that have been extended from the local NGS network. This system serves to orient the meridians and subsequent lines and stations. Assuring the correct orientation of the meridians is an important task, since it is on this true north-south line that all surveying and mapping depends. Most surveying textbooks[3–5] describe astronomical checking of baseline meridians and should be consulted as to proper procedures.

The greatest likelihood of meridian error occurs when the meridians are extended underground. A great deal of care should be exercised during this phase of the survey since once the meridian is below ground, it is impossible to perform an astronomic check of its orientation. As with the establishment of vertical control underground, the extension of meridians underground is most troublesome in the case of shaft entry systems. Staley[5] and Frush[6] both provide discussions of several methods for carrying meridians underground. Two of these methods are coplaning and north-seeking gyroscopic techniques.

Coplaning uses two carefully stabilized plumb bobs suspended in the shaft to establish a reference line. The azimuth of this line on the surface is assumed to be the same underground. The azimuth of the plumb line above ground is measured with the instrument aligned with the plumb line, as shown in Fig. 2-13. The B.S. used to orient the instrument is ideally a permanent monument on the mine site tied to the state plane coordinate system. The angle between the B.S. and the plumb line is established using the multiple-angle-right technique described earlier. Once the azimuth of the plumb line is determined, the instrument can be taken underground. As on the surface, the instrument is aligned with the plumb line and the initial stations 1 and 2 established.

The principle of *gyroscopic* techniques is that the spinning gyroscope

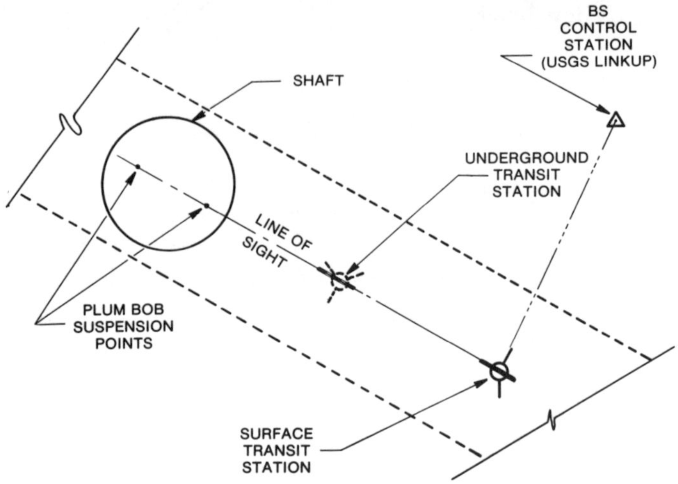

Figure 2-13 Transfer of meridian down shafts: two-wire method.

is influenced by the spinning of the earth such that the axis of the gyroscope aligns with and oscillates about the geodetic meridian. Several gyro attachments are commercially produced which provide a ±20″ standard deviation, and some special commercial types provide errors as low as a few seconds.[3]

Using the gyro-theodolite, the meridian underground can be established very quickly without the awkward and time-consuming procedures required of the plumbing operation. The only limitation that adversely influences the common adoption of the gyro-theodolite is the higher cost of the attachment. An added advantage which somewhat offsets the cost is the fact that the gyro-theodolite can operate anywhere underground. Thus the meridian can be checked or reestablished in locations away from a portal, which in many mines may be as much as 10 miles.

Underground traversing. Once the transit or theodolite has been oriented to the reference meridian, the traversing of the surface and underground property can begin. Backsights are taken on previous stations from the occupied traverse station to reorient the instrument following each new setup. The angle right is turned to the forward station and either the azimuth bearing or angle right is recorded.

Underground traversing differs from surface methods in that the instrument is set up *under* rather than *over* the station. The survey stations used underground are called *spad locations.* All spads are theoretically located in the centerline of the entry and are permanently positioned in the mine roof. The spad is installed by first drilling a $\frac{1}{2}$- to 1-in. hole in the top and securing a wooden plug in the hole. The spad (see Fig. 2-14), together with an identification tag, is then driven into the plug. The methods of spad designation are discussed in Chapter 4.

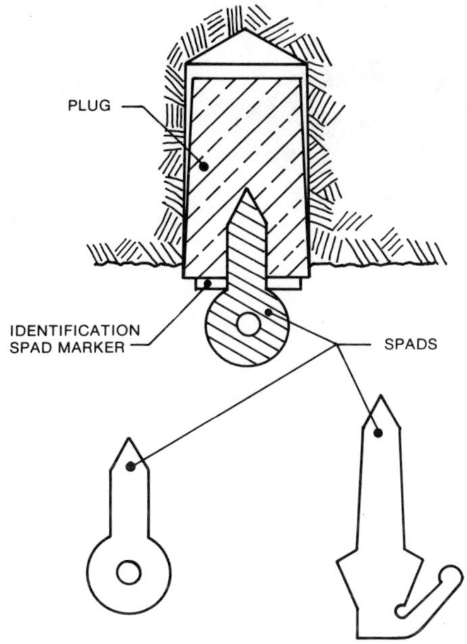

Figure 2-14 Typical spad installation.

2.4 PHOTOGRAMMETRY

Photogrammetry is a system of combined vertical and horizontal control using aerial photographs to obtain the needed measurements. Photogrammetry has been used primarily for topographical mapping of large areas and is well suited for mine site reconnaissance mapping. Photogrammetry, the subject of many texts,[7,8] is discussed briefly in Chapters 4 and 6.

2.5 INSTRUMENTATION

The fundamental design principles of the surveying instruments have changed little over the past years (with the exception of EDM equipment); however, the reliability and accuracy have been significantly improved. The principal instruments used for surveying are the steel tape, the transit, the theodolite, the level, the leveling rod, electronic distance measuring equipment, and field data processing equipment.

Steel Tape. Historically, measuring distance by a steel tape has been the only reliable method. Although newer developments have advanced additional methods, taping remains the most common distance-measuring method. Figure 2-15 shows a typical 100-ft steel tape. Tapes vary in material composition, accuracy, and the system of units (SI or English) used.

Transit. The engineer's transit is the traditional instrument used for

Figure 2-15 A 100-ft steel tape.

determining angular orientation. Figure 2-16 shows a typical engineer's transit, consisting of a telescope suspended between two standards, permitting the telescope to be tilted through the vertical plane. The standards are mounted on a circular base plate. The upper plate is permitted to rotate within an additional lower plate, thus describing a horizontal circle. The lower plate, which can also be made free to rotate, is attached to the leveling head. The leveling head supports the instrument, attaches to the tripod, and enables the surveyor to adjust the level of the instrument. Several types of transits are available with differing accuracies, magnifications, eyepiece and crosshair configurations, and so forth. Additional references and manufacturers' literature should be consulted before selecting a transit suitable for your purposes.

Theodolite. The theodolite is the newest innovation in the evolution of the transit, resulting from advances in optical theory and manufacturing. The theodolite is a highly accurate optical transit which permits the reading

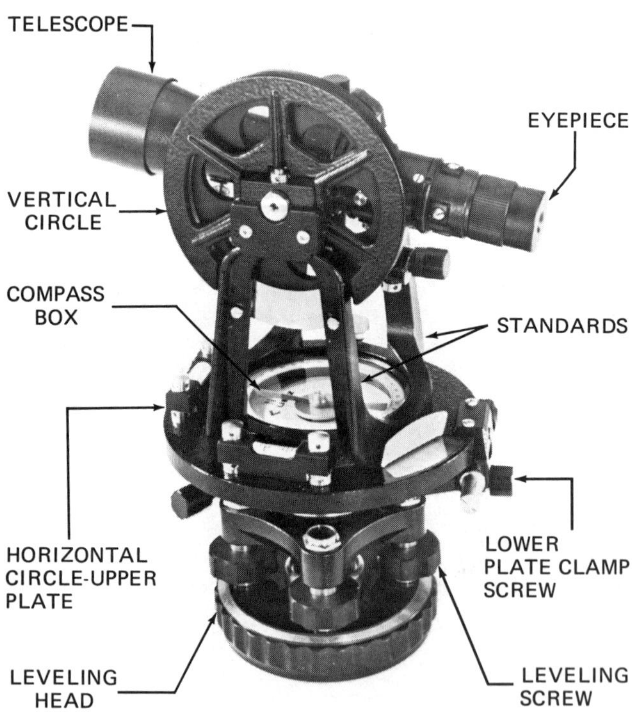

Figure 2-16 Engineer's transit.

of both horizontal and vertical circles through a side-mounted eyepiece. A typical theodolite used for mine surveying is shown in Fig. 2-17. The advantage of the theodolite over the transit is that the instrument is more accurate, more compact, lighter, easier to set up (since most theodolites are equipped with self-leveling compensators and have an optical plummet, replacing the need for plumbing), totally enclosed, and easier to read (all readings are made through the same eyepiece). Additionally, theodolites can be tied to electronic data-processing and recording equipment. Because of these advantages, most underground and surface mining operations today use the theodolite for their surveying needs.

Level. Various types of levels are available. The most recent advancement in leveling instruments has been the automatic level, shown in Fig. 2-18. An internal compensator tied to the direction of gravitational force automatically maintains a horizontal line of sight. The advantage of such a device is that the speed of setup is greatly enhanced. All that is required is a rough leveling of the instrument and the freeing of the compensator, which brings the instrument into a true horizontal plane.

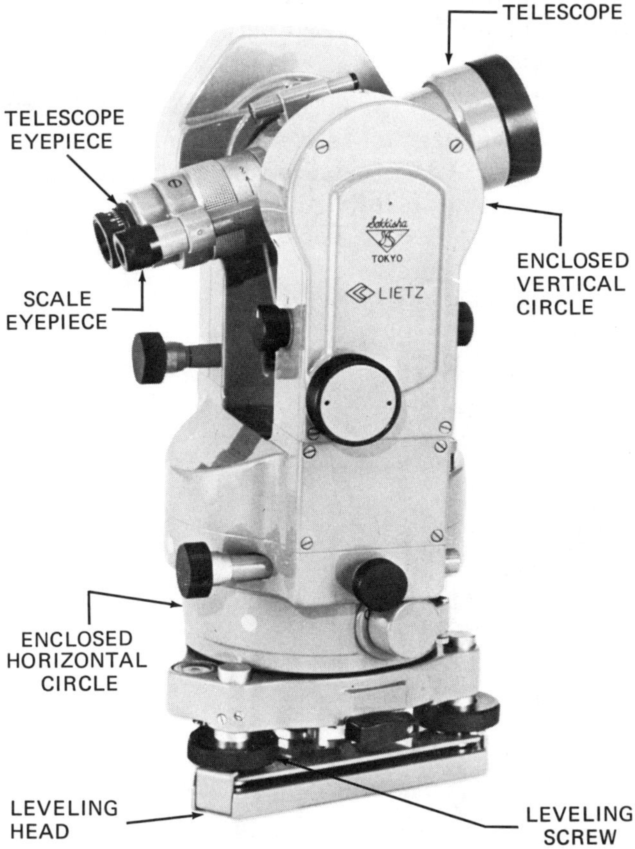

Figure 2-17 Theodolite.

Figure 2-18 Automatic level. (Courtesy of Lietz.)

Leveling Rod. Modern leveling rods are essentially unchanged except that lighter and more durable materials have increased the useful life of a typical rod. Figure 2-19 shows one type of a leveling rod. Like steel tapes, leveling rods can be purchased in a variety of graduations, colors, and materials. Manufacturers' literature should be consulted before selecting the rods that best meet your particular needs.

Electronic Distance Measurement. EDM equipment is perhaps the most revolutionary advance that has been made in surveying since the development of the first reliable transit. EDM is based on the principle that electromagnetic radiation (whether radio waves, light waves, or heat waves), broadcast over some distance and reflected back, will exhibit waveform

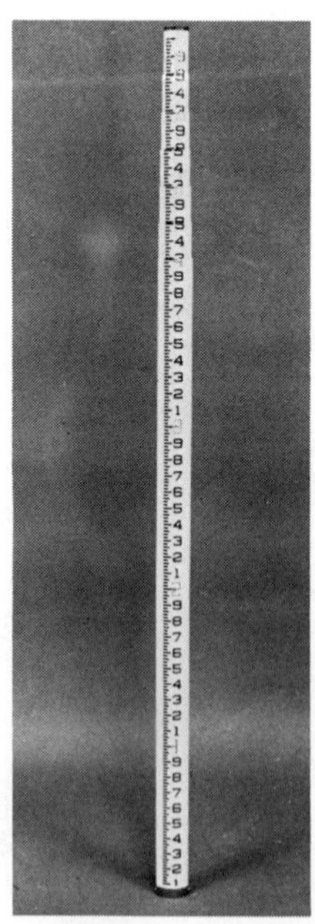

Figure 2-19 Lightweight leveling rod. (Courtesy of Lietz.)

phase displacement. The difference between the waveform being broadcast and that being reflected is proportional to the distance traveled. The EDM equipment compares the two waveforms, makes adjustments for the environmental factors and for the doubling of the distance, and automatically computes the distance measured. Figure 2-20 depicts a typical EDM accessory mounted on a theodolite. The setup of the EDM requires less than 2 minutes; thus even for short distances a measurement can be taken in the same amount of time that correct taping would require. Over long distances, and across rivers, lakes, or valleys, the EDM has no equal.

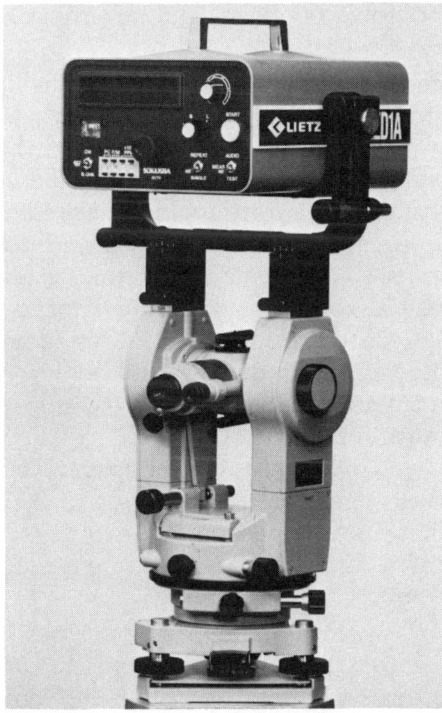

Figure 2-20 Electronic distance measuring equipment. (Courtesy of Lietz.)

Field Data Processing Equipment. Another significant advancement in surveying procedures has been the introduction of field electronic data processing (FEDP equipment. FEDP can be tied to both the EDM equipment and the theodolite to compute, store, and display a variety of surveying data in the field. The same data once required hours of laborious office computation. One such device uses the vertical angle, sloped distance, and horizontal angle to yield horizontal distance, vertical distance, and even station coordinates. Some manufacturers are providing cassette recorders, which record the raw surveying data. The tapes are then taken to the office, where data analyzers process the data. These processors can be interfaced with printers to provide a hard copy and even tied to plotters to generate maps. The application of such equipment is bounded only by the imagination of those using it.

2.6 NOTE TAKING

Regardless of the advent of FEDP equipment, the bases of all computations and subsequent mapping operations are the field notes. All notes taken in field, whether for vertical, horizontal, or side feature control, must meet certain standards. Other than correctly setting up the instrument and executing the proper procedures, the most critical task of the survey leader is to prepare field notes which are:

1. *Intelligible* without the need for extensive oral or written explanations. This requires that the notes be clear, legible, concise, and in a standard format.
2. *Valid* in representation of field findings. Validity is based on the accuracy of the record and the truthfulness of the survey leader. Notes must not be created, only recorded. If data are missing, the notes should reflect this. When mistakes are made they should be crossed out, never erased, and if necessary a side note included explaining the error. All data that may be significant should be logged and documented. Never depend on recollection as a means of storing valuable information. Pocket engineering calculators have greatly enhanced the accuracy and speed of field calculations. However, field notes should always be subjected to detailed office checking once the survey has been completed.
3. Contain correct *identification.* Identification of the survey party members includes the transit operator (identified by the symbol ⩚), the rodman (identified by ϕ), the notekeeper (identified by N), and the chainmen (HC for head chainman and RC for rear chainman).[9] Additional information that may be needed is the location or site identification; day, month, and year; time of survey; environmental data; and instrument identification.
4. Correct use of *representation.* The various types of representative format include (a) verbal representation—written descriptions, (b) graphical representations—drawings and maps, (c) symbolic representations—the use of special configurations to represent a real-world feature or occurrence, and (d) mathematical representation—the use of numbers and mathematical formulas to communicate data.

2.6.1 Vertical Control Notes

Differential leveling field notes consist of compiling a tabulation of the measurements taken *and* of field computations made regarding the elevations. An example of the format commonly used is shown in Fig. 2-5. As a minimum these notes always contain the station or turning point identification, the backsight (B.S.) measurement, the height of instrument (H.I.) calculation, the foresight (F.S.) measurement, and the elevation (Elev.) calculation of the point of interest.

Underground elevation notes differ from surface notes only in the method of calculating H.I. and elevation. Since the stations are on the roof and the rod is inverted, the H.I. is the B.S. plus the B.M. and the elevation of the forward station is the H.I. minus the F.S. Figure 2-6(a) illustrates underground elevation notes.

2.6.2 Horizontal Control Notes

Field data required to establish horizontal control include station identification code (letter, numeration, etc., of the occupied station to forward or rear station designation), bearing angle of this designated line, and the horizontal distance of the line (or the sloped distances and vertical angle measurements). An example of this minimum set of data is illustrated in Fig. 2-21. The horizontal control notes underground do not differ substantially from this format except that they usually denote rib and crosscut locations between stations by side notes.

STA	AZIM.	BEARING	VERT ANGLE	SLOPE DIST.	HORIZ DIST.	STA

Figure 2-21 Traverse field data.

2.6.3 Underground Side Notes[5,6,10]

In addition to the minimum field notes, data, and methods previously discussed, the mine surveyor must include certain *side notes* which are essential to accurate layout of mine excavations. Side notes include excavation features, seam notes, and notes about faults, slips, contacts, anomalies, and slope dimensions.

The most important side notes that must be kept are those fixing and locating the positions of ribs, corners, crosscuts, and other structural features relative to the theoretical entry centerline. Several methods exist for doing this, including the offset method and angle right.

Offset side notes are illustrated in Fig. 2-22. The first line of these notes[11] can be interpreted as follows. "2-8" is the centerline from spad number 2 to spad number 8 measured at an azimuth of 53°20′, a vertical angle of 0°18′, and a sloped distance of 138.18 ft. The second line is interpreted as follows: "+4" means a distance of 4 ft along the centerline starting at station 2 and moving (in) toward station 8; "R6 L9" means perpendicular

BUFF - ⊤ ROGERS - ϕ JOHNSON - H.C.			3-15-79		JENKINS ADVANCE PTS		38
2-8	53°20′		0°18′	138.2	H=138.2		
	+4	R6	L-		COR	R+	
	+8	R6	L7		COR	LT	
	+27	R5	L7				
	+58	R7	L6	+76	R7	L6	XCT R+/LT
	+112	R7	L6				

Figure 2-22 Example of underground survey notes taken in offset form.

to the centerline there exists a pillar wall or rib to the right 6 ft away and a rib to the left 9 ft; "Cor Rt" is a corner formed by the intersection of a crosscut and entry and means a rib corner to the right, while to the left is a point on the rib. The third line, "+8 R6 L7 Cor Lt," means that 8 ft from spad 2 moving (in) toward spad 8 at a location 6 ft to the right and 7 ft to the left there exists a corner on the left and a rib on the right. Figure 2-23 shows how these notes are plotted on the mine map.

This form of side noting, locating the rib from the entry centerline, is called offset side noting and is the most popular method of detailing room and pillar excavations. Other formats and methods of side noting include map designation of offsets and the angle-right method. With the *map desig-*

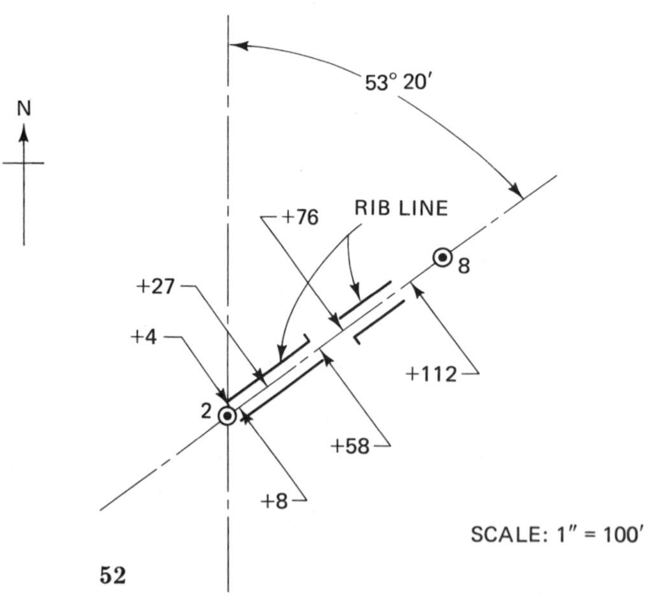

Figure 2-23 Map plotting using offset rib notes. (1 in. = 100 ft.)

nation format, the offset details are placed directly on an idealized room and pillar map, as shown in Fig. 2-24.[12] With the *angle-right* method, the transit is turned from a B.S. to predominant features on the rib. By recording the angle right and the distance from the occupied station to the feature, the necessary rib features can be located. This method is illustrated in Fig. 2-25.

STA.	AZ.	VERT ∡	SLOPE DIST.
100-101	282°30'	-0°36'	136.92
101-102	284°45'	-0°14'	142.97
102-103	284°26'	-0°03'	143.03
103-104	285°18'	-0°31'	135.89
104-105	282°41'	-0°01'	138.17
105-106	285°52'	-0°08'	145.18
106-107	280°30'	-0°19'	138.88
107-108	284°46'	-0°31'	135.85
108-109	284°20'	+0°24'	141.66

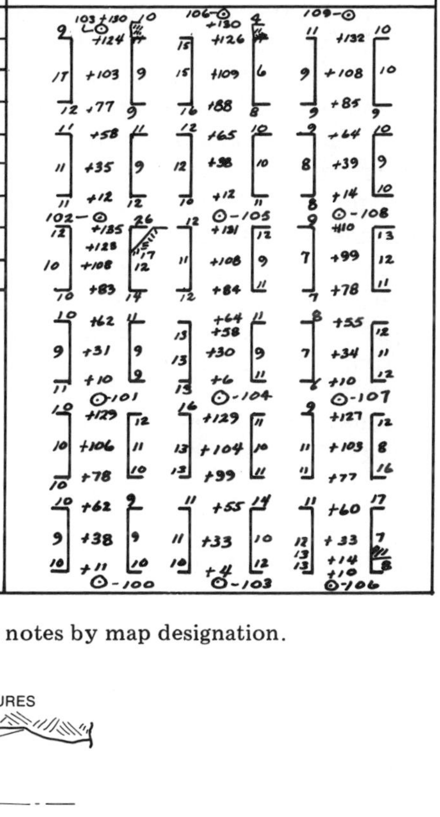

Figure 2-24 Offset notes by map designation.

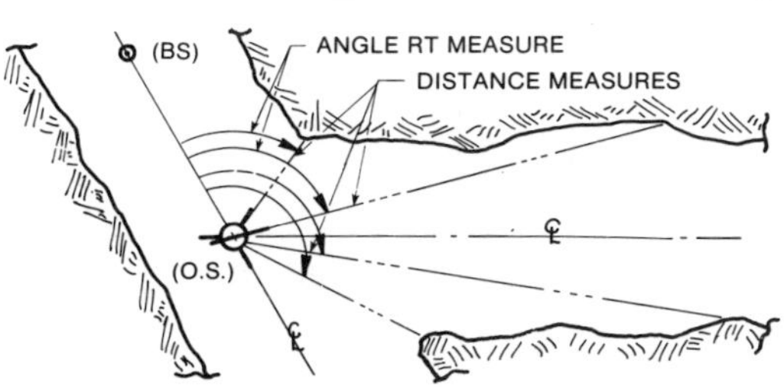

Figure 2-25 Angle-right measurement of rib positions.

2.6.4 Engineer's Notebook

Each engineer, survey team leader, lead technician, or other technical professional should maintain a personal record of daily activities, ideas, and computations. Such a notebook is a daily logbook of what was done during the course of the project and serves as a historical record, providing both ethical and legal documentation (provided that the notes have been properly recorded, witnessed, and dated).

The types of data that are included in each entry vary depending on whether the specific entry is a logging of general activities, recording of an idea or invention, computations for a specific project, or whatever. In general those data that are always included in an entry are (1) title of the entry, (2) date of entry, (3) a sketch and title of sketch, (4) assumptions or reference materials, (5) the entry (computation, activities, etc.), and (6) statement of recommendations. Additionally, ideas, inventions, and certain computations should be properly witnessed by an unbiased party who has the technical expertise to comprehend the entry.

Technical computations supporting other entries constitute the most numerous and important recordings found in the notebook. To facilitate the recording of the entry and the possible future review of these calculations, most engineers use a format similar to that shown in Fig. 2-26. The entry should be *TITLED*, *DATED*, and *SIGNED*. The data that were measured or otherwise obtained in field, or which were previously known, should be condensed and recorded under the title "*Given*." "*Sketches*" of the situation should be included and significant factors highlighted. Next, the information that must be determined is enumerated under the general heading "*Find*." Next the assumptions, referenced data, and equations should be

Figure 2-26 (a) Computation format for engineer's notebook. (b) Idea disclosure format for engineer's notebook.

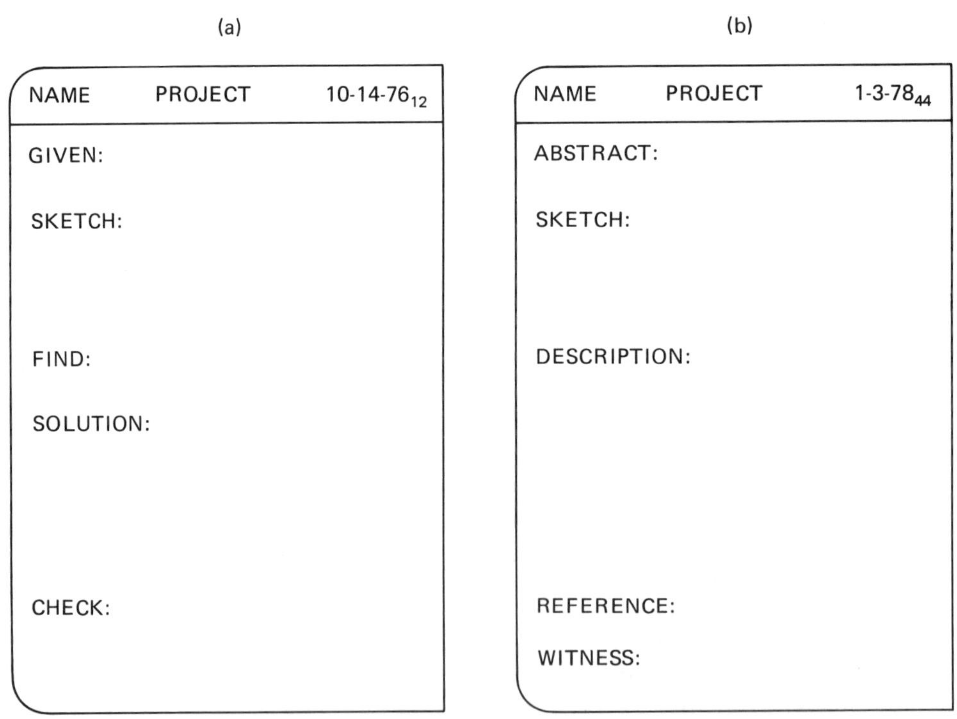

noted under "*Reference.*" Next, the calculations of "*Solution*" should be compiled. All calculations and steps should be included. Finally, the recalculation or "*Check*" should confirm the previous calculations. It should be noted that this check is not a mere confirmation of mathematical operation but is rather an alternative method of solution being applied to these data. (For example, an analytical solution may be checked using a graphical solution.)

Most standard logbooks used as the engineer's personal notebook are hard-bound with nondetachable pages. This type of notebook is best for daily activity recordings and for legal purposes (since pages cannot be removed without being noticeable). However, many engineers prefer the convenience of loose-leaf notebooks, since the calculations can be grouped according to project or type of calculation (e.g., hydraulic, surveying, mechanics, etc.) and separated from daily activities schedules, which must be kept according to chronological progression. Regardless of the type of notebook used, the entries must be kept clear, concise, legible, well documented, and as needed, witnessed. To prepare for this future role in the engineering profession, you should begin to compile all homework assignments in a single notebook using either the format shown or another suitable format that will organize your efforts.

2.7 MEASUREMENTS AND DATA

Regardless of the care with which instruments have been set up or the years of experience of the recorder, *no measurement taken in the field will be exact.* Many types of errors contribute to the inaccuracy of the reading, including instrument errors and environmental influences (e.g., temperature, pressure, humidity, variances in gravitational force, etc.), contributing to recording error and human error. The first two types of errors, instrument and environmental errors, are treated in detail by most standard surveying texts. However, your understanding of human error is of importance to detailed mapping and mine drafting.

Human recording error is not concerned with human mistakes, that is, mistakes that can be reduced through better attention to detail, better instrument procedures, better setup of procedures, and so forth, but is that variation that exists between two people reading the same setting. The standard surveying textbooks all include chapters treating the subject of accuracy and error and should be referred to for their more detailed presentations. The importance to subsequent mapping is that those methods of control that treat each point independently from adjacent points should be used, thereby avoiding the accumulation of error resulting from constant high or low readings. For surveying, these methods include use of the UTM or state plane coordinate systems, and for cartography and drafting, these methods include rectangular coordinate positioning.

2.7.1 Significant Figures

Since the introduction of electronic pocket calculators in the early 1970s, the use of data containing five or more known digits has presented little or no problem. Most calculators operate with a capacity of eight significant figures (those digits whose value is known) and automatic floating decimal, and therefore the rounding or truncation error (that error that results when the calculator automatically rounds or drops those numbers containing more than eight significant figures) is quite small in comparison to the error attributable to instrumental, environmental, and human causes.

Unless a computer is employed, the rule of thumb is to use or carry computations to one more significant figure than the data or one more place than can be accurately used. For example, an average draftsman can accurately scale a dimension to ± 0.01 in.; thus a computed dimension of 0.49218 in. need be carried only to the 0.xxx place, three significant figures, or 0.492 in.

The example above also illustrates the procedure of "rounding off" data. Rounding off depends on the use of the data. That is to say, in drafting, machining, and other engineering activities, the use of even integers is preferable to the use of odd numbers. Thus for engineering calculations the following rule is used to determine the significant figure in the Nth place:

> Those values in the $N+1$ place that are less than 5, round to the next lower value; those values in the $N+1$ place that are greater than 5, round to the next higher number; and those numbers in the $N+1$ place that are exactly 5, round to the nearest (up or down) *even* number.

Example 1

(a) 684.431 (to five significant figures) = 682.43
(b) 9.12643 (to three significant figures) = 9.13
(c) 8.525 (to three significant figures) = 8.52

2.7.2 Types of Computations[3–5,8]

Various types of computations are needed to reduce survey field notes into data suitable for mine mapping and design. These include traverse calculations and calculations of areas and volumes.

Traverse calculations. After the field notes have been gathered, the next step in the engineering process is to reduce these notes to data that are more amenable to mapping and design. Specifically, the coordinates of the survey stations and spads must be determined. To calculate the coordinates most practitioners use a traverse computation sheet similar to that shown in Fig. 2-27. This sheet is merely a convenient chart for recording the field and calculated data. On this sheet the station or vector between stations appears

Vol. and Page	Station Corner	HOR. ANGLE		Slope Distance	Meas. Angle	Hor. Distance	LATITUDE		DEPARTURE		CO-ORDINATES		D.M.D.	DOUBLE AREAS	
		Azimuth	Quadrant				N	S	E	W					

COMPUTED BY ______ DATE ______ ACCOUNT/LOCATION ______

CHECKED BY ______ DATE ______ TYPE OF COMPUTATION ______

Figure 2-27 Example of traverse computation sheet.

in the first column, followed by the azimuth, bearing, vertical angle, slope distance, and the first calculated value, horizontal distance (assuming, of course, that the horizontal distance was not measured directly). Figure 2-28 shows a typical sloping profile, the sloped distance, the vertical angle, and the desired horizontal distance.

The calculation of the horizontal distance follows:

$$\text{H.D.} = \text{S.D.}(\cos\theta) \tag{2-7}$$

where H.D. = horizontal distance
S.D. = sloped distance
θ = vertical angle

Thus for the first line in the traverse (Fig. 2-29);

$$\text{H.D.} = 344.66(\cos 1^\circ 30')$$
$$= 344.54178$$

Latitudes and Departures. Following the calculation of the horizontal distance, the next step in this data-reduction process is to calculate the latitude and departure of the line. A *latitude* of a line is the projection of that line on the reference meridian. The *departure* of a line is the projection of the line on an east-west line perpendicular to the meridian. Figure 2-29 shows a diagram of the first line in the traverse. By convention, latitudes to the north are assumed positive, those to the south are assumed negative; departures to the east are assumed positive, those to the west are assumed negative.

The calculation of the latitudes and departures of lines in a traverse are mathematically summarized as follows:

$$\text{latitude} = (\text{horizontal distance})\,[\cos(\text{bearing angle})] \tag{2-8}$$

$$\text{departure} = (\text{horizontal distance})\,[\sin(\text{bearing angle})] \tag{2-9}$$

For the first line in the traverse:

$$\text{lat.} = 344.54(\cos 25^\circ 30') = 310.97824$$

$$\text{dep.} = 344.54(\sin 25^\circ 30') = 148.32902$$

Linear Error of Closure and Precision. The next computation (for a closed traverse) is to determine the extent of the *linear error of closure* (LEC). For an ideal or perfect (closed) traverse the sums of the latitudes and departures of all lines will equal zero. For this example, however, an error of 0.53 in the departure and 3.96 in the latitude is calculated. The linear error of closure is the straight-line difference between the origin and closing points of the traverse and is calculated from the Pythagorean theorem as

$$\text{LEC} = \sqrt{(\Sigma\ \text{lat.})^2 + (\Sigma\ \text{dep.})^2} \tag{2-10}$$

where LEC = linear error of closure
Σ lat. = summation of latitudes
Σ dep. = summation of departures

STA. NO.	AZIM. ANG.	BEAR. ANG.	VERT ANG.	SLOPE DIST.	HOR. DIST.	LATITUDE		DEPARTURE		COORDINATES			
						N	S	E	W	LAT.		DEP.	
1-2	334°30′	N25°30′W	+1°30′	344.66	344.54	310.98			148.33	+310.98			−148.33
2-3	287°05′	N72°55′W	−2°1′30″	255.315	255.16	74.96			243.90	+385.94			−392.23
3-4	200°13′	S40°13′W	+6°20′	339.34	337.27		257.54		217.77	+128.40			−610.00
4-5	161°58′	S18°02E	0°0′	291.14	291.14		276.89	89.97			−148.49		−520.03
5-1	74°29′	N74°29′E	−3°58′	541.55	540.25	144.53		520.56			−3.96	0.53	
							3.96	0.53					

COMPUTED BY: ______________ DATE: ______________ PROJECT

CHECKED BY: ______________ DATE: ______________ ACCOUNT

Figure 2-28 Example of traverse commutation sheet.

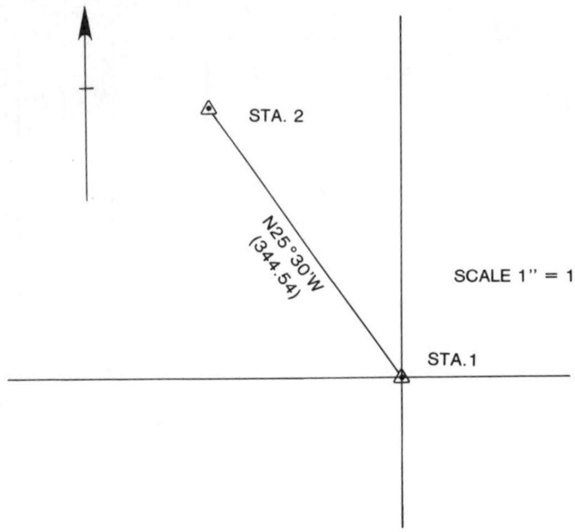

Figure 2-29 Plot of time 1–2 from Fig. 2–28 traverse.

The ratio of the LEC to the total length of the traverse provides a measurement of the precision of the survey. For the example above, a LEC is equal to 3.995 ft measured over 1768.36 ft, or a precision of 1/442.6. This may or may not be an acceptable error. If it is not, a resurvey is in order. If it is acceptable, the traverse should be balanced.

Balancing. Traverse balancing is a process of distributing the error throughout the traverse and assumes that the error randomly and equally occurred during each of the individual measurements made during the course of the traverse. The effect of balancing, a distortion treatment, is to make the summation of latitudes and departures in a closed traverse equal to zero. If the surveyor noted problems during the survey, then instead of equal distribution of error, the error distribution may be more accurately weighted on those measurements.

Several distortion treatments exist for traverse balancing, but the simplest is the *compass rule* method. The compass rule assumes that the correction made to individual latitude or departure measures is equal to the ratio of the corresponding closure to the perimeter multiplied by the length of the side. Mathematically,

$$C_{1\text{L}} = \frac{\Sigma \text{ lat.}}{\Sigma \text{ H.D.}_i} (\text{H.D.}_1) \tag{2-11}$$

where $C_{1\text{L}}$ = correction in the latitude of the first measurement
Σ lat. = total latitude closure error
Σ H.D.$_i$ = summation of the individual horizontal distances
H.D.$_1$ = horizontal distance of the first measurement

Similarly, the correction for the departure in the first measurement is stated as

$$C_{1\text{D}} = \frac{\Sigma \text{ dep.}}{\Sigma \text{ H.D.}_i} (\text{H.D.}_1) \tag{2-12}$$

For the example, then,

$$C_{1L} = \frac{3.96}{1768.36}(344.54)$$
$$= 0.7715$$
$$C_{1D} = \frac{0.53}{1768.36}(344.54)$$
$$= 0.1032$$

Care must be taken to ensure that corrections in latitude are subtracted from southerly measurements and added in the case of northerly measurements, since the total error is to the south. Similarly, departures are corrected by subtracting corrections from easterly measurements and adding corrections to westerly measurements.

Coordinates. The next step is to compute the value of the x-y coordinates of each survey station. The first station will either have known coordinates or will act as an assumed origin with coordinates of $x_1 = OY_1 = O$. To determine subsequent station coordinates, the following can be generally stated:

$$X_2 = X_1 \pm \text{departure of line 1-2} \tag{2-13a}$$
$$X_i = X_{i-1} \pm \text{departure of line } (i-1)-i \tag{2-13b}$$

or

$$Y_2 = Y_1 \pm \text{latitude of line 1–2} \tag{2-14a}$$

or

$$Y_j = Y_{j-1} \pm \text{latitude of line } (j-1)-j \tag{2-14b}$$

Again, for a closed traverse the last point should have the same coordinates as the starting point.

Areas and volumes. The next stage of computation usually entails calculations of the areas and volumes defined by the survey. The simplest method available for determining the area defined by the survey traverse is computation by coordinates. The procedure for applying this method requires that the plot be a closed traverse bounded by straight lines. The procedure is defined by Eq. 2-15.

$$A = \tfrac{1}{2} \Sigma\, Y_i(X_{i+1} - X_{i-1}) \tag{2-15}$$

where A = area

Y_i = ordinate or latitude coordinate of station i, where i = 1, 2, 3, 4, 5, and 6

X_{i-1} = departure coordinate of a station in advance of i

X_{i-1} = departure coordinate of back sight of station i

Example 2

For the six-station plot shown in Fig. 2-30, Eq. 2-15 is written

$$A = \frac{1}{2}\left(Y_1(X_2 - X_6) + Y_2(X_3 - X_1) + Y_3(X_4 - X_2) + Y_4 (X_5 - X_3) + Y_5(X_6 - X_4) + Y_6(X_1 - X_5)\right)$$

Substitute:

$$A = \frac{1}{2}\left(\cdot\ 0(517.2 - 41.8) + 234.7(1114.6 - 0) + 53.5(980 - 517.2) + (-342.9)(403.6 - 1114.6) + (-501.4)(41.8 - 980.3) + (-314.2)(0 - 403.6)\right)$$

$$= 563766.65 \text{ ft}^2$$

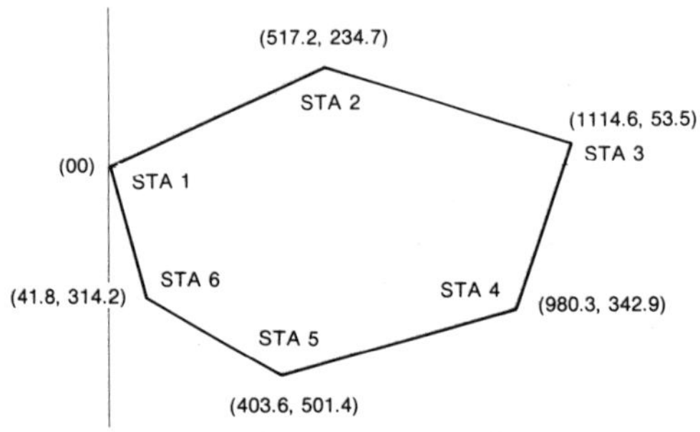

Figure 2-30 Closed traverse.

Another less analytical, although more involved, method for computing areas entails the subdivision of the plot into familiar geometric shapes such as squares, triangles, rectangles, trapezoids, and so forth. Using the known distances and angles, the area of the traverse can be determined by summing the areas of such subdivision:

$$A = \Sigma A_i \tag{2-16}$$

where A = total area

A_i = individual areas.

Such a method is illustrated in the following example (Fig. 2-31).

Example 3

$$A = A_A + A_B + A_C + A_D + A_E + A_F + A_G$$

$$= \frac{1}{2}(148.33)310.98 + 392.23(310.98 + \frac{1}{2}(392.23)(74.97) + \frac{1}{2}(217.77)(257.54 + (127.80)(128.40) + \frac{1}{2}(89.97)(276.89) + \frac{1}{2}(520.56)(144.53)$$

$$= 254{,}268.2 \text{ ft}^2$$

Often, the plot is not bounded by straight lines. For these plots two methods are commonly employed for computing the area contained within the traverse, the trapezoidal method and Simpson's one-third rule. The *trapezoidal* method, also called the offset method, requires that the property be divided into a series of "trapezoids" of width b, as shown in Fig. 2-32.

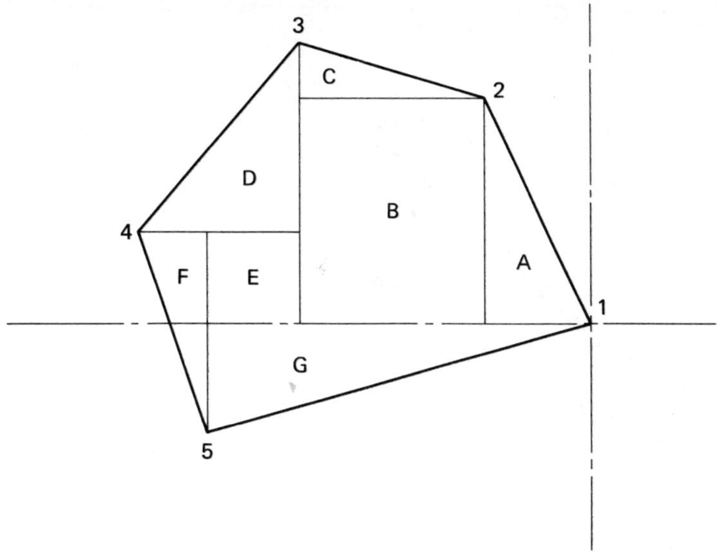

$$A\ TOTAL = A_A + A_B + A_C + A_D + A_E + A_F + A_G$$

Figure 2-31 Computing the area of Fig. 2-11 traverse.

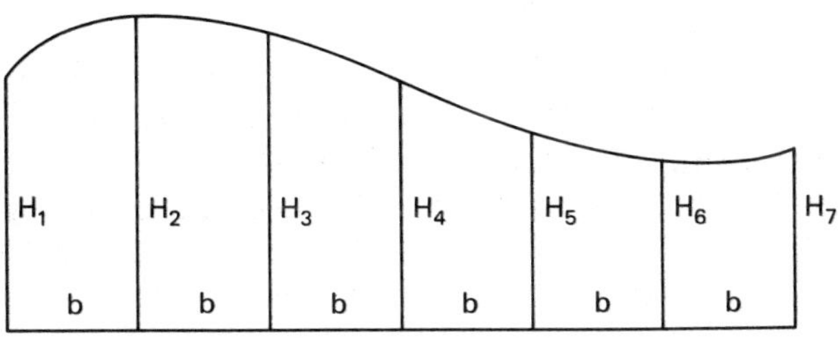

$$A\ TOTAL = b\left[\frac{H_1 + H_7}{2} + (H_2 + H_3 + H_4 + H_5 + H_6)\right]$$

Figure 2-32 Division of an irregular area into trapezoids.

The rationale assumes that b is small enough so that the variation of the end lines from a straight line drawn through the centerline of the curve is insignificant. The area of the first trapezoid is $b(h_1 + h_2/2)$, the second $b(h_2 + h_3/2)$, and so forth. These can be summed to determine the total area according to Eq. 2-17:

$$A = b\left[\left(\frac{h_1 + h_n}{2}\right) + (h_2 + h_3 + \ldots + h_{N-1})\right] \tag{2-17}$$

A final method presented here for determining areas is by *planimetric* measurement. A planimeter, shown in Fig. 2-33, is a device that mechanically integrates the area within a closed traverse. The traverse is first plotted to scale, then the planimeter is used to trace the outline of the traverse. (Consult manufacturers' literature for proper procedures of instrument set-up.) The area is usually within 1% of the actual. The principal advantage of

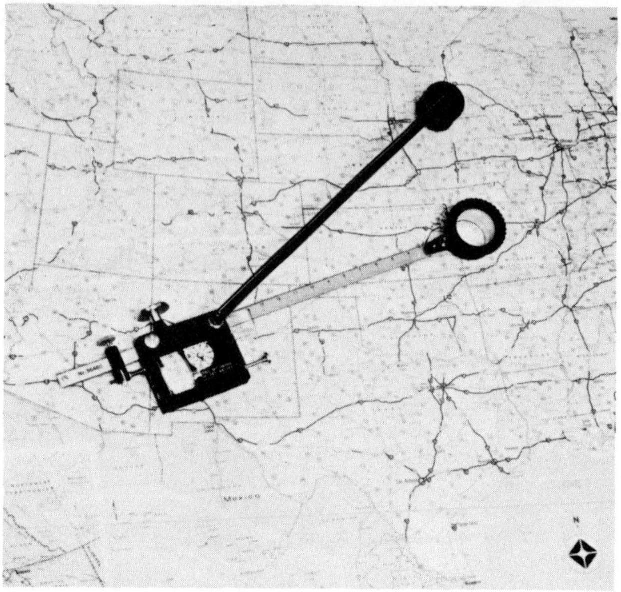

Figure 2-33 Planimeter for area measurement. (Courtesy of Teledyne Post.)

the planimeter is that extremely irregular shapes, such as coal properties, can be determined quickly and accurately.

Bedded deposits, such as coal, are often regular and assumed to be of constant thickness. The property volumes in these cases are calculated by multiplying the area times the thickness (derived through core sampling or other methods; see Chapter 6). A common practice in coal exploration is to first plot the coal property data on a 7.5-minute topographic quadrangle. Then determine the area of the property using planimeter measurement. Finally, the volume is simply the product of the area times the known thickness.

Cases where the thickness of the seam is variable require special treatment. One method is to partition the property into smaller, more individually uniform parcels, and to use an average thickness multiplier for each individual parcel. The total volume would then be the summation of the individual volumes. Computer methods may also be used but are considered beyond the scope of this presentation.

REFERENCES

1. Federal Mine Safety and Health Act of 1977, Public Law 91-173 as amended by Public Law 95-164, U.S. Government Printing Office, Washington, D.C. 1977.
2. Coal Mine Safety Law of 1971, State of West Virginia, W.Va., Dept. of Mines, Charleston, W.Va., 1971.
3. Milton O. Schmidt and William Horace Rayner, *Fundamentals of Surveying*, 2nd ed., D. Van Nostrand Company, New York, 1978.

4. Raymond E. Davis, Francis S. Foote, and Joe W. Kelley, *Surveying Theory and Practice*, 5th ed., McGraw-Hill Book Company, New York, 1968.
5. W. W. Staley, *Introduction to Mine Surveying*, Stanford University Press, Stanford, Calif., 1964.
6. Charles G. Frush, "Mine Surveying," in Ivan A. Given, ed., *SME Mining Engineering Handbook*, Vol. 2, Society of Mining Engineers, AIME, New York, 1973.
7. Francis H. Moffitt, *Photogrammetry*, 2nd ed., International Textbook Co., Scranton, Pa., 1967.
8. Paul R. Wolf, *Elements of Photogrammetry*, McGraw-Hill Book Company, 1974.
9. Russell C. Brinker, *Elementary Surveying*, 5th ed., International Textbook Co., Scranton, Pa., 1969.
10. T. Carl Shelton, Jr., and Stephen E. Merritt, "Surveying and Mapping," in Samuel M. Cassidy, ed., *Elements of Practical Coal Mining*, Society of Mining Engineers, AIME, New York, 1973.
11. Thomas Nunnally, unpublished lecture Notes.
12. H. Dotson Cather, unpublished lecture Notes.

PROBLEMS

1. Complete the field notes below for a surface survey:

Station	B.S.	H.I.	F.S.	Elev.
BM-A	6.066			913.266
TP 1	11.052		14.778	
TP 2	9.144		7.506	
BMB	19.98		11.736	
TP 3	0.776		10.638	
TP 4	16.002		0.576	
BMC			2.07	

2. Determine the underground elevations of floor and roof. Assume a seam thickness of 62 in.

Station	B.S.	H.I.	F.S.	Elev.	Floor
BMA	1.22			914.6	
1	2.29		2.78		
2	2.64		2.36		
3	2.47		0.77		
4	2.4		1.22		
5	1.2		2.02		
BMB			1.2		

3. Determine the elevation of the spads, assuming that the elevation of spad 48 is 916 ft, the H.I. is 914 ft, and uniform thickness of the seam.

Station	Azimuth	Vertical angle	Slope distance (ft)
48-49	67°15′	-0°32′	115
48-47	337°30′	+2°15′	160
49-50	67°15′	-0°40′	82
49-55	157°	-2°	164
48-52	157°30′	-2°30′	168
52-56	157°30′	-1°55′	155

Station	Elevation
46	
47	
48	916
49	
50	
52	
55	
56	

4. Change the following azimuths to bearings (make a sketch of each).

Azimuth	Bearing
148°25′02″	
36°14′15″	
231°03′14″	
125°12′20″	
320°36′15″	

Change the following bearings to azimuths.

Bearing	Azimuth
S41°34′20″E	
N82°38′14″E	
N18°24′50″W	
S62°54′12″W	

5. Set up a traverse sheet for the following and plot the coordinates at a scale of 1:4000. Assume that the coordinates of A are (0, 0).

Station	Azimuth	Vertical angle	Slope distance (meters)
A			
A-B	350°30′	0°00′	689.32
B-C	303°05′	3°30′	510.63
C-D	236°13′	0°00′	678.68
DE	177°58′	-10°30′	582.28
EA	90°29′	0°00′	1082.71
A			

6. Balance the data in Problem 6 and replot using units of feet.
7. Set up a traverse sheet for the data in Problem 4 and plot using a scale of 1 in = 100 ft. Assume the coordinates of station 48 to be (250, -100).
8. Assume that the spads in Problem 4 are located on the centerline of each entry. Assuming the entry width to be 20 ft, plot the rib locations and write a set of offset notes for each measurement.
9. Use the coordinate method to calculate the area of the following data.[11]

Station	Azimuth	Vertical angle	Slope distance (ft)	Coordinates	
1				38,921.33	82,241.91
1-2	134°34′10″	14°28′48″	336.61		
2-3	168°21′10″	15°00′22″	641.32		
3-4	146°07′40″	9°08′24″	385.92		
4-5	256°56′00″	0°28′	275.14		
5-6	66°23′40″	6°13′	118.07		
6-7	352°09′00″	7°04′	156.50		
7-8	336°44′20″	10°40′33″	496.86		
8-9	348°48′10″	17°18′27″	384.29		
9-1	328°03′40″	15°49′	251.54		
1					

3

Overview of Drafting and Cartography

3.0 INTRODUCTION

Like mathematics and language, drafting and cartography are methods of communication. Your ability to use graphical techniques to communicate effectively depends on mastering the principles, skills, and standard practices of the field. Drafting principles, skills, and practices, unlike those of mathematics, are subject to change or interpretation depending on the requirements of the situation. This chapter presents the fundamentals of drafting and cartographic design and cannot therefore discuss all possible variations. Only two concepts will be presented as invariant: (1) to be acceptable, drafting work must be neat, clean, legible, and free of obvious erasures, rips, blots, smudges, or other mistakes; and (2) any drafting principle or standard practice may be violated or suspended by the draftsman if and only if the change is necessary to effect better communication. The value of any drawing is always judged by these two criteria: neatness and the ability to communicate.

3.1 SCOPE

The field of drafting can be partitioned into various categories, based on the governing principles or methods employed, as illustrated in Fig. 3-1. Four major classes of drawings are delineated as comprising the field of drafting: (1) object projecting, (2) mapping (cartography), (3) relational diagramming, and (4) mathematical diagramming.

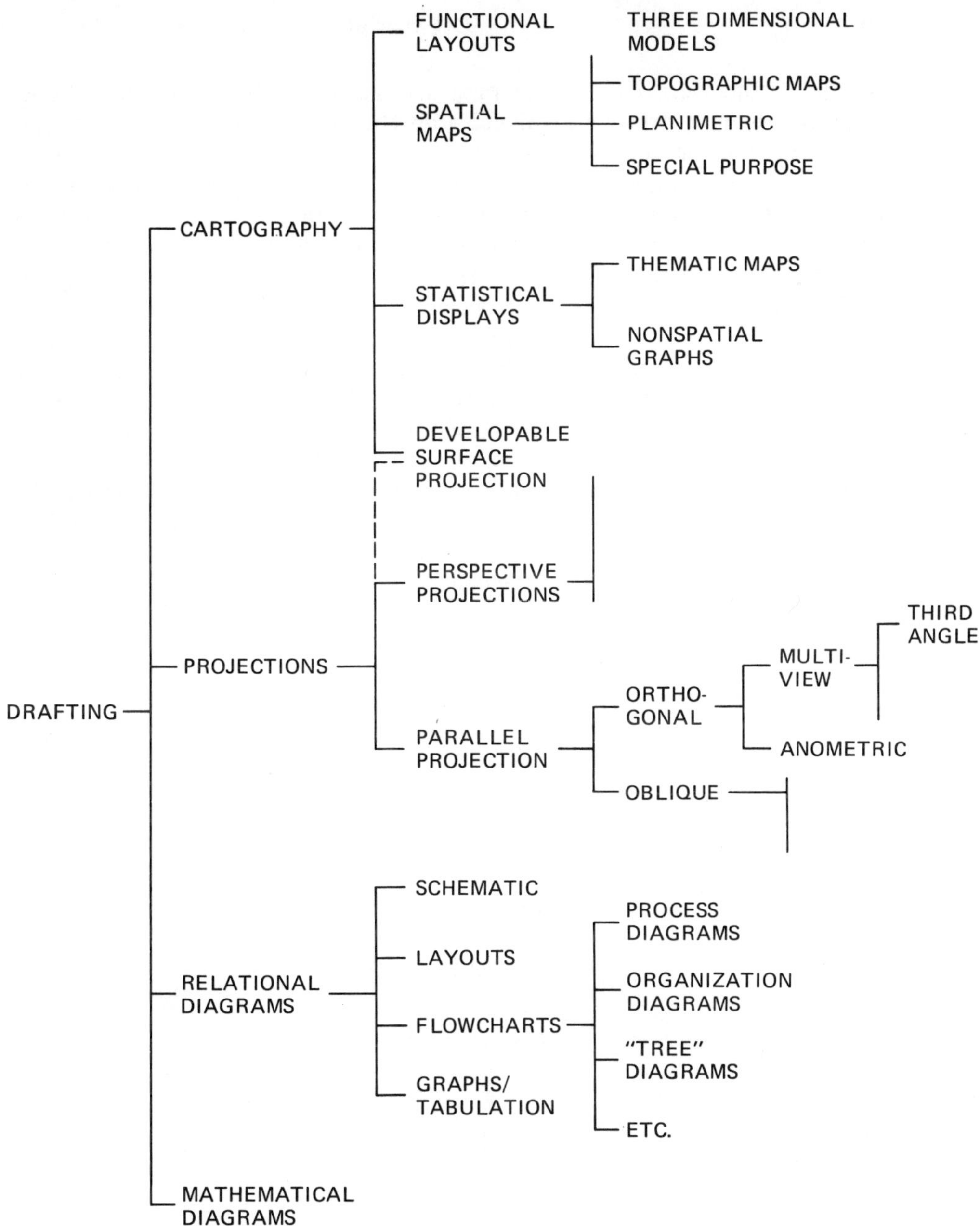

Figure 3-1 Hierarchical diagram of drafting techniques.

3.1.1 Projections

The class of drawings known as object projecting can be subdivided into three distinct groups: perspective projection, developable projection, and parallel projection. *Perspective* projection assumes the observer to be located a finite distance from an object, such that the lines of sight or projections from the viewer to the object form a cone expanding outward from the observer to the object. The plane upon which the object is to be projected is between the viewer and the object as shown in Fig. 3-2(a). The second class of projecting is *developable surface* projection, in which two-dimensional and mathematic surfaces are used to translate from spherical to plane geometry. Developable projection is the basis for global cartography. Developable surfaces are exemplified by plane, conical, and cylindrical projections, as discussed in Chapter 1. The third group of projections, *parallel* projection, assumes that the observer is located an infinite distance away from the object. The result of such an assumption is that the projectors become parallel to one another, as shown in Fig. 3-2(b). Parallel projections are preferred to perspective projections in most engineering disciplines. Parallel projections can be further subdivided, as shown in Fig. 3-1, into oblique and

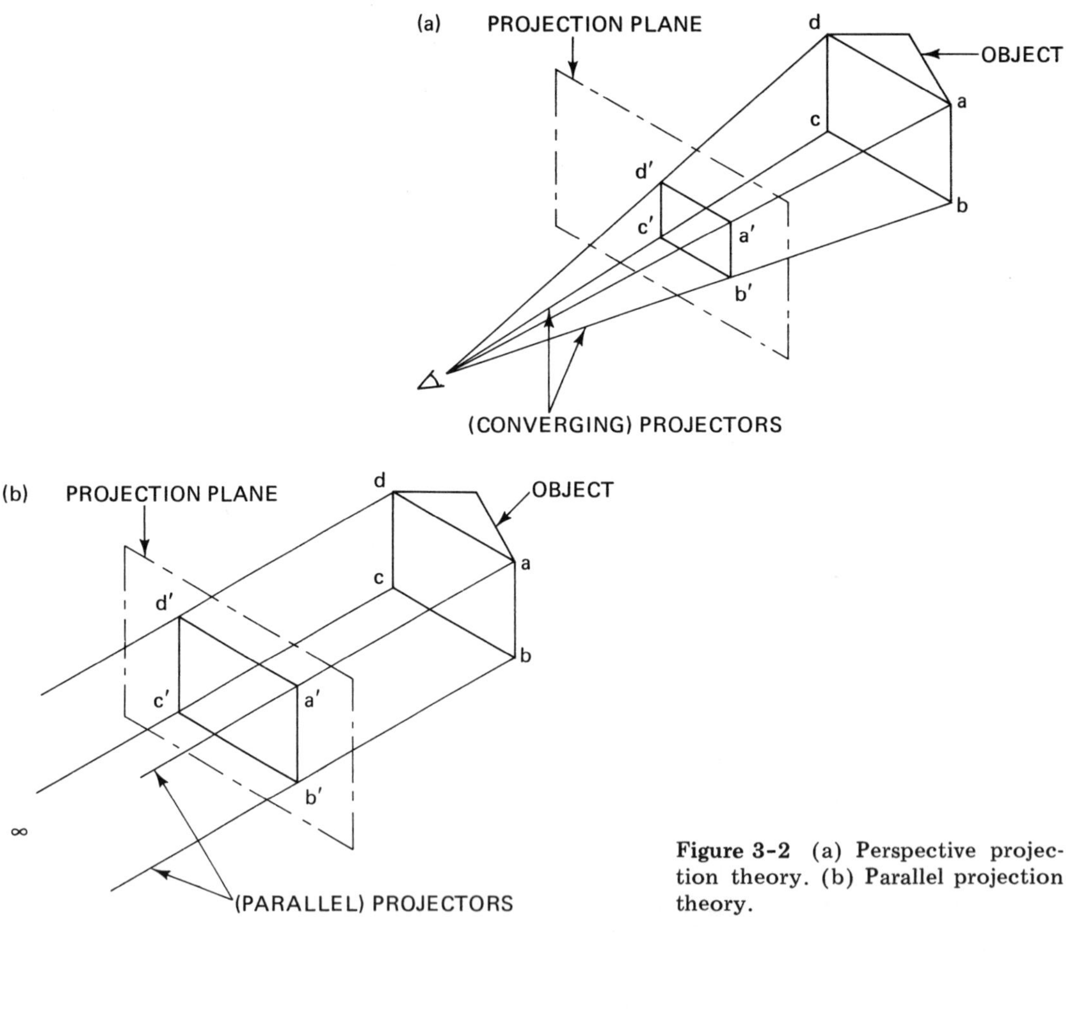

Figure 3-2 (a) Perspective projection theory. (b) Parallel projection theory.

orthogonal parallel projections. The application of multiview orthogonal projections, using techniques of descriptive geometry for resolving certain three-dimensional problems, is discussed in more detail in Chapter 6.

3.1.2 Cartography

The second major category of drafting is cartography, the art and science of mapmaking. Although cartography employs most of the principles contained within the category of projections, it constitutes the largest portion of mine drafting time and applications and has therefore been shown as a separate grouping in order to detail component parts. Cartography is subdivided into four principal groups: spatial maps, statistical displays, functional layouts, and models. *Spatial maps* are maps which are basically intended to portray the horizontal and vertical control of features in a specific area. Topographic and planimetric maps are examples of such maps within the spatial category. Topographic maps are described in detail in Chapter 4.

Statistical displays are drawings that emphasize the statistical account of some topic contained within a region. Two types of statistical displays are thematic maps, which show the spatial distribution of a topic, and statistical diagrams, which are not concerned with the spatial qualities of the topic but account only for the quantity. Types of thematic maps are the dot map, the choropleth map, and the isopleth map. Thematic maps are described in Chapter 6. Examples of statistical diagrams are line graphs, bar graphs, scatter plots, and so on. Statistical diagrams are developed in detail in this text.

The third subdivision of cartography, *functional layouts*, portray the spatial positions of components or activities constituting some functional system. A functional system is any organized collection of subsystems that operate to produce a desired outcome, process, or service. In mining the functional systems that we will describe include the ventilation system, the transportation system, the drainage control system, and the electrical distribution and service system. These functional layouts are discussed in Chapters 7 through 10.

The last category of cartography includes paradigmatic or descriptive *models* and probative or experimental models. These models may be physical, such as three-dimensional globes, or abstract, such as a computer simulation. Because of the complex mathematics and model-building theories involved in cartographic modeling, this category is not discussed in this text.

3.1.3 Relational Diagrams

The third major category of drafting is called relational diagrams. All drawings that fall within this grouping show interrelationships between two or more quantities. Figure 3-1 is an example of a relational diagram depicting drawing classifications within the drafting field. Major subdivisions of the category include flowcharts, schematic diagrams, sectional layouts, and line graphs or tabulations. Flowcharts show the elements and relationships that exist within processes, systems, organizations, and so forth. Flowcharts are normally quite simple, representing an element with a square, rectangle, or other common geometric symbol and showing the direct relationships between elements with a line. Flowcharts are discussed in Chapter 6 and are used in various sections in Chapters 7 through 10. Schematic diagrams are similar to flowcharts, differing only in the use of more descriptive symbolization and in complexity. The only types of schematic diagrams discussed in this book are electrical service diagrams detailed in Chapter 10. Sectional layouts show the internal features of an object by simultaneously displaying multiple cross-sectional views of the object. These views may be of different cross sections taken at critical locations or they may be cross sections taken at the same location but at different times. Sectional profiles are described in Chapter 4. The final subdivision of relational diagrams is line graphs or tabulations. These drawings are similar to the statistical diagrams presented under the cartography category, but the data or topics of interest are not related to land forms or areas. These are beyond the scope of this text.

3.1.4 Mathematical Diagrams

Mathematical diagrams are diagrams that provide graphical relationships which can be manipulated to yield new forms of data. These diagrams include Venn diagrams, matrix diagrams, nomographs, graphical integration or differentiation techniques, and so forth. These diagrams are also beyond the introductory scope of this text.

3.2 METHODS OF DRAFTING

Drafting can be subdivided into two general methods of drawing: (1) the freehand or technical sketch, and (2) the instrument drawing. The *technical sketch* is perhaps the most valuable tool the engineer or engineering technologist can and should develop. The advantage of the technical sketch is that it requires only the aid of paper, scale, and pencil to prepare an adequate drawing which can be used and almost universally understood. The freehand sketch, when done properly, has *all* of the characteristics and follows *all* of the same drafting practices as an instrument drawing. Therefore, the freehand sketch is *just* as *accurate* and *just* as *informative* as an instrument drawing. *Instrument drawings* differ from technical sketches only in the use of mechanical tools or instruments, which greatly aid and usually accelerate the drafting process. Instruments, although initially cumbersome for the novice, supplement the skill of the draftsman, thereby permitting him or her to concentrate on the technical details of the drawing. For this reason, the teaching of instrument drafting usually precedes that of technical sketching. This text will present instrument-assisted drawing, leaving the more difficult principles of sketching for subsequent courses.

3.3 SUPPLIES AND INSTRUMENTS

Drafting proficiency depends, in part, on the proper selection of supplies and use of various instruments. These, in turn, depend on the draftsman's knowledge of the capabilities and limitations of each. The proper selection of the "best" supplies or instruments will greatly facilitate the speed and accuracy of any draftsman. The axiom that "no one ever regrets buying the best" is particularly true of drafting supplies and instruments.

3.3.1 Supplies

The following supplies are necessary for mine drafting and map construction.

Drafting pencils. Drafting pencils (Fig. 3-3) can be purchased in a variety of constructions and types, ranging from the common cedar-laminated

Figure 3-3 Typical drafting pencils and heads: (a) cedar shaft, (b) metal lead holder, (c) leads for metal holder, (d) 0.05-mm leads, (e) 0.05-mm mechanical pencil holder, (f) 0.03-mm leads, (g) 0.03-mm mechanical pencil holder.

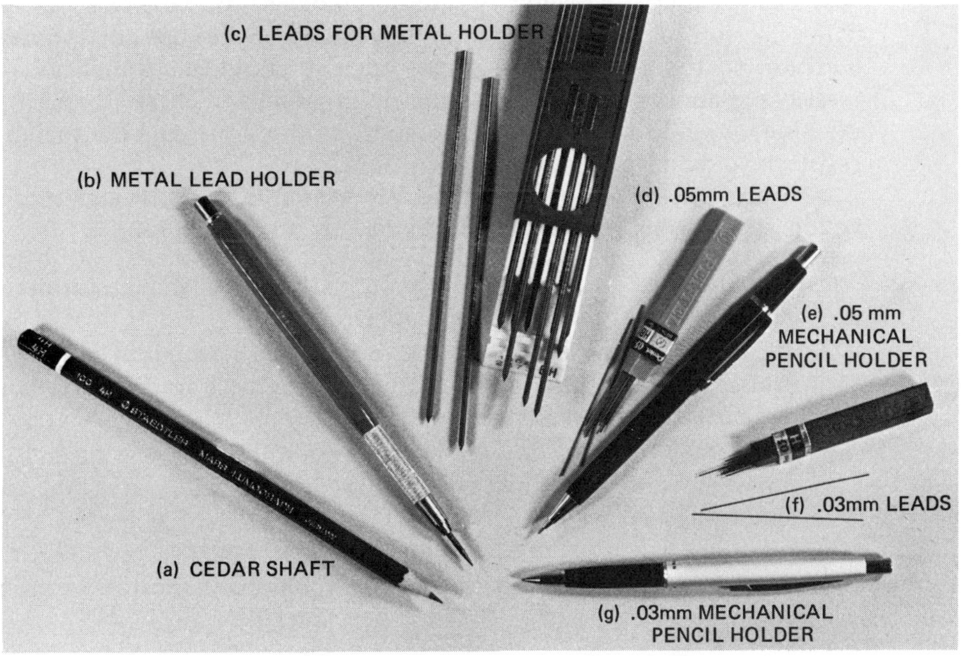

graphite pencil to the newer "thin lead" (0.3 mm) mechanical pencils, which do not require sharpening and thereby contribute to faster drafting.

More variable than the pencil's construction and more important to an acceptable drawing is the ranking of the graphite lead. Leads are ranked according to the graphic composition and amount of binder. Figure 3-4 shows the relative ranking of leads. For most drafting work the student will find that 6H lead provides the "best" preliminary construction line, 4H the best dimensioning or secondary object line, and 2H or H the best general object line. The softer grades (HB to 6B) are usually reserved for technical and artistic sketching.

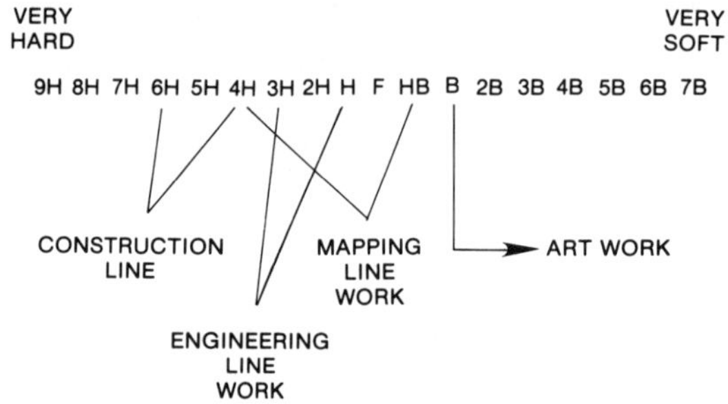

Figure 3-4 Relationship of lead hardness and application.

The student should be careful to purchase a high-quality set of drafting pencils and avoid the common No. 2 type of pencils. These more common leads do not have the quality control found in drafting leads; thus the student will experience hard and/or soft spots even within the same pencil.

Drafting paper. Several types and grades of drafting paper are available. The selection of the best grade is dependent on the use for the drawing. In addition to the end use, the paper selected should be dimensionally stable (resist expanding and contracting with environmental changes), cleanly erasable, suitable for various types of reproduction (e.g., microfilming), and resistant to damage (e.g., tearing, crinkling, etc.).

Four types of paper are generally available for drawing: tracing bond, vellum, polyester film (Mylar), and cloth.

1. *Tracing bond* is the least costly paper, but is most susceptible to dimensional instability and general damage.
2. *Vellum* is a better-quality rag bond paper which is in more general drafting use. Vellum is also slightly susceptible to dimensional changes, depending on environmental variations (e.g., humidity, temperature).
3. *Polyester film* (e.g., Mylar) is the most stable drafting surface available, provides good reproduction quality, and is highly resistant to damage and aging. In general, you will find this type of paper to be the most suitable surface for preliminary and working maps.
4. *Cloth* is thin linen impregnated to facilitate cohesiveness. Because of

the lower cost and greater availability of other drafting surfaces, cloth surfaces are rarely used except for special inking projects and "finished" maps.

Drafting paper can be purchased in rolls (24 to 54 in. in width) for large-scale, large-area cartography or in standard sheet sizes which are identified by letter designations, A through F. [These standards, like most used throughout this text, are specified by the American National Standards Institute (ANSI).] The available sizes are shown in Table 3-1.

TABLE 3-1 ANSI DRAWING SIZES

Letter designation	Size (in.)	Margin (in.)
A	$8\frac{1}{2} \times 11$	$\frac{3}{8} \times \frac{1}{4}$
B	11×17	$\frac{3}{8}$
C	17×22	$\frac{1}{2}$
D	22×34	$\frac{1}{2}$
E	34×44	$\frac{1}{2}$
F	28×40	$\frac{1}{2}$

In addition to the various grades and sizes, drafting paper can be purchased with different cross-sectional patterns (e.g., $\log_{10}$-$\log_{10}$ paper), which may or may not reproduce when the drawing is reproduced. This type of paper is often used for map or true positioning drafting, since the grid pattern facilitates coordinate positioning.

Drafting erasers. Several types of erasing materials exist, varying in composition, pliability, and softness. Some of the more common types of erasers are displayed in Fig. 3-5, including the art gum, the soft pink, the vinyl white, the harder red, and specialty erasers. Each type of eraser is useful for different tasks.

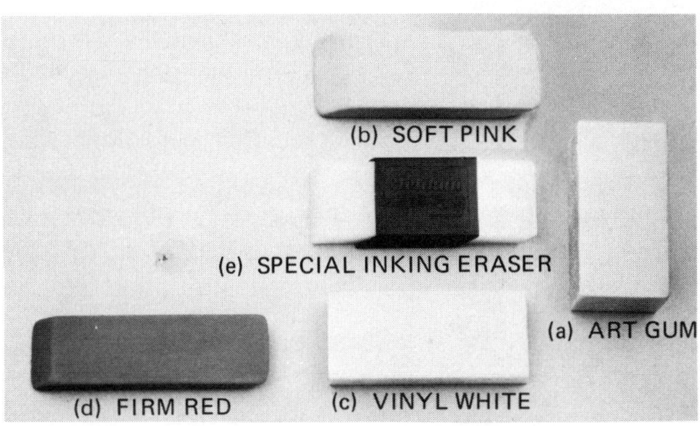

Figure 3-5 Common erasers: (a) artgum, (b) soft pink, (c) vinyl white, (d) firm red, (e) special inking eraser. (Courtesy of Teledyne Post.)

1. Artgum is a very soft eraser that has a low amount of binder and therefore erases cleanly, although less effectively than some other types.
2. Soft pink is the most useful and common eraser available for general drafting. It normally provides clean erasures without excessively damaging the paper. (Like most erasers, the soft pink will collect oils and dirt from normal handling and therefore the edge to be used should be cleaned by making test wipes on a separate sheet before attempting to use it on an actual drawing.)
3. The vinyl white eraser is a light-colored vinyl eraser which tends to provide cleaner and more complete erasures on difficult surfaces such as vellum and tracing papers.
4. The firm ruby eraser is a slightly harder grade of eraser than the soft pink, providing clean erasure of harder markings; however, the firm ruby tends to abrade the paper and is not recommended for less-wear-resistant tracing paper.
5. Speciality erasers include (a) ink erasers, which have a harder composition and binder; (b) dusting pads, which contain gum fragments in a loose weave bag; and (c) special configurations made for erasing machines, as shown in Fig. 3-6.

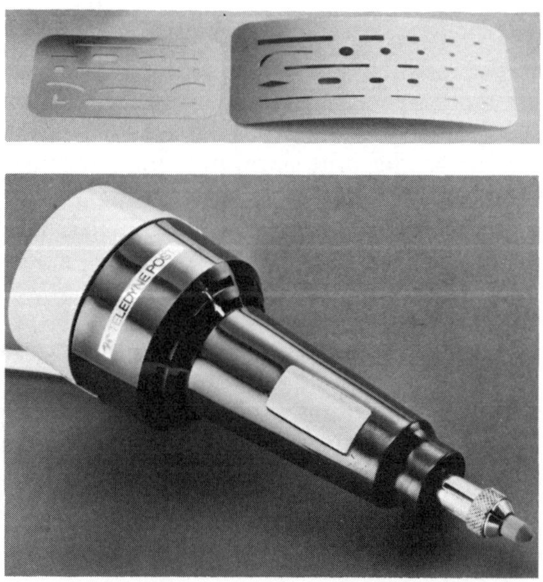

Figure 3-6 Electrical erasing machine. (Courtesy of Teledyne Post.)

3.3.2 Instruments

Drafting surfaces. Various surfaces depicted in Fig. 3-7(a)-(c) are used to aid the draftsman, including (1) the drawing board and T-square, (2) the parallel arm board, (3) the drafting table and machine, and (4) speciality surfaces (not shown).

(a)

(b)

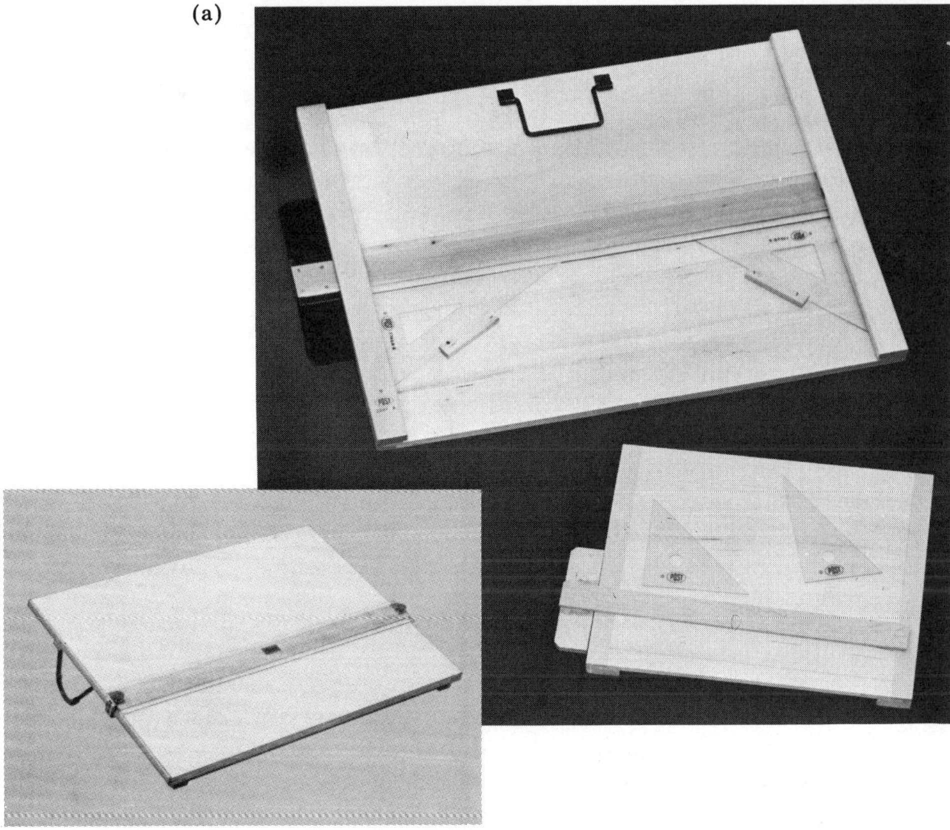

Figure 3-7 (a) T-square. (b) Parallel arm board. (c) Drafting table and machine with indexing head. [(a) and (c), Courtesy of Teledyne Post; (b), courtesy of Pickett Industries.]

(c)

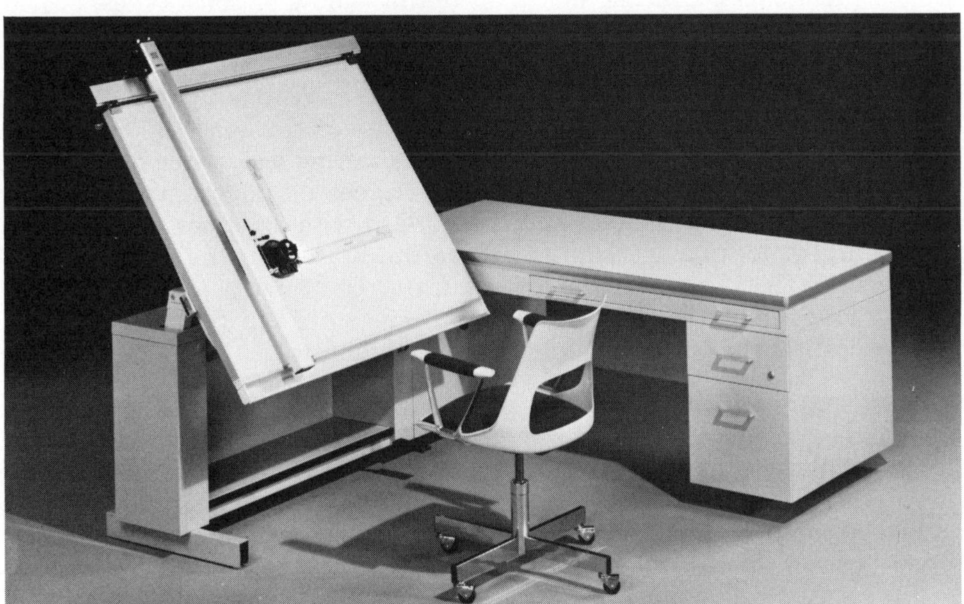

Drawing Board and T-Square. Still the most basic surface for drafting is the drawing board and T-square. Boards can be obtained in a variety of grades and sizes. An adequate board should have a perfectly flat smooth surface and have metal strips on either side to ensure reproducible tracking of the T-square head. The T-square, which is used to provide a parallel horizontal surface, consists of a head and arm which must be mutually perpendicular and secured in such a way that their relationship never varies. The arm ideally has bottom and upper edges which are inlaid with clear plastic.

The head of the T-square is held firmly against the left edge of the board (assuming a right-handed draftsman) by the draftsman using his left hand while he manipulates the drafting pencil with his right hand. Vertical lines are drawn using right triangles which are held firmly against the edge of the T-square arm with the left hand, while drawing the necessary lines with the right hand. (You should never attempt to hold the T-square against the bottom or top of the board for constructing vertical lines, since these edges are not always perpendicular to the side edge.)

Parallel Arm Board. The advantage of parallel arm boards is that the draftsman no longer needs to worry about keeping the head of the T-square firmly against the edge of the board, thereby freeing his left hand for other tasks. The parallel arm board maintains the horizontal orientation throughout the vertical range of travel and is used for more accurate work than the T-square board.

Table and Machine. Most professional drafting today is done using a drafting table mounted with a drafting machine. The drafting table provides the draftsman with a choice of board tilt angles and heights to suit his preferences. Drafting machines (either elbow or track types) provide a vertical and horizontal translating head which maintains the same orientation at any position across the surface of the board.

The control head can mount two scales, one vertical and one horizontal. These scales have the characteristics of free scales and thereby eliminate the need for manipulating a separate drafting scale while drawing. The indexing head is usually calibrated in degrees (although grads, mils, and other divisions are available) and is equipped with a veriner for accuracy of reading. A locking mechanism allows the draftsman the ability to lock in any angle with just a press of a button. Drafting tables and machines save a considerable amount of time and are quite reliable; however, their initial cost for the occasional draftsman is prohibitive.

Speciality Surfaces. Certain accuracy requirements demand surfaces that are highly specialized. *Coordinatographs* and automated drafting machines are examples of these speciality surfaces. These systems permit the recording of necessary scales, scale ratios, coordinates, and line weights into a computer. The computer then controls the translation of the drawing head.

Drafting scales. Scales shown in Fig. 3-8 may be purchased in a variety of sizes, configurations, and graduations in both English and SI units. The two basic types of scales are the *architect's scale*, fractionally segmented in English units, and the *engineer's scale*, segmented in decimal units. Scales are divided into the most commonly used segments. For example, a common triangular architect's scale (English units) is divided into the following scales:

$\frac{3}{32}$-in. scale and $\frac{3}{16}$-in. scale

$\frac{1}{8}$-in. scale and $\frac{1}{4}$-in. scale

$\frac{3}{8}$-in. scale and $\frac{3}{4}$-in. scale

$\frac{1}{2}$-in. scale and 1-in. scale

$1\frac{1}{2}$-in. scale and 3-in. scale

A typical triangular engineering scale (English units) is divided into decimal units, where each side of the scale is equal to a decimal part of an inch as follows:

10th scale = 0.1 of an inch

20th scale = 0.05 of an inch

30th scale = $0.03\overline{3}$ of an inch

40th scale = 0.025 of an inch

50th scale = 0.02 of an inch

60th scale = $0.01\overline{6}$ of an inch

The engineering type of scale has gained a wide following in most sectors of drafting, but in a field like architecture, many draftsmen still prefer using the architect's rule. If the student were to purchase only one scale, the triangular engineering scale would provide the greatest future service.

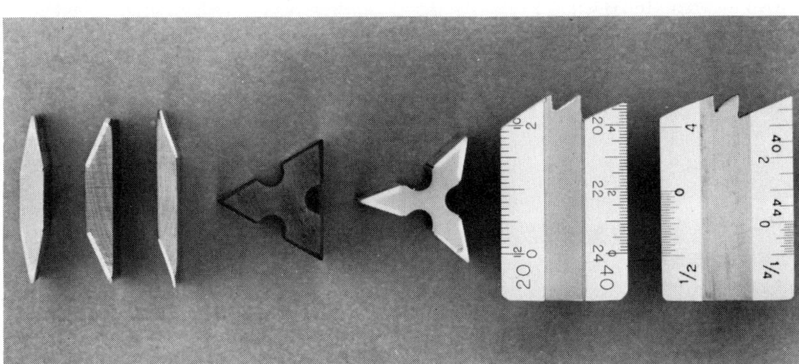

Figure 3-8 Various scale configurations and graduations. (Courtesy of Teledyne Post.)

In practice the scale is rarely used to directly lay off distance on the tracing. Instead, a set of dividers are extended along the scale to the desired length and then this length is physically transferred to the drawing. The exception to this practice occurs with drafting machines, which have scales mounted on the control head.

Dividers and compasses. Dividers (Fig. 3-9) are one of the most useful drafting instruments. They are used to transfer lengths (either from the scale to the tracing or between views on the tracing) and to divide distances into a prescribed number of equal parts.

Proper handling of the dividers is important to ensure the speed and accuracy desired. Dividers are held by the thumb and index finger, and the second and third fingers are used to steady or adjust the instrument.

Compasses, shown in Fig. 3-10, are used to draw circles and arcs of constant radii. Bow compasses are used to draw small circles; larger radii can be obtained by adding leg extensions or using beam compasses. The handling of the compass follows that used to manipulate the dividers.

Triangles. Triangles (Fig. 3-11) greatly facilitate drafting procedures. The two essential triangles are the 45° and the 30°–60° triangles. The triangles purchased should have uniform straight edges and, most important, a true 90° angle. These two triangles, coordinated with the T-square, should enable the draftsman to produce easily a series of perpendicular or parallel lines at 0°, 15°, 30°, 45°, 60°, 75°, and 90° angle orientations. Additionally, any two triangles can be used to generate lines parallel to any given line.

Other instruments. Other instruments may be used to aid the drafting procedure.

Templates. A template is a plastic sheet that has been stamped or molded with standard symbols or common configurations which allow for quicker, often more precise drawing of these shapes. Figure 3-12 illustrates common types of templates.

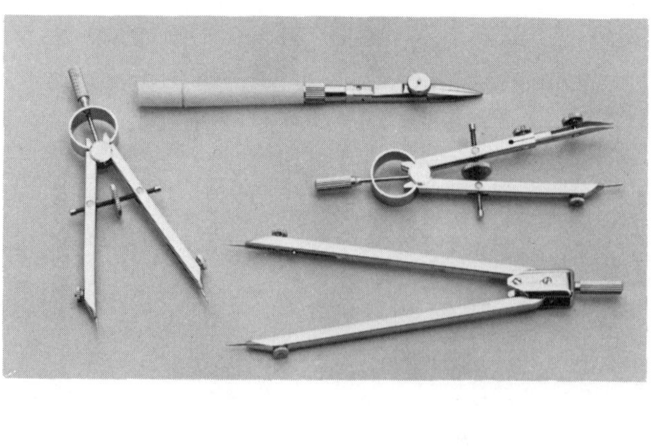

Figure 3-9 Dividers. (Courtesy of Teledyne Post.)

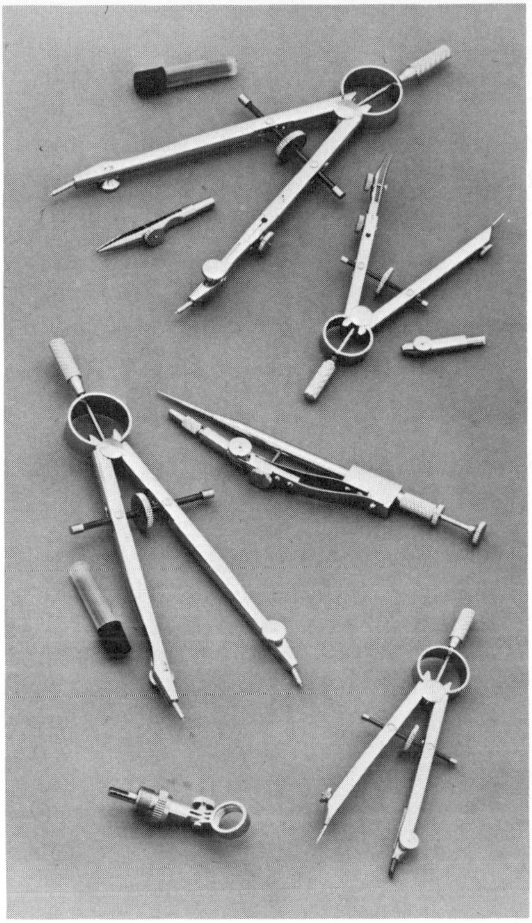

Figure 3-10 Compasses. (Courtesy of Teledyne Post.)

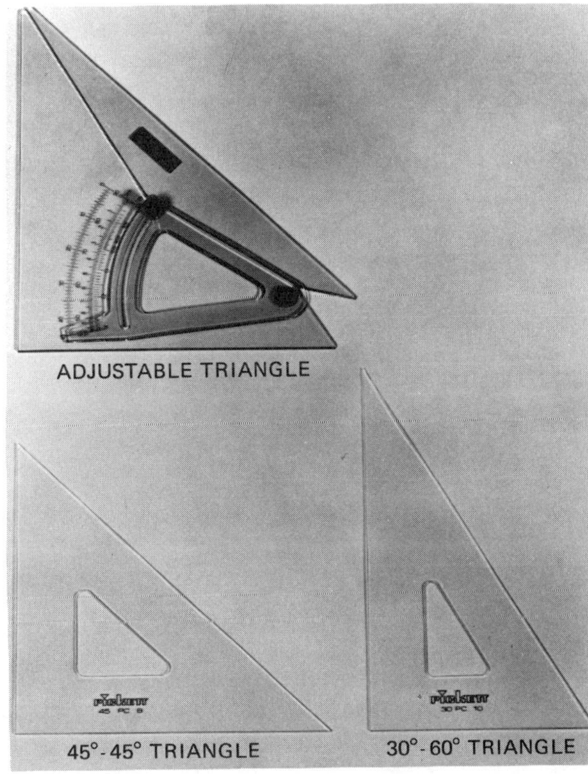

Figure 3-11 Triangles. (Courtesy of Pickett Industries.)

Figure 3-12 Mapping and layout templates. (Courtesy of Pickett Industries.)

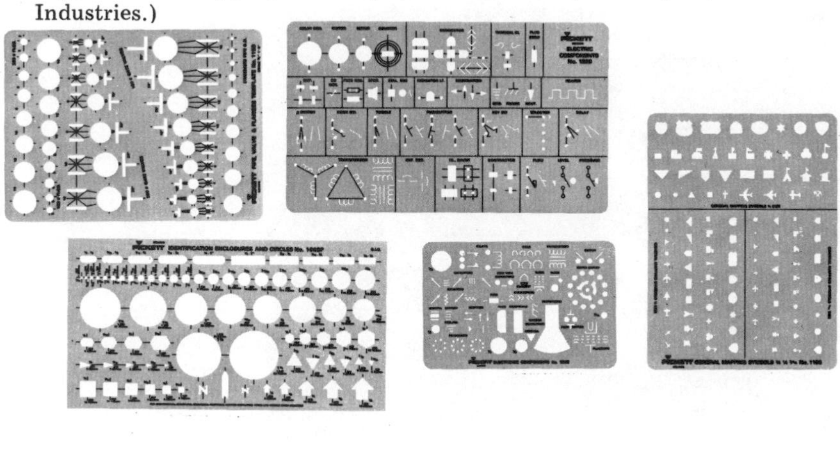

Protractors. A protractor (Fig. 3-13) is a circular or semicircular plastic sheet which divides a circle into degree graduations relative to a center. Protractors used to lay off intermediate angles are essential instruments when drafting machines are not available.

Assorted Combination Curves. French curves (Fig. 3-14) are used to smoothly connect several known points with a combination of curves to differing radii.

Adhesive Aids. Various symbols and crosshatching overlays are available with pressure-sensitive adhesive backing that can be placed or moved on the drawing any number of times without requiring redrawing. Once the

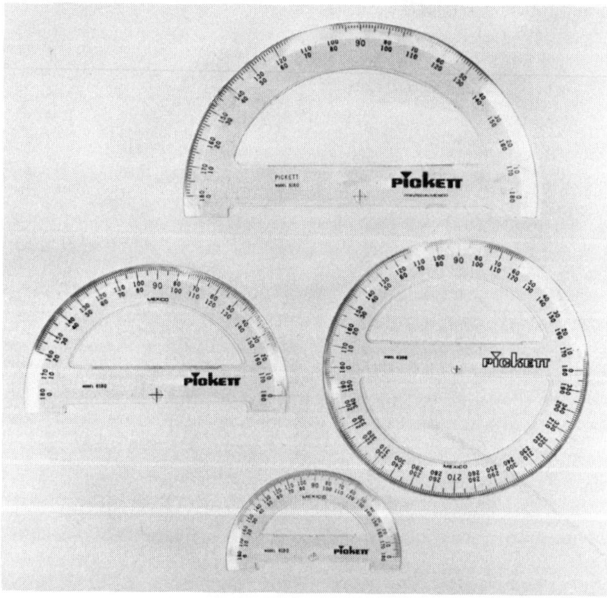

Figure 3-13 Protractors. (Courtesy of Pickett Industries.)

Figure 3-14 French curves (Courtesy of Pickett Industries.)

location has been fixed, reproductions may be made without further modification or treatment.

Proportional Dividers. Proportional dividers (Fig. 3-15) are used to directly divide any specified length into any number of equally spaced lengths.

Electric Erasers. Erasing machines (Fig. 3-6) are used to speed the erasing process and improve the accuracy and cleanliness of erasing.

Erasing Shields. Erasing shields (Fig. 3-6) are usually constructed of thin stainless-steel sheet stamped with various shapes, allowing the draftsman to erase unwanted lines while covering and thereby protecting lines in close proximity.

Inking Instruments. Twenty years ago drafting and inking were somewhat synonymous terms. Presently, inking drafting is practiced only on finished maps, while pencil or lead drafting is used for preliminary and working drawings. A typical set of inking instruments needed for finished maps is shown in Fig. 3-16.

Scribing Instruments. Scribing is a process of creating a photographic negative using a metal scribe to cut or remove an opaque coating from a transparent base film. Once the coating is removed, photomechanical or photographic reproduction methods are used to prepare positive prints. Originally, glass was used for the base material; however, the introduction of vinyl and polyester films has improved the cost competitiveness. Scribing is the most accurate method, providing the cleanest, sharpest, and most stable reproductions. A typical instrument set is depicted in Fig. 3-17.

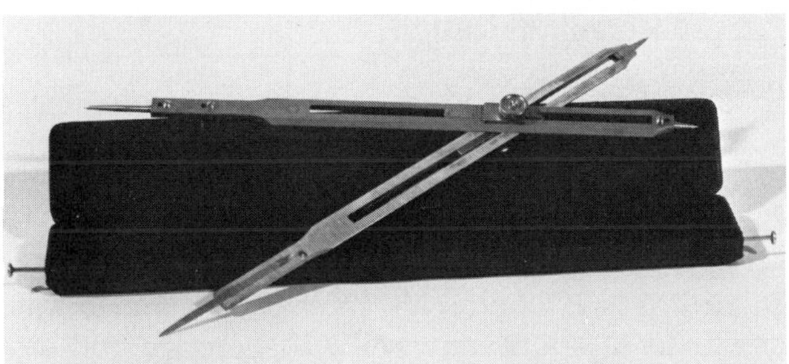

Figure 3-15 Proportional dividers.

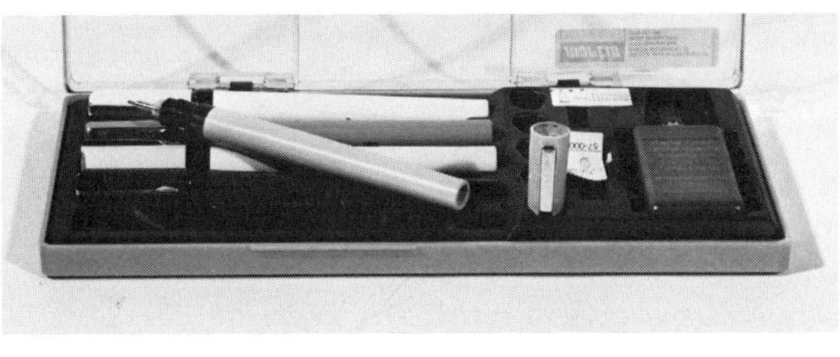

Figure 3-16 Inking instruments.

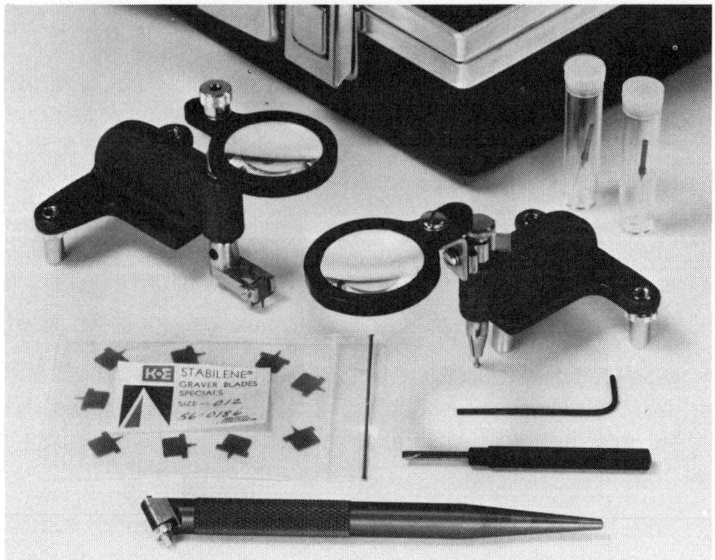

Figure 3-17 Typical etching instruments. (Courtesy of Keuffel and Esser Co., Morristown, N.J..)

3.4 DRAFTING ROOM PRACTICES

The two irrevocable rules governing standard drafting room practices were stated in the introduction to this chapter.

1. Drafting work must be neat, clean, legible, and free of obvious erasures, rips, blots, smudges, or other mistakes.
2. Any variation from standard practices must be made with the intent and result of effecting better communication.

Two other concerns occasionally offered as drafting guidelines are speed and accuracy. Both speed and accuracy are important, particularly in industry; however, both are somewhat dependent on the draftsman's familiarity with the equipment, techniques, and problems. Initially, speed and accuracy are at odds with one another. As one increases speed, drafting accuracy decreases. In general, neatness is conducive to accuracy and frequent practice contributes to speed. Both traits improve with practice and attention to detail.

3.4.1 General Preparations

To begin drafting work, all instruments should be cleaned and leads sharpened. Instruments should be arranged in orderly groups to the right of the drafting surface (for right-handed draftsmen). The paper should be aligned with the horizontal control instrument (T-square, parallel arm, etc.), and securely taped at each corner. The vertical control should be checked using a triangle to ensure perpendicularity with the horizontal. When an old drawing is being revised, the existing horizontal and vertical lines depicted on the drawing should be used to align the paper with the horizontal control instrument, thereby ensuring proper parallel conformance of old and new content.

The optimum paper-lead combination should be selected. The lead needed will vary with the type of paper, humidity, and temperature. For example, humidity tends to soften the paper; thus a softer pencil may be needed to avoid cutting the surface. In general, 5H or 6H is used for initial construction and layout, and 2H or H is used for object outlining.

The first lines on a paper should be the view boundaries or control coordinates needed for true positioning and map drafting. The lines that follow are generally those needed for initial construction and layout. If properly constructed, these lines will be so light that they will not reproduce, thereby eliminating the necessity of having to erase them from the final form.

3.4.2 Line Conventions and Use

The standards of line convention used in North America have been compiled by the American National Standards Institute[1] in their standard ANSI Y 14.2. Three relative weights of line have been specified as heavy, medium, and light. Further, ANSI has specified an alphabet of lines. The name, weight, and dimensional standards of each line are shown in Fig. 3-18.

Figure 3-18 ANSI standards of lines.

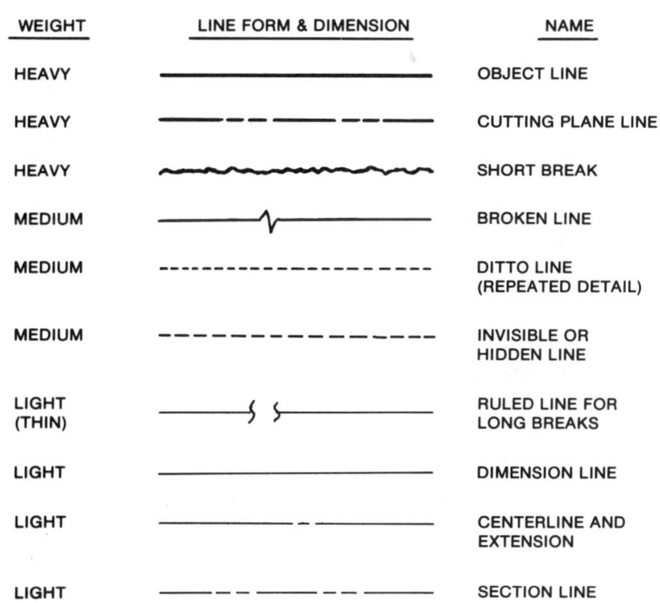

Standard map line conventions, particularly for mine mapping, have never been uniformly agreed upon; however, Fig. 3-19 is a compilation of the major lines used most frequently on functional layouts. Recognize that tints and colors, as dictated by the USGS standards,[2] cannot be used. Thus Fig. 3-19 is a modification of USGS guidelines.

FEATURE TYPE		DESCRIPTION	SYMBOL	ADDITIONAL
ROADS	CLASS 1	PRIMARY HIGHWAY	HEAVY .030-.040	NAME OR NUMBER DESTINATION
	CLASS 2	ALL WEATHER	HEAVY .030-.040	
	CLASS 3	LIGHT DUTY	MEDIUM .020	
	CLASS 4	UNIMPROVED	MEDIUM .02	
	CLASS 5	TRAIL	LIGHT	
BOUNDARY		NATIONAL	HEAVY	
		STATE	MEDIUM	
		COUNTY	MEDIUM	
		INCORPORATED TOWN	MEDIUM	
		PARCEL OWNERSHIP	LIGHT	
RAILROADS		STANDARD GAUGE	MEDIUM	
		NARROW GAUGE	LIGHT	
		ABANDONED	LIGHT	
		MULTIPLE TRACK	MEDIUM .02	
		SPUR	MEDIUM	
UTILITIES	POWER	<13.8 KVA	LIGHT	
		69.0 KVA	LIGHT	
		>69.0 KVA	LIGHT	
INDUSTRIAL PIPE LINE		GAS WITH SHUT-OFF VALVE	LIGHT	
		LIQUID OR SURRY	LIGHT	
ORIENTATION		(PRIME MERIDIAN)	MEDIUM	
		MERIDIAN PARALLELS	LIGHT	
		HORIZONTAL CONTROL & MONUMENT	LIGHT	
		VERTICAL CONTROL	BMX 824	

Figure 3-19 Proposed line standards for a mine map (1 in. = 500 ft).

Techniques. When lead drawings are being produced, the most common point configuration is a conical point, produced by rotating the pencil over a file, sandpaper pad (Fig. 3-20), or by using a standard pointer. A "needle-sharp" point should be avoided, since the point will easily break or require frequent, time-consuming sharpening. After sharpening, the loose graphite is removed from the point using a clean cloth, clay ball, or Styrofoam pad. The pencil should be held at an angle (approximately 60°) to the paper and drawn along the edge of the triangle or T-square as illustrated in Fig. 3-21. To ensure constant line weight and to increase the point life, the pencil

Figure 3-20 Sandpaper pad and pointers.

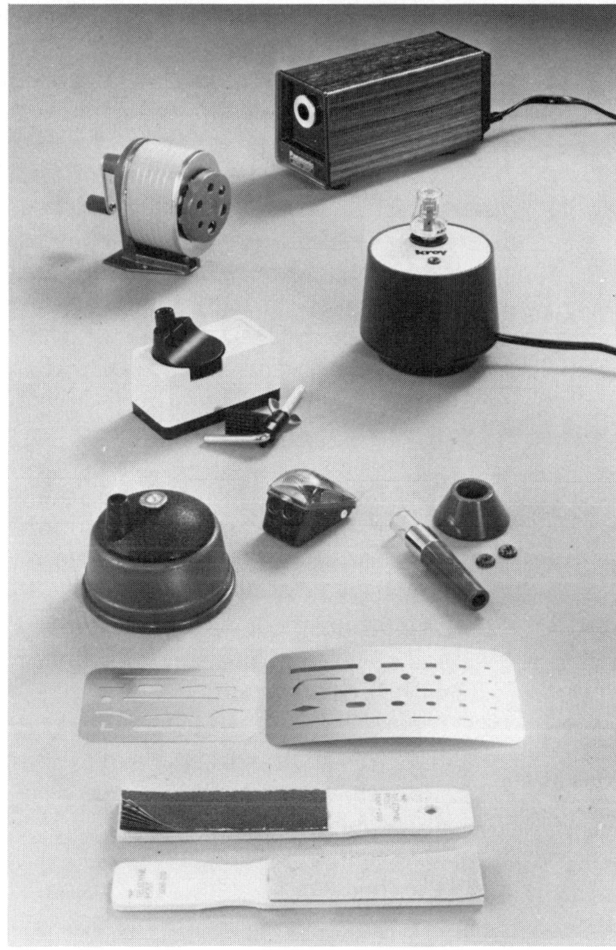

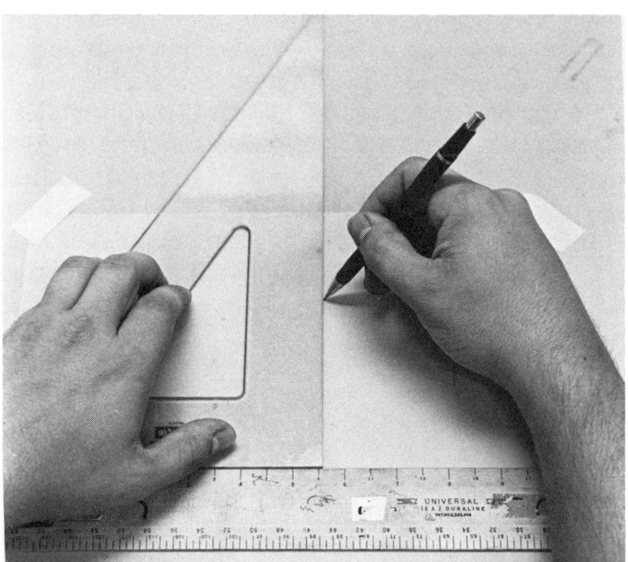

Figure 3-21 Proper positioning of pencil and left hand for instrument manipulation.

should be rotated slowly and uniformly as the line is drawn. Line weights between 0.0067 and 0.0125 in. are most frequently used for unspecified lines.

Line weight plays a major role in the interpretation of the drawing by others. Lines, other symbolization, and their arrangement depict the view desired by the cartographer. The heavier the line, the greater the importance attributed that feature. Although the line weights shown in Fig. 3-19 are presented as standard, the line weights eventually used are dependent on the degree of generalization. Generalization is somewhat a function of map scale, the relative importance of one feature compared to another, and the end use of the map. Thus the line weight selected fits into the same classification of data treatments discussed in Chapter 1:

1. *Simplification.* Features are ranked according to relative importance, the most important being proportionately heavier than the less important features. Features of little or no importance are correspondingly not displayed.
2. *Enhancement.* Features are set off from the surrounding features by adding color, line shading, line continuity, or size not in proportion with the map scale.
3. *Distortion.* Features that would draw attention from the region can be shown with less weight even though they are of equal value (e.g., heavy property lines depicting one area, contrasted with lighter surrounding property lines).

3.4.3 Lettering

One of the most critical controls over the appearance and acceptability of a drawing is the quality of the lettering that accompanies the graphics. Good lettering *can* make faulty linework appear acceptable, but poor lettering *will always* make good graphics *unacceptable.* Lettering quite literally will make or break a drawing, report, or other assignment that it accompanies.

Like line weight, lettering can be used to simplify, enhance, or distort the content being depicted. The presence of printed letters or numbers, such as titles, names, and code numbers, alone has the effect of setting off the content so labeled. The relative importance of features depicted can be affected by letter style, form size, color, contrast (weight and darkness), position, and quality.

Letter style (the design of the character) is one of the most important factors affecting the interpretation of the map. Various standard lettering includes Gothic, Roman, and Old English or text, shown in Fig. 3-22. All styles are used, but the Gothic is used more for engineering and scientific work.

GOTHIC LETTERING STYLE

ROMAN LETTERING STYLE

ITALIC LETTERING STYLE

OLD ENGLISH LETTERING STYLE

Figure 3-22 Examples of various lettering styles.

Letter form refers to the use of capital and lowercase letters, the degree of incline, and the height-to-width ratios that are used. For all preliminary drafting work you should use only capital letters of the same form. When a finished map is required, different forms can then be selected without bias.

Letter size strongly affects the attention of the viewer. For most drafting and preliminary mapping, the size of the letter, like the form, should remain constant. A typical size range would be between $\frac{3}{32}$ and $\frac{1}{4}$ in. high.

The color of the letter is somewhat influenced by the feature being described. For consistency the following features and color of print are usually associated:[3]

1. Hydrographic features: blue
2. Cultural features: black and red
3. Surface relief features: brown and black
4. Surface vegetation features: green, yellow, and brown
5. Special symbols: black

Of course, when preliminary work is mapped, black is the only color used.

The *contrast of letters* refers to the weight and darkness of the lettering compared to other letters and to other map symbols. In general, darker, more contrasting letters attract more viewer attention than do the lighter, less contrasting letters. Contrast on preliminary tracings should be constant.

The *position* of the lettering will also tend to focus viewer attention to the form being titled. Positioning helps to delineate the extent of the feature. For preliminary drafting work the position of the lettering should be such that the viewer can read the drawing from either the bottom (south) or right-hand (east) side of the tracing. If needed, leaders can be used to locate and identify specific features otherwise too small to locate.

The *quality* of lettering refers to the neatness, consistency, and uniformity with which the letters have been drawn (or placed) upon the drawing. No excuse is accepted nor should any be offered for poor lettering. Freehand lettering is done with single strokes of the hand. All good letters use lettering guidelines (Fig. 3-23), which consist of two horizontal lines

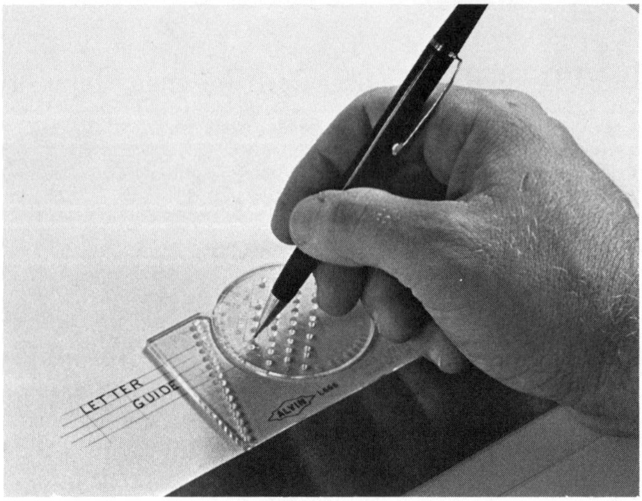

Figure 3-23 Use of the lettering guide.

drawn at the appropriate height for the desired letters; vertical guides are randomly placed across the space to be lettered to aid in vertical control. A series of evenly spaced horizontal lines is then placed to facilitate letter spacing.

3.5 MAPPING SYMBOLIZATION

Maps are representations of an original; therefore, all marks, points, and figures on a map symbolize the original features. In general, symbols should be (1) *simple* in composition (size, outline, clarity), (2) *obvious* in what they are intended to represent, and (3) *easily drawn*. Based on these guidelines, symbols are not details of the specific feature but are generalizations of the nature or form of the feature. As the scale factor increases, the degree of generalization increases. This change in generalization results in a corresponding change in the symbol, thus complicating any efforts to compile an "absolute" set of standard symbols. Symbol generalization is illustrated in Fig. 3-24.

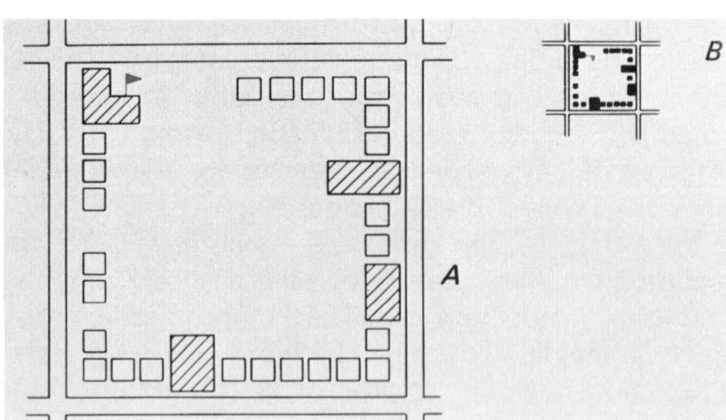

Figure 3-24 Generalization of symbols due to scale changes. (Courtesy of USGS.)

Like linework (Fig. 3-19), map symbolization is dependent upon the map scale, use, and the relative importance of features. These three factors determine the degree of simplification, enhancement, and distortion of the features on the map. The extent to which the symbols have been standardized depends primarily on the frequency of use of the symbols.

Major topographic features and the corresponding symbols are divided into five categories:

1. Hydrographic features
2. Cultural (human-made) features
3. Surface relief (elevation) features
4. Vegetation or agricultural features
5. Special features (e.g., mining)

The first four groups of features have been used fairly extensively in past mapping efforts, whereas special features such as those used to represent or detail mining operations have not been extensively used in any uniform manner. A brief review of the features follows.

Hydrographic features. Hydrographic features are those defining and locating water-related information, such as rivers, lakes, bay areas, and channels. Special thematic and functional maps include the hydrograph, which depicts the features of runoff; isochronal maps, which depict characteristics of the watershed; and aquifer maps, which show water-bearing geologic features. Some of the more useful hydrologic features for mine cartography are shown in Fig. 3-25.

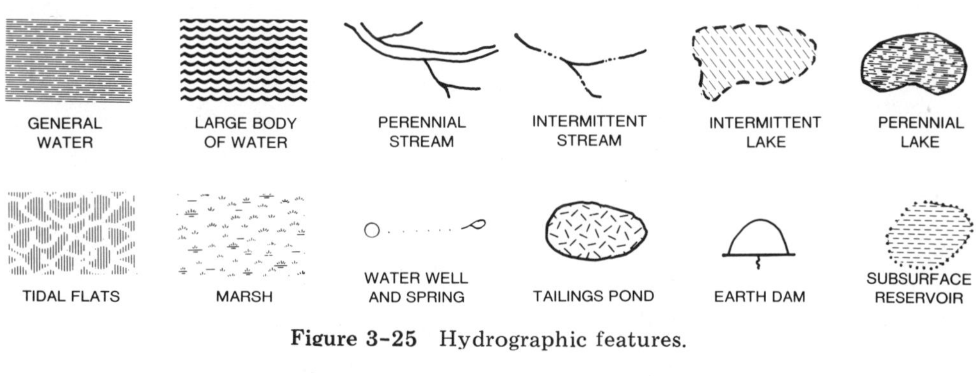

Figure 3-25 Hydrographic features.

Cultural features. Cultural features are any human-made feature, such as buildings, roads, railroads, and so forth. At times these features also fall within the definition of the other groupings; for instance, dams, canals, and reservoirs are also hydrographic features. Although mine excavations, ventilation controls, haulage and transportation features, and the like are all human-made features, they are discussed below under specialty features.

Surface relief. The vertical control of a map defines the topographical relief or changes in elevations of the surface. Several methods are shown in Figs. 1-13 through 1-17 for representing topographic relief. The most common and useful method is the use of contour lines. Figure 3-26 depicts the major form of contour lines. Contour mapping is discussed in detail in Chapter 4.

Vegetation features. Surface woodlands, grasslands, and other forms of vegetation are rarely depicted on working drawings for a mine. Exceptions are revegetation and environmental impact studies connected with surface mining. Usually, these studies use conventional crosshatching to represent vegetation types. When more specific symbolization is required, transparent overlays are available, as depicted in Fig. 3-27.

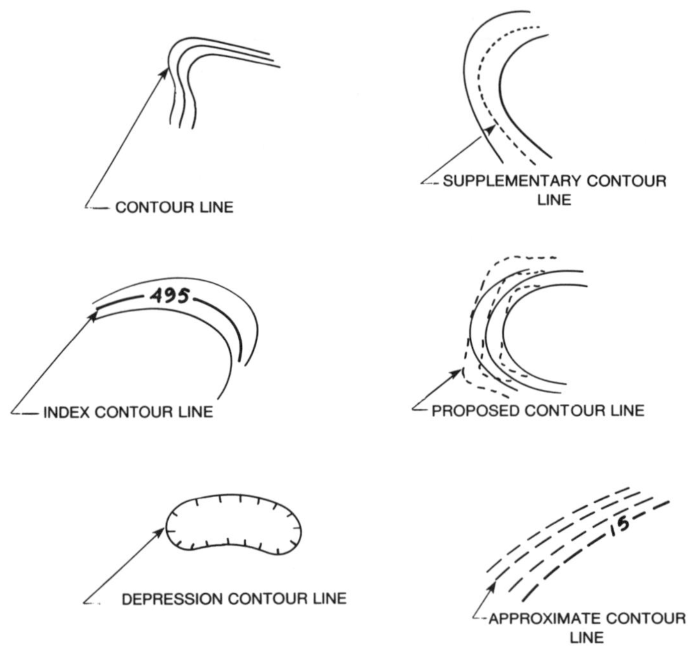

Figure 3-26 Common types of contour lines.

Figure 3-27 Vegetation features.

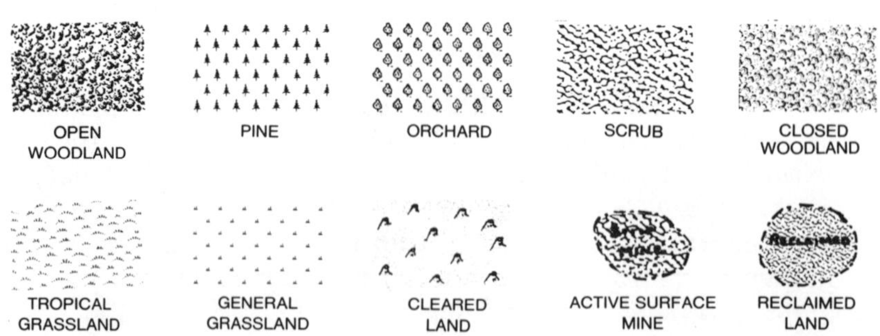

Specialty symbolization. The speciality feature class covers those features not previously described. For our purposes, we are most concerned with geologic features, oil and natural gas features, and mining features.

Geologic features that need to be symbolized are those principally concerned with describing the subsurface structural character of a region. These features include the type of subsurface material, (e.g., limestone and sandstone) and the structural anomalies (e.g., faults and discontinuities; Figs. 3-28 and 3-29).

Oil and natural gas symbolization is concerned primarily with designating the nature and type of well site, as shown in Fig. 3-30. When the field extent is needed, a dashed line is commonly used.

Mining feature symbolization is the least standardized of the three. One reason for this is the complex nature of mining activities. A second reason is the many different approaches taken to accomplish the same task. Another is the desire on the part of the mining firm to keep plans and activities secret (one way is by using special symbolization codes). A fourth reason is a lack

Figure 3–28 Geologic materials symbols.

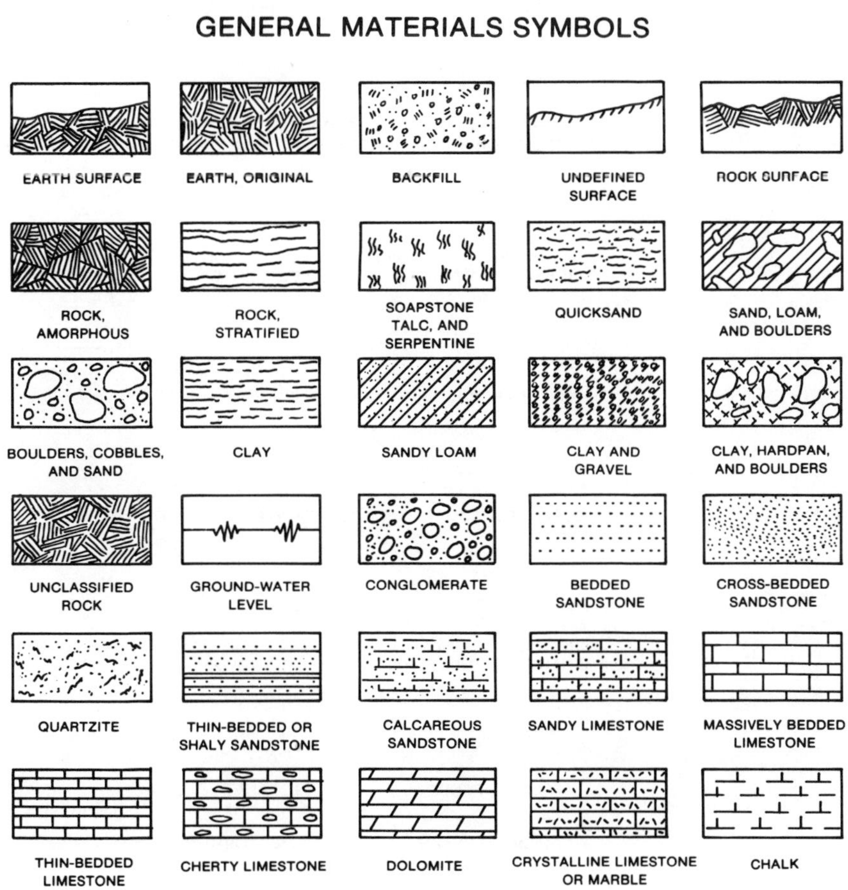

Figure 3-28 (continued)

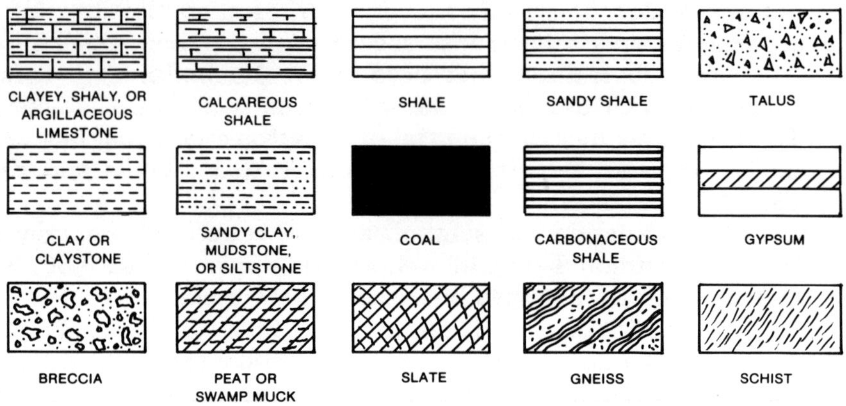

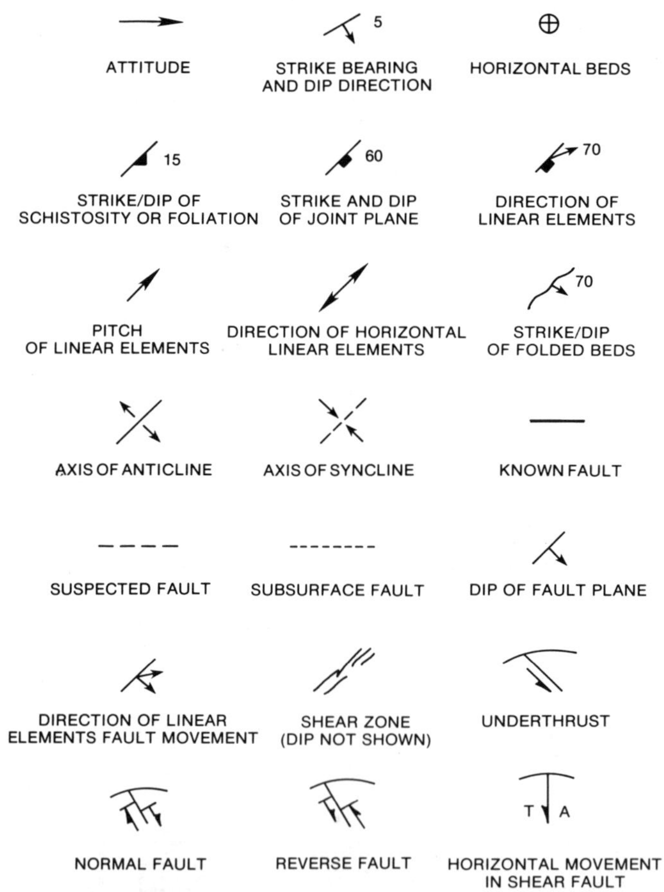

Figure 3-29 Standard geologic map symbols.

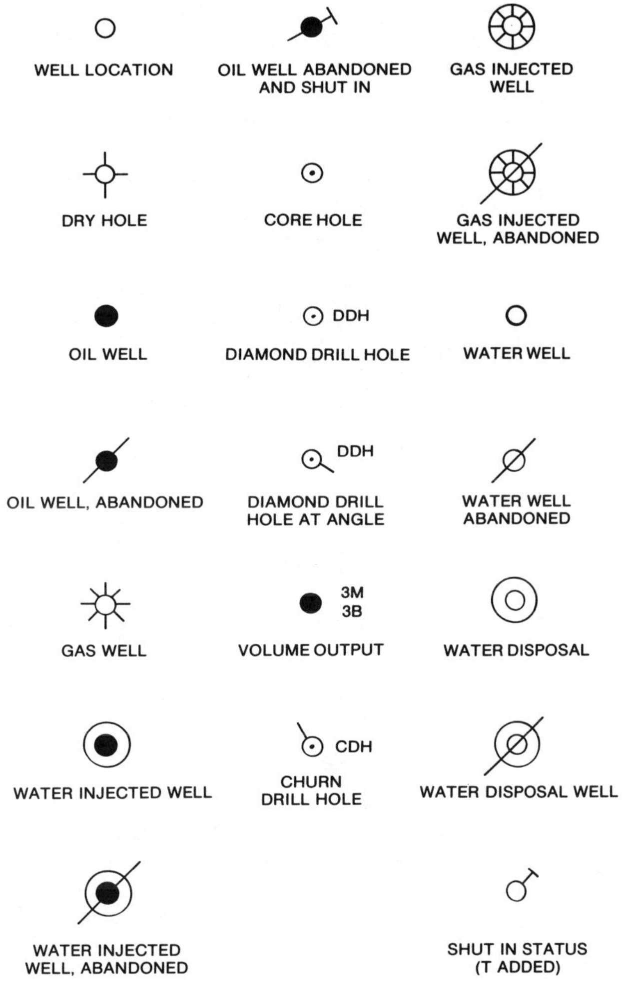

Figure 3-30 Standard petroleum and bore hole symbols.

of interest on the part of the larger and older firms to standardize their activities with those of others. Additional reasons can be stated, but the reality of the situation is that standardization will occur either by agreement of operators or through government mandate.

Various features are needed to describe properly the various phases (prospecting, exploration, exploitation, closing) and functions (e.g., ventilation, haulage, drainage, electrical, preparation, and roof control) of mining operations. The Appendix is an initial effort toward standardizing the major functional activities. Figure 3-31 shows the more common mine mapping features, and Figs. 3-32 and 3-33 depict the more common flowsheet symbols that may be needed.

Figure 3-31 Proposed mine map symbols (1 in. = 100 ft). (*, 1 in = 500 ft: **, 1 in. = 20 ft.)

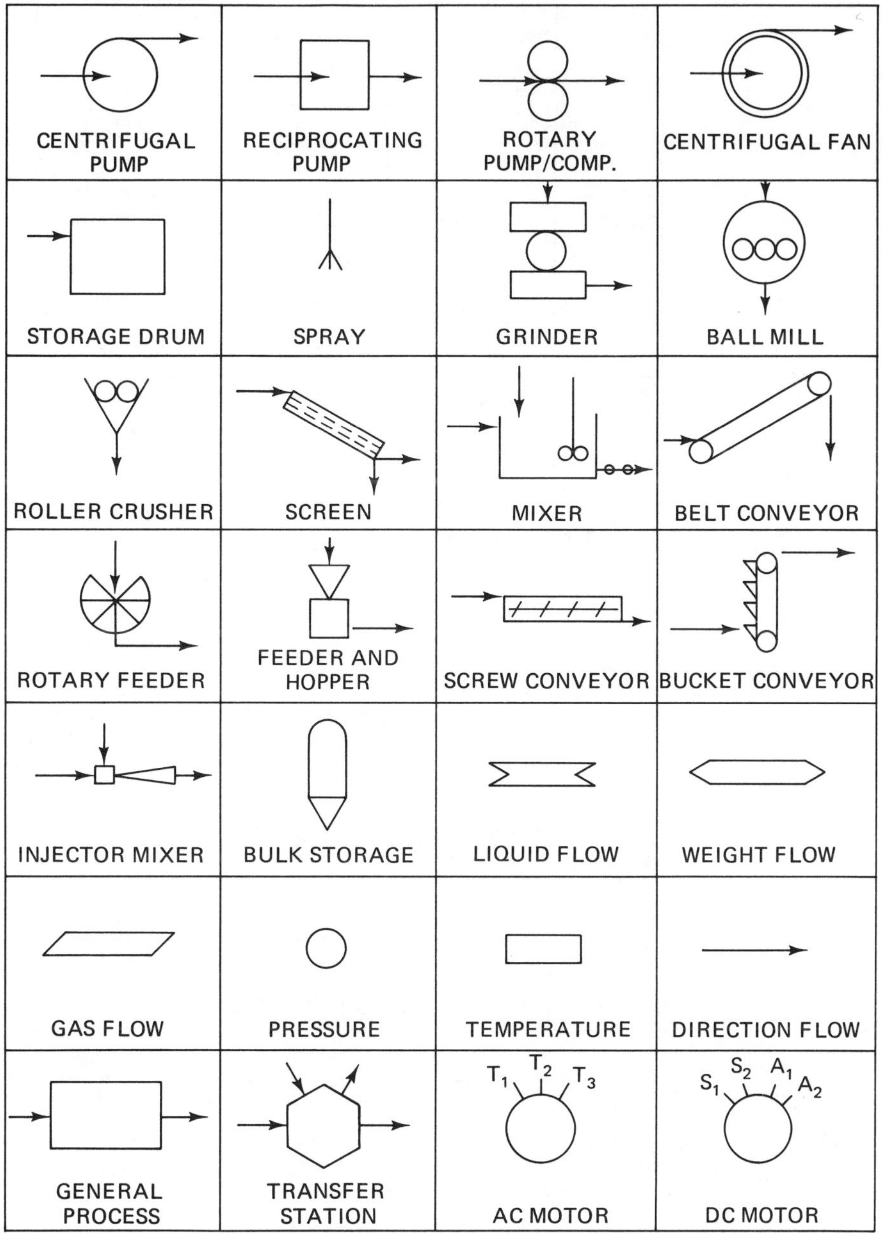

Figure 3-32 Flowsheet symbols.

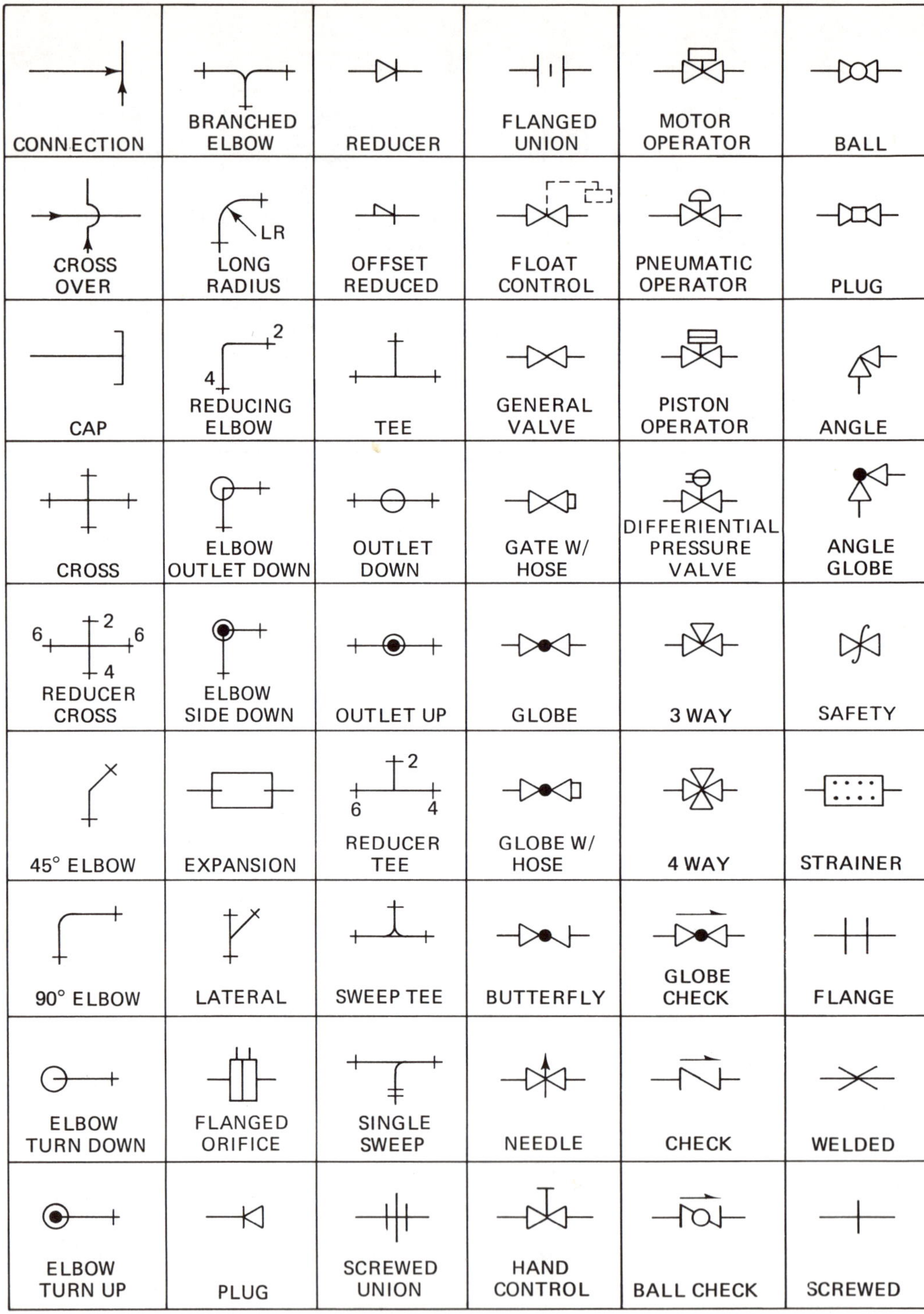

Figure 3–33 Schematic pipe symbols.

REFERENCES

1. Standards Y 14.1 through Y 14.5, American National Standards Institute, Inc., New York, 1957.
2. Morris M. Thompson, *Maps for America*, U.S. Dept. of the Interior, U.S. Geological Survey, U.S. Government Printing Office, Washington, D.C., 1979.
3. Topographic Maps, U.S. Dept. of the Interior, U.S. Geological Survey, U.S. Government Printing Office, Washington, D.C., 1976.

PROBLEMS

1. Prepare a relational diagram showing the operations of a mining company (including administrative functions).
2. Prepare a flowchart of the operations and activities needed to construct a mine map.
3. What types of statistical displays might be needed prior to opening a mine?
4. Using only 45°-45° and 30°-60° triangles and a T-square, prepare the following angles: 15°, 30°, 45°, 60°, 75°, 90°, and 105°.
5. Prepare a series of parallel lines $\frac{1}{4}$ in. apart at a 15° angle to the horizontal.
6. Draw any line *AB* and use your dividers to divide the line in (a) half; (b) thirds; (c) fifths.
7. Using the same line weight and thickness, draw a square 32 ft $\times$ 32 ft (scale 1 in. = 20 ft). Within the square, draw a 32-ft-diameter circle, and within the circle draw a hexagon. Hold the drawing up to a light and observe the variations in line density.
8. Redraw the map in Fig. 1-13 at a size four times that shown.

4

Procedures of Mine Map Construction

4.0 INTRODUCTION

The major portion of the mine drafting effort is directed toward producing, updating, and revising mine maps. Many types of maps are needed to effectively plan, design, open, and operate a modern mine. The two most important maps are the property topographical map and the general mine map. Several auxiliary types of maps and diagrams that are needed to support these two major documents are described in later chapters.

The major purpose of this chapter is to present the basic methods of producing topographic and general mine maps. It is important that you be aware of general mining practices and limitations which most affect the layout of mine operations. Section 4.1 presents an overview of underground mining practices and nomenclature. Section 4.2 abstracts the major requirements for mine maps as mandated by federal and state regulations. Section 4.3 presents general information needed for the initial preparation of any map. Section 4.4 describes procedures of contour construction, profiling, and topographic mapping. Section 4.5 presents the development of mine maps using field survey notes, and lastly, Section 4.6 introduces projection and functional layout.

4.1 UNDERGROUND LAYOUTS AND PRACTICES

4.1.1 Classification by Entry

Underground coal mines are classified using a wide variety of criteria. The simplest classification scheme is the method of access from the surface to the coal seam. Three distinct methods are employed: (1) drift entry, (2) slope entry, and (3) shaft entry. A *drift entry*, sometimes called a *punch mine*, is a horizontal set of tunnels driven into the earth, often directly following the outcropped coal seam, as shown in Fig. 4-1(a). A *sloped entry* system is one in which access to the seam is gained by driving a set of inclined tunnels from the surface to the seam, as depicted in Fig. 4-1(b). A *shaft entry* system is one in which one or more vertical tunnels are driven from the surface to one or more coal seams lying below the entry point, as shown in Fig. 4-1(c).

Figure 4-1 (a) Drift entry. (b) Slope entry. (c) Shaft entry.

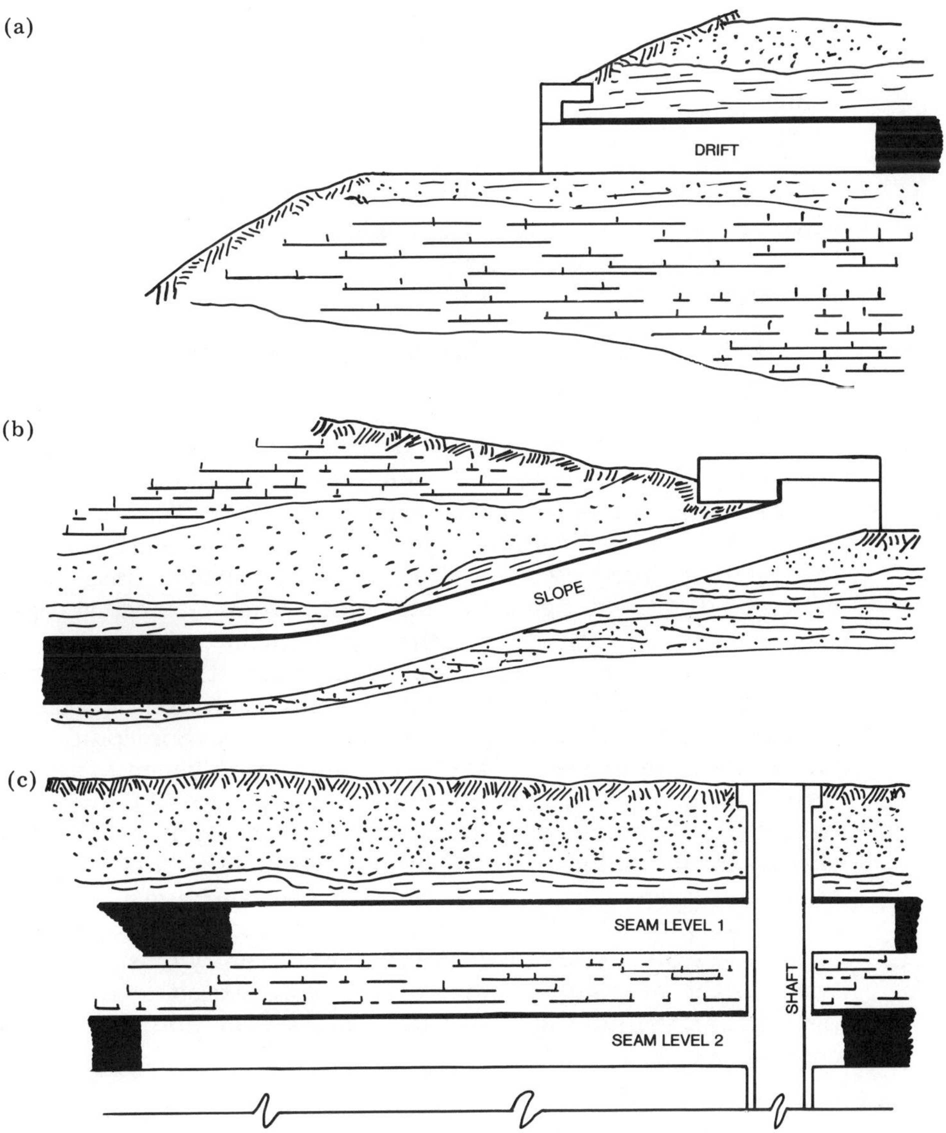

The individual tunnels that constitute the entry system are commonly called *main entries* or simply "mains." The main entries differ from one mine to another in terms of entry width, configuration, number, and direction.

Width. The width of the entry is influenced by (1) the thickness of the coal seam, (2) the methods and equipment needed for mining, (3) the stability of the immediate top and surrounding strata, and (4) factors related to haulage.

As the thickness of the coal decreases, the width of the entries must increase to permit proper airflow capacities and velocities. As the methods of coal extraction become more mechanized and as the equipment becomes larger, the width of the entries must correspondingly increase to accommodate proper operation and maneuverability of machinery. The greater the stability of the immediate top, the wider the entries can be; conversely, the more susceptible the immediate strata is to various rock failures, the shorter must be the entry width. Finally, various haulage factors affect the required width of the main entries. Principal among these is haulage or production capacity. In general, larger entry widths are needed to accommodate the larger equipment and increased traffic frequencies required to meet higher production needs.

Configuration. The configuration of the entry refers to the geometric cross section of the entry. The geometric shape varies from rectangular openings with square corners to fully circular openings. The factors affecting the resultant entry configuration are (1) the type of equipment used to mine the entry, (2) the stability of the immediate top (and longevity of the entry), and (3) the thickness of the coal seam.

The major factor influencing the shape of the main entries is the type of equipment being used to drive each tunnel. When continuous miners are being used, the configuration can range from rectangular openings with square corners (for ripper or drum-type miners) to oval entries (for boring-type miners). If the tunnels are being bored, the resulting configuration would be circular. The stability of the top is adversely affected by square-cornered configurations. Square corners such as those produced by ripper miners result in zones of "stress concentrations" which often cause premature failures. Thus, if the immediate strata is "weak" (particularly in shear strength), the configuration selected should be characterized by more gentle changes in the wall geometry. The thickness of the coal seam usually determines the height of the entry. Since the cross-sectional flow area ($w \times h$) needed for proper ventilation and required production capacities are predetermined or otherwise fixed, the width of the entry must be increased as the seam height decreases. Within limits, more exaggerated rectangular configurations are found in "low coal" mines than would be found in "high coal" mines.

Number. The number of mains constituting an entry system depends primarily on the present and future needs of the mine. These needs include (1) the types of haulage systems and number of entries needed for men, equipment, supplies, and coal and the number of individual entries required for each; (2) the size of the mine, number of individual working sections, and the number of ventilation intakes needed to supply the fresh air requirements within the mine; and (3) the size and geometry of the coal reserve and the number of submains required to subdivide the property properly to allow for "balanced" mining. As the number of submains increases, the total number of mains will increase proportionately.

Location. The location of the entry portals on the surface and the direction of the mains depend on (1) the geological stability of the overburden, (2) the location of surface support systems, and (3) factors affecting mining balance. Many geologic factors influence the stability of the ground. The more significant factors are the degree of faults, fractures, and joint discontinuities within the subsurface structure. Opening portals will be located at those sites surrounded by a minimum number of geologic discontinuities. Surface support systems include surface transportation, electrical power, preparation plants, and the like. The positions of these needed systems relative to the locations of stable subsurface points help to determine the optimum entry points for the mains. Finally, the direction the main entries asume is determined by subdividing the mine property according to several possible criteria:

1. *Area of reserve.* Bisect the property so that half of the property lies on either side of the main and balances the production capacity.
2. *Coal analysis.* Divide the property so that half of the Btu/lb, percentage of sulfur, and percentage of ash content, or other qualitative characteristic of the coal will be on either side of the main, balancing the quality of coal extracted.
3. *Roof (ground) stability.* Divide the property so that problems associated with poor top can be equally distributed on either side of the main, thus balancing the ground control conditions.
4. *Environmental variables.* Divide the property so that the same problems of methane liberation, water intrusion, and so forth are balanced on either side of the bisecting main entries.

4.1.2 Classification by Stoping Practices

A second scheme used to classify underground operations was developed by the U.S. Bureau of Mines and is based on the nature of ore extraction and the methods used to support the underground stopes or openings.[1] This scheme is partitioned according to:

1. Naturally supported stopes (or openings)
2. Artificially supported stopes
3. Caved stopes
4. Combination stopes

Naturally supported stopes. The "naturally supported" category includes (1) open, self-supporting stopes and (2) open stopes with pillar supports. The strata surrounding most coal seams is usually quite weak and discontinuous and for the most part cannot be considered within the self-supporting category. Thus most underground coal mining employs small openings with pillars of coal left intact between these entries to support the overburden (and tectonic) loads, as shown in Fig. 4-2. These alternating rooms and pillars form the bases for most coal mining operations today.

Figure 4-2 Typical room and pillar mining. (Courtesy of Joy Machinery Co.)

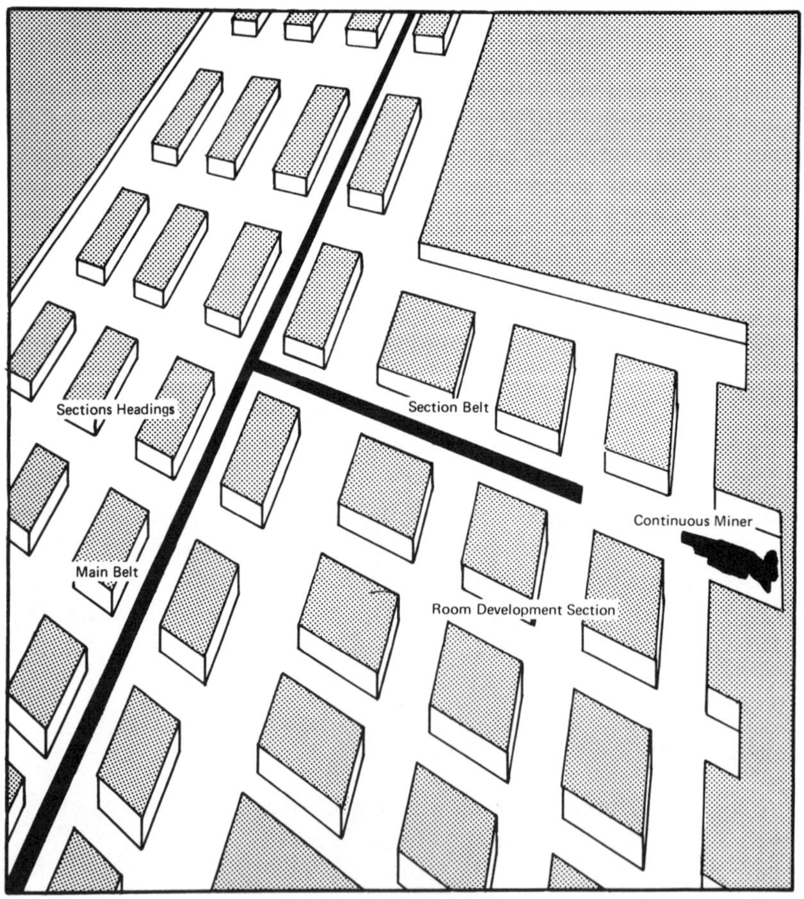

Artificially supported stopes. The pillars left during the room and pillar method of mining support the overburden and tectonic loads. However, the volume of material immediately above each entry, usually called the *minor load*, is "supported" by the strength of the immediate stratum. Since this stratum is often weak and unpredictable, additional mechanical support is usually required. This support may be timber sets, roof bolts, cribbing, yieldable rings, or other artificial mechanical support systems. Thus in most room and pillar mining, the entries are maintained using both natural pillar and artificial support.

Caved stopes. Caving is a mining practice in which the top is permitted to collapse following the extraction of the ore. Underground coal mining employs three variations of mining processes that use caving as a strategy: (1) pillaring or retreat mining, (2) longwall mining, and (3) shortwall mining. *Pillaring* is the systematic extraction of support pillars once the development mining of one region of the property has been completed. Extraction usually begins at the property boundary and retreats toward the main entries. The top is allowed to fall along a predetermined fall line after each pillar is removed, thereby relieving some of the overburden weight from pillars that remain. Pillaring is illustrated in Fig. 4-3.

Longwall mining is a system whereby a large block of coal (from 500 to 850 ft in face width and up to 10,000 ft in length) is mined by alternately cutting a 6- to 36-in. width of coal from the face. The system may be of either an advancing or a retreating type. With the advancing method, support entries (supplying power, intake air, return air courses, haulage, and the like) are driven slightly ahead of the mining of the block. In the retreating method depicted in Fig. 4-4, the support entries are completed prior to the mining of the longwall block. In both cases, the longwall face is protected and supported by artificial supports (either shields or chocks) which can be advanced as the mining of the face advances. As the supports are advanced, the immediate top behind them is left unsupported and is therefore allowed to fall.

Shortwall mining procedures are similar to those of longwall mining, except that the face is usually less than 250 ft wide and a continuous mining machine is used to extract the coal from the face instead of the more complex plows or cutting machines associated with the longwall system. A typical shortwall panel in the process of being mined is shown in Fig. 4-5.

Combination stopes. Underground coal mining employs all of the methods described above. Often, one mine will use longwall, shortwall, pillaring, and room and pillar development simultaneously. The use and location of each depends on (1) the location of property boundaries. (2) the stability of the top, (3) the location and density of gas or oil wells, (4) the surface status (extent of man made development) and factors affecting subsidence,

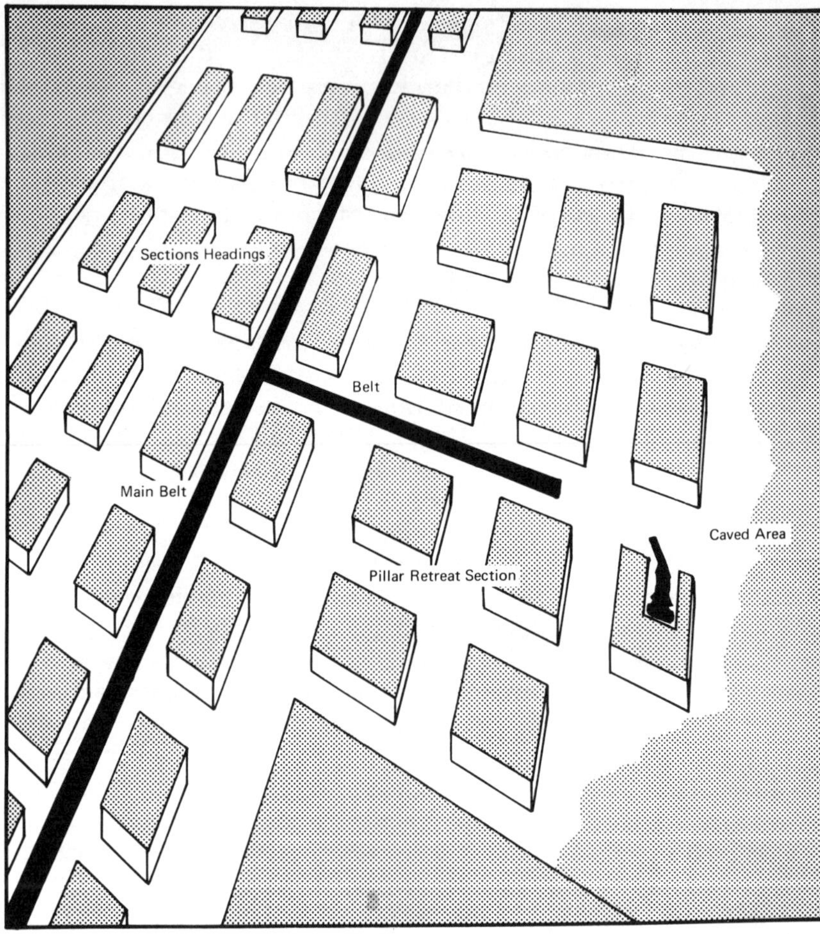

Figure 4-3 Pillar recover section. (Courtesy Joy Machinery Co.)

(5) the stability and nature of bottom, and (6) economic, marketing, and cash flow factors. The "optimum method" requires a careful assessment of the goals and limitations of the mining concern and a blending of the methods most suitable to the circumstances.

4.1.3 Nomenclature[2]

Before discussing the specific methods that can be used to identify and describe locations underground, we should first agree on the terms and definitions commonly used to describe underground workings. The terms can be broken into groups: (1) those describing a section of the mine, (2) those defining the types of pillars, and (3) those identifying the purpose of the opening.

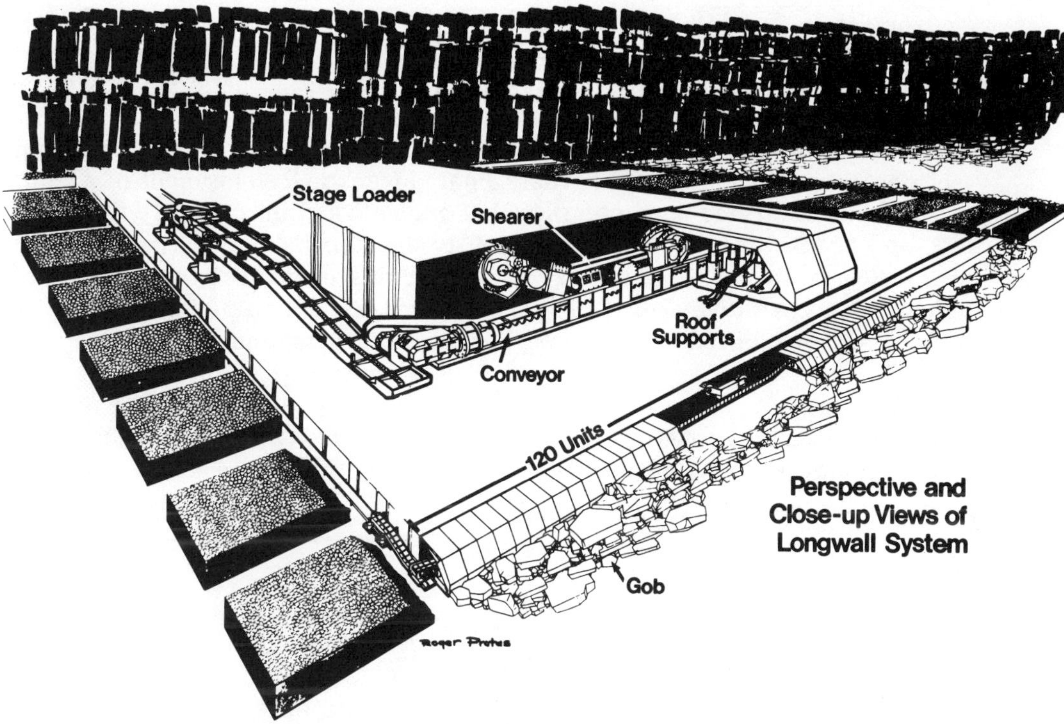

Figure 4-4 Layout of a retreat longwall mining section. (Courtesy of Joy Machinery Co.)

Figure 4-5 Layout of a shortwall mining section. (Courtesy of Joy Machinery Co.)

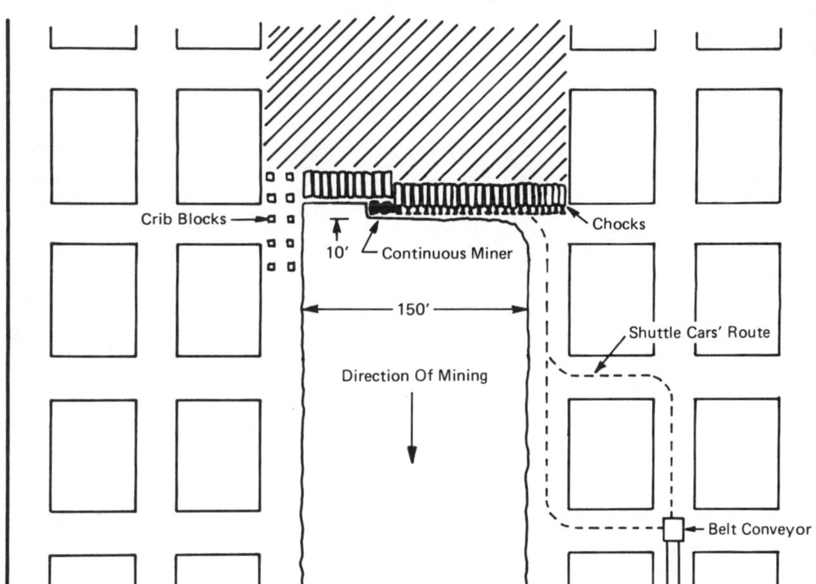

Sections. Sections of the mine can be described as working, development, retreating, longwall, shortwall, or abandoned. All types of sections, except abandoned sections, are considered *working* sections as long as workers are mining coal. *Development* sections are those working sections that advance openings into "virgin" or unmined coal. *Retreating* or pillaring sections extract the coal pillars that remain after the development of a room and pillar section has been completed. *Longwall* section is simply the active face and entries used to mine a previously developed longwall panel. *Shortwall* section is the active face and adjacent entries needed to mine a shortwall panel. Finally, an *abandoned* or idle section is any region of the mine that is not actively being used or is "closed" or sealed to personnel.

Pillars. Four types of pillars are shown in Fig. 4-6 and can be defined in most underground workings: (1) chain pillars, (2) barrier pillars, (3) safety pillars, and (4) speciality pillars. *Chain* pillars are those regularly

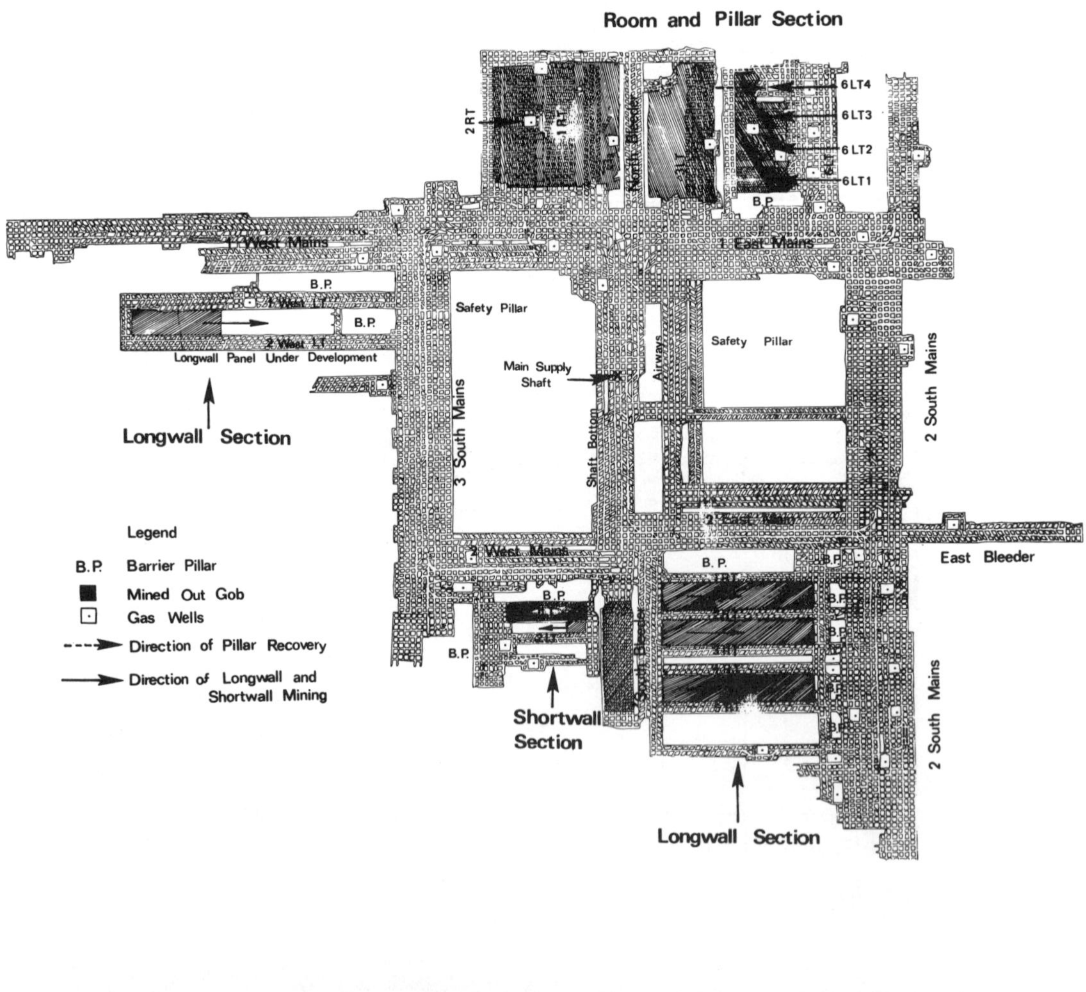

shaped pillars that remain to support the major loads after development entries and crosscuts have been driven. Chain pillars are commonly 80 ft long but may be as much as 200 ft long if roof stability necessitates greater support. *Barrier* pillars are usually much longer and wider than common chain pillars and are left in strategic locations to provide greater ground stability or to divide one region of the mine from another. Barrier pillars are left to protect the longwall headgate and tailgate entries and to isolate extremely gassy sections. *Safety* pillars are blocklike pillars, usually larger than barrier pillars, which are most commonly placed around shafts or main entry systems to ensure maximum ground stability. *Speciality* pillars are pillars left to separate miners from potentially dangerous conditions. For instance, the law requires that a 200 ft × 200 ft speciality pillar with a gas well in the center be left to separate the operations of mining from those of gas recovery.

Entries. The identification of entries is most often attributed to the use or purpose of that opening. Before denoting the various types, you should be aware of the various terms that are used to describe underground openings, including entries, headings, tunnels, drifts, slopes, room, cuts, stopes, courses, passageways, crosscuts, and breakthroughs. To eliminate some of the confusion that so many synonyms can cause, *entry* will describe all general openings used for major haulage, ventilation, and transportation, and *crosscut* will refer to those openings that span between entries and are used to facilitate ventilation and face transportation.

The various types of entries can be distinguished by the following terms: main entries (mains), cross mains (submains), section entries, room (panel) entries, rooms, bleeder entries, intakes, and returns. Figure 4-7 illustrates the layout relationship between these terms. *Mains* are entries that link the coal seam with the surface. Mains provide the initial bisection or division of the coal reserve. *Submains* or *cross mains* are driven off the mains at intervals up to 1500 yards and are used to provide access from the mains to the various regions of the coal reserve. *Section entries* driven perpendicular to submains at intervals up to 500 yards are used to subdivide the property into smaller regions or longwall panels. *Panel* or *room entries*, usually driven on 80-ft centers, establish blocks of coal that will be developed by room and pillar methods and provide access to the rooms. *Rooms* are driven perpendicular to the room entries and are often wider than the previously mentioned entries since the longevity of a room is significantly less than that of access entries.

The entry types can also be described in terms of their ventilation function. *Bleeder entries* or bleeders are driven in advance of mining to liberate methane gas from virgin coal and surround caved regions remaining after longwall, shortwall, or pillar mining, supplying a set of air courses for the gob ventilation. Bleeder entries thus provide a method by which methane gas can be controlled within a mine. *Intakes* are entries through which fresh air is

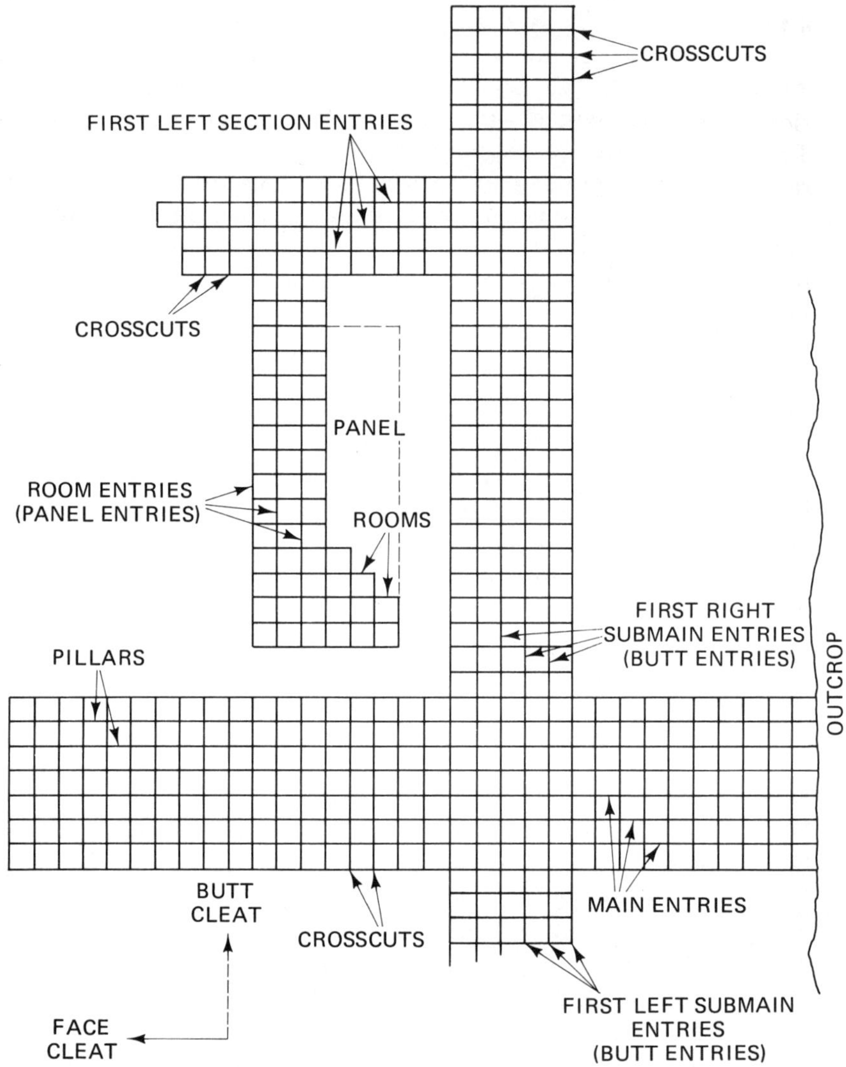

Figure 4-7 Entry orientation and nomenclature. (Courtesy of W.V. Mine Extension Service.)

directed to working or other sections. *Returns* are entries through which stale, gaseous, or dusty air is coursed away from active sections and out of the mine. (See also the Appendix.)

4.1.4 Entry and Section Identification

As the mine expands, the haulage, transportation, ventilation, communication, drainage, and other subsystem operations become more complicated. To lessen the severity of problems, the mine is often partitioned into separate quadrants or sections that operate independently of one another.

To facilitate rapid communications regarding transportation, supply, safety, or other problems, a system of region identification is needed. Also needed is a similar system by which mine survey teams and mappers can identify the location of any point within the mine.

One method usually accepted for meeting both needs is to use a system of identifying areas of the mine using *survey stations.* Survey stations are located along the heading centerline and are driven into the strata above the heading. In this way the survey station can be permanently located where it will not become easily dislodged. Survey stations, called *spads*, are numbered or coded in a systematic fashion and can thus be used for positioning identification (both physically in the mine and on the mine map). The symbol for a spad location is shown in Fig. 2-15 and is discussed in detail in the Appendix.

Various types of coded numbering systems can and have been used for spad identification, each applicable for a specific set of circumstances. Some different types are described below.[3,4]

Sequential system. The sequential numbering system, used mostly for smaller mining operations, numbers each spad sequentially according to the time the spad was installed. Thus spad 172 would follow the installation of spad 171 and precede the installation of spad 173. The greatest advantage of such a system is that it is simple, fast, and requires little advance planning. The disadvantage is that spad 172 may not be located outby (the portal side) spad 171 or even in a completely different quadrant of the mine; thus no *control* or consistency can be obtained. Although such a system does not hamper mapping of the mine, it does inhibit systematic communications or control system based on the spad number and is therefore limited in applicability.

Position-coded-sequential system. The position-coded-sequential system uses a three-part number to locate the general area of the workings. This system is used in more moderately sized mining operations. For example, with a spad number of 4-NW-172, the "4" can represent the level or entry point of the mine, the "NW" the northwest quadrant, and the "172" the sequential spad number located in that quadrant. Although this system is better than the sequential system, it still has certain obvious communication limitations.

Position-coded-distance system. The position-coded-distance system assumes that the mine map positioning control uses a rectangular coordinate system for positioning. With this system each spad is identified with either a three- or a four-element number. For example, the number 4-W220-N448 would be deciphered as follows: the "4" indicates level or main entry offtake, and the spad is located at coordinates (220, 448) in the NW quadrant of the mine. The obvious advantage, apart from significantly facilitating the mapping process, is that the spad numbers can be used to identify accurately any section of the mine. The greatest disadvantage of this system of identification is that it requires the survey team to calculate the exact location relative to the origin and to inscribe the spad plate with that information. However, with the advent of the surveying microprocessor carried by the instrument man or interlocked with the survey instruments, this becomes little more than reading, recording, and inscribing the coordinates furnished by the digital output of the instrument.

4.2 LAWS AND REGULATIONS

The federal government and most state governments, recognizing the critical importance of uniform mine maps, have issued regulations governing the preparation of mine maps and the information or content they must depict. These requirements are not all-encompassing, but they do provide mine operators and drafting personnel with certain guidelines that were not present prior to their enactment. The federal regulations covering mine maps are contained in "Code 30 Federal Regulations—Mine Safety and Health Act of 1969 as Amended in 1977."[5] Parts 75 and 77 of this act affect the mapping, layout, and general operation of underground coal mines. Although an attempt as been made to coordinate the requirements of the various state and federal agencies, some differences still exist. Table 4-1 summarizes the federal regulations compared with the West Virginia 1971 Coal Mine Safety Law, Section 22-2-1.

The greatest differences between the two documents are the more specific survey and mapping requirements of West Virginia. These are the use of a rectangular coordinate system for both surveying and mapping, the minimum specification (three) of survey monuments on the mine property, and the required use of either the USGS or the U.S. Coast and Geodetic Survey system for vertical control. Each state has similar enactments which are intended to ensure the safety and health of those working within the mining industry and to control the impact (environmental, economics, etc.), of the mining operation on the adjacent region. You should contact the Depart-

TABLE 4-1 SUMMARY OF STATE AND FEDERAL LAWS AND REGULATIONS GOVERNING MINE MAPS*

I. Regulations Governing the Surveying of the Mine Property and Representation of Those Notes.

1. Mine maps must show the scale and orientation of the map. (75.1200-1(b)), ((22-2-1 (33)))
2. The scale of mine maps submitted to the Secretary shall not be less than 100 or more than 500 ft to the inch. (75.1200-2(a)), ((22-2-1 (29-30)))
3. (a) Mine traverses shall be advanced by closed-loop methods of traversing or other equally accurate methods of traversing. (75.1200-2(b))
   (b) Mapping shall be done by the rectangular coordinate traversing method and meridians carried through and tied between at least two parallel entries of each development panel and panels or workings adjacent to mine boundaries or abandoned workings. ((22-2-1 (144-151)))
4. (a) The location and description of at least two permanent baseline points coordinated with the underground and surface mine traverses, and the location and description of at least two permanent elevation bench marks used in connection with establishing or referencing mine elevation surveys. (75.1200-1(h)), ((22-2-1 (81-86)))
   (b) These surveys shall originate from at least three permanent survey monuments on the surface of the mine property. The monuments shall be clearly referenced and described in the operator's records. Elevations shall be tied to either the U.S. Geological Survey or the U.S. Coast and Geodetic Survey bench mark system, be clearly referenced, and described on such map. ((22-2-1 (151-158)))
5. Contour lines passing through whole-number elevations of the coal bed being mined. The spacing of such lines shall not exceed 10-ft elevation levels, except that a broader spacing of contour lines may be approved by the District Manager for steeply pitching coal beds. Contour lines may be placed on overlays or tracing attached to mine maps. (75.1200-2(m)), ((22-2-1 (55-61)))
6. Such maps shall be made or certified by a registered engineer or a registered surveyor of the State in which the mine is located. (75.1201), ((22-2-1 (4-9)))
7. Mine maps shall be revised and supplemented at intervals of not more than 6 months. (75.1202-1, ((22-2-1 (101-108)))

II. Regulations Governing the Information Content Represented by the Mine Map

Cadastral/Cultural (not directly related to mine workings)

1. Name and address of mine. (71.1200-1(a)), ((22-2-1 (32)))
2. Adjacent mine workings within 1000 ft. (75.1200(h)), ((22-2-1 (70-72)))
3. Mines above or below. (75.1200(i)), ((22-2-1 (87)))
4. Either producing or abandoned oil and gas wells located within 500 ft of such mine and any underground area of such mine. (75.1200(m)), ((22-2-1 (91-93)))
5. The property or boundary lines of the mine. (75.1200-1(c)), ((22-2-1 (34)))
6. The location of railroad tracks and public highways leading to the mine, and mine buildings of a permanent nature with identifying names shown. (75.1200-1(g)), ((22-2-1 (76-98)))
7. All drill holes that penetrate the coal bed being mined. (75.1200-1(d)), ((22-2-1 (38-39)))
8. The location of all high-pressure pipelines, high-voltage power lines, and principal roads. ((22-2-1 (94-95)))
9. Where overburden is less than 100 ft, occupied dwellings. ((22-2-1 (99-100)))

Table 4-1 (continued)

Geological/Hydrology

1. Contour lines of all elevations. (75.1200(d)), ((22-2-1 (55-61)))
2. Dip of the coal bed. (75.1200(f)), ((22-2-1 (40)))
3. Water pools above. (75.1200(j)), ((22-2-1 (88)))
4. The location of any body of water dammed in the mine or held back in any portion of the mine; provided, however, that such bodies of water may be shown on overlays or tracings attached to the mine maps used to show contour lines as provided under paragraph (m) of this section. (75.1200-1(i)), ((22-2-1(73-77)))
5. The elevation of any body of water dammed in the mine or held back in any portion of the mine. (75.1200-1(k))), ((11-2-1 (78-79)))
6. The outcrop of the coal bed within the bounds of the mine property. ((22-2-1 (41-42)))
7. The location of the principal streams and bodies of water on the surface ((22-2-1 (89-90)))

III. Mine-Related Information

1. The active workings. (75.1200(a))
2. All pillared, worked-out, and abandoned areas, except as provided in this section. (75.1200(b)), ((22-2-1 (80)))
3. Entries and air courses with the direction of airflow indicated by arrows. (75.1200(c)), ((22-2-1 (64-65)))
4. Elevations of all main and cross or side entries.
5. Escapeways. (75.1200(e)), ((22-2-1 (118))) and ((22-2-1 (69)))
6. Such other information as the Secretary may require. Such maps shall identify those areas of the mine that have been pillared, worked out, or abandoned, which are inaccessible or cannot be entered safely, and on which no information is available. (75.1200(k-1)), ((22-2-1 (80))) (partial) ((22-2-1 (101-102))) (partial)
7. All shaft, slope, drift, and tunnel openings and auger and strip-mined areas of the coal bed being mined. (75.1200-1(e)), ((22-2-1 (35-37)))
8. The location of all surface mine ventilation fans; the locations may be designated on the mine map by symbols. (75-1200-1(f)), ((22-2-1 (66-68)))
9. The elevations of tops and bottoms of shafts and slopes, and the floor at the entrance to drift and tunnel openings. (75.1200-1(j)), ((22-2-1 (43-45)))
10. The elevation of the floor at intervals of not more than 200 ft in:
    (1) At least one entry of each working section, and main and cross entries.
    (2) The last line of open crosscuts of each working section, and main and cross entries before such sections and main cross entries are abandoned.
    (3) Rooms advancing toward or adjacent to property or boundary lines or adjacent mines. (75.1200-1(k1,2,3)), ((22-2-1 (46-54)))
11. Temporary notations shall include:
    (1) The location of each working face of each working place. ((22-2-1 (48-49)))
    (2) Pillars, mined, or other such second mining ((22-2-1 (62-63)))
    (3) Permanent ventilation controls constructed or removed, such as seals, overcasts, undercasts, regulators, and permanent stoppings, and the direction of air currents indicated. ((22-2-1 (114-178)))
    (4) Escapeways designated by means of symbols. ((22-2-1 (69)))

*( ), reference for the Code of Federal Regulations[4]; (( )), reference for the 1971 West Virginia State Law.[5]

ment of Mines, Department of Natural Resources, or equivalent agency in your region to obtain applicable copies of the mine mapping regulations.

In addition to the regulations covering general mine maps outlined in Table 4-1, similar regulations mandate the information content of certain functional maps and diagrams that must accompany permits and mining applications. The functional maps and appropriate federal regulations governing their content include:

1. Electrical layout (75.508)
2. Ventilation layout (75.316-1)
3. Escapeway map (75.1704-2d)
4. Sealing plan to abandoned areas (75.330)
5. (Surface) ground control plan (77.1000)
6. Roof control plan (75.200)
7. Slope/shaft permit (77.1900)
8. Surface mine map (77.1200)
9. Underwater mining permit (75.1716-3)
10. Impoundment/waste pile plan (77.216-2)

These regulations are discussed as appropriate in later chapters.

4.3 GENERAL MAPPING PREPARATIONS

4.3.1 Scale

Once the field data have been obtained and treated, the first operation in preparing a map is to select the most appropriate scale. For mine mapping, large scales are employed. By law, mine map scales must vary between 1 in. = 100 ft and 1 in. = 500 ft; however, larger and smaller scales are often employed for design and layout work. In general, 1 in. = 100 ft is the smallest practical scale for depicting any excavation detail. In the case of roof control layouts, the scale varies between 1 in. = 5 ft and 1 in. = 40 ft (75.200-5f[5]). Scales smaller than 1 in. = 100 ft result in drafting errors which are significantly greater than errors incurred in the collection and treatment of field data. As a rule, state and federal regulations will usually yield to needed accuracy. Therefore, select the scale that best satisfies your needs.

4.3.2 Paper Selection

The next stage in map preparation is to determine the most suitable paper size for portraying the information. The selection of paper size is dependent on (1) the total latitude and total departure of the traverse, (2) the scale ratio, and (3) the availability of paper types and sizes (see Table 3-1). The total length and width of paper needed to construct the map is then determined by multiplying the total latitude and departure (found by subtracting the smallest coordinate from the largest coordinate) by the scale ratio. The paper size (selected from Table 3-1) would then be the size of paper that will accommodate these dimensions plus an appropriate margin. Let us assume that for a certain survey the total latitude spans 4100 ft, the total departure covers 6200 ft, and that we need to show this to a scale ratio of 1 in./500 ft. The total scaled length in latitude is the product of 4100 ft and 1/500, or 8.2 in; the scaled departure is equal to 12.4 in. Allowing a minimum 1-in. margin, the minimum paper size would be 9.2 × 13.4 in. From Table 3-1, the closest standard paper size would be B, 11 × 17 in.

4.3.3 Orientation of Data

The third stage is to orient the paper so that the plot will be where the draftsman wishes. This is accomplished by positioning the horizontal control points, principally the points of origin, baseline, and meridian orientation.

The point of origin on the paper will usually coincide with the field point where the survey began and to which all bearings and distances refer. To place the origin on the paper, determine the latitude and departure distances from the origin to those points corresponding to maximum or minimum points on the traverse and use the scale factor to lay off this point on the plot. The meridian orientation on the paper should be such that the points of maximum departure and latitude remain within the confines of the paper. (Most often the meridian spans top to bottom.)

4.3.4 Horizontal Control

The horizontal positioning of subsequent points along the traverse can now be established. Four methods of horizontal control can be used: (1) protractor and scale plotting, (2) tangent plotting, (3) cord plotting, and (4) coordinate plotting. The first three differ only in their method of determining the map angle that corresponds to the horizontal angle of the traverse lines. These three methods are described only briefly, as they are less appropriate for accurate mine mapping than the fourth method, coordinate plotting.

The *protractor and scale* method uses a protractor or automatic dividing head to lay off the appropriate angle relative to the initial meridian. An engineering scale is used to lay off the line distance relative to the origin. The tangent method uses the mathematical tangent of the needed angular departure to establish a second point, which lies on the line. With two points determined along the line (one being the origin), the angle is established, and

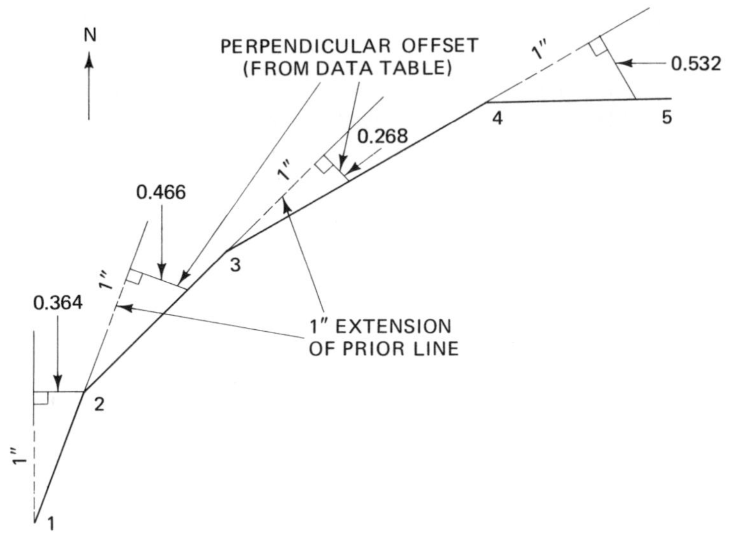

TANGENT METHOD PLOTTING TABLE/DATA

LINE	BEARING	DIST.	ANGLE	TANG'T	ADJ.	OPP.
1-2	N20° 00'E	100.0	20.0°	0.364	1	0.3640
2-3	N45° 00'E	150.0	25.0°	0.4663	1	0.4663
3-4	N60° 00'E	225.0	15.0°	0.2679	1	0.2679
4-5	N88° 00'E	1400	28.0°	0.5319	1	0.5319

Figure 4-8 Traverse plotting using tangent methods.

an engineering scale is used to lay off the distance. This method is illustrated in Fig. 4-8. The third method, *cord* plotting, is similar to the tangent method except that an arc of known radius is laid out from each successive station, cutting an extent in the line drawn from the station along the previous bearing. From the point where the extension line is cut, a chord of calculated length is laid off. The calculated length of the cord is equal to twice the sine of the half-angle between the adjacent bearings times the length of the arc. This method is illustrated in Fig. 4-9. Sloane and Montz[6] suggest constructing plotting tables for both tangent and chord plotting, similar to those shown in the example figures.

The *coordinate* method is the most accurate, popular, and in some states the only legally acceptable plotting method. The mapper first establishes a set of rectangular coordinates. This procedure is quite simple, particularly with the availability of a wide variety of coordinate lined paper. The reference latitudes and departures should be lightly denoted near or within the margin of the paper. Once the coordinate system has been established, the location of each point on the paper depends only on the corresponding *x-y* coordinates, which are usually a part of the survey computations. Figure 4-10 shows a simple coordinate plot of traverse calculations originally shown in Fig. 2-27.

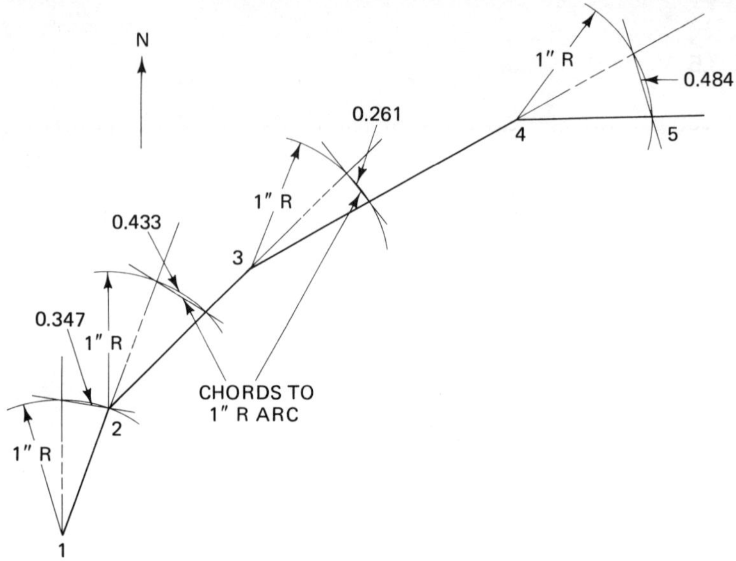

CHORD METHOD – PLOTTING TABLE/DATA

LINE	BEARING	DIST.	½ ANGLE	SINE	RAD.	CHORD
1-2	N20° 00′E	100.0	10.0°	0.1736	1″	0.3473
2-3	N45° 00′E	150.0	12.5°	0.2164	1″	0.4329
3-4	N60° 00′E	225.0	7.5°	0.1305	1″	0.2610
4-5	N88° 00′E	140.0	14.0°	0.2419	1″	0.4838

Figure 4-9 Traverse plotting using chord methods.

Figure 4-10 Plot of previous traverse, Fig. 2-28, using coordinate methods.

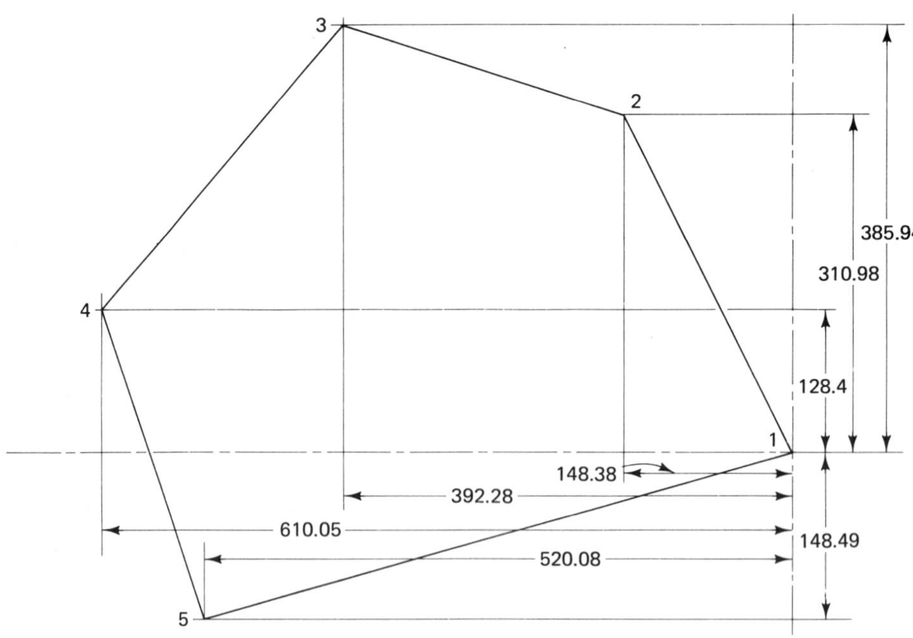

4.3.5 Vertical Control

After the horizontal control for the various points has been represented on the drawing, the vertical control should be (correspondingly) mapped. This requires that the elevation of each traverse point previously laid off and any additional elevations be recorded on the drawing. Detailed vertical control could then be established by drawing the contour lines. This procedure is developed in the following section.

4.4 TOPOGRAPHIC MAPPING

A topographic map is a line and symbol representation of natural and selected human-made features of a part of the earth's surface plotted to a definite scale. A distinguishing characteristic of the topographic map is the portrayal of the shape and elevation of the terrain by contour lines. The topographic map is a simple contour map, on which is plotted other physical and cultural characteristics of the region as determined by cadastral, hydrographic, and engineering surveys. The construction of a topographic map is a staged process usually beginning with the development of a contour map, followed by the overlaving of those other topographic data considered significant by the mapper.

4.4.1 Contour Maps

A *contour* is an *imaginary* line on the earth's surface that connects all *adjacent* points having the *same elevation* above or below a reference plane. Thus a contour is the intersection of a horizontal plane and some other surface. A *contour line* is an actual line drawn on a plan view map which *represents* the imaginary contour. The *vertical distance* between any two *adjacent* contours is called the *contour interval.* The contour interval is *constant* for a given contour map and is arbitrarily selected by the cartographer. Contour lines have certain characteristics which influence their construction and use. The cartographer drawing contour lines should always conform to the following guidelines:

1. Contour lines are never-ending and eventually close on themselves (Assuming that the paper and survey are large enough).
2. Contour lines never cross, except in cases of vertical walls or overhangs. In such cases the lower contours should be dashed lines.[5]
3. Contour lines that intersect streams or rivers always appear to cross perpendicular and straight to the centerline of water flow. Ideally, the contour line should not be drawn across the stream but should disappear at the shoreline.[5]
4. Contour lines never split.

5. Contour lines are always perpendicular to drainage (dip) and parallel to the horizontal (strike).
6. The contour interval never changes on a map.
7. Every fifth contour line is darkened to facilitate identification.
8. Contours are always measured in whole numbers (integers).
9. Existing contours are shown as solid lines and proposed contours are shown as dashed lines.

The importance of contour maps to mining operations cannot be overemphasized, particularly when drainage or transportation are considered. As noted above, the contour line is always perpendicular to the direction of water flow or dip. The *dip* is the greatest angle formed between the surface (or the stratum being considered; see Fig. 4-11) and a true horizontal plane. Thus the dip is always the direction of natural drainage. The *strike* is the bearing angle of a horizontal line on the surface of the earth or of the seam stratum and is always perpendicular to the dip when viewed in the plan or top view. Thus the tangent to the contour line is the strike. Strike and dip, when constant, are symbolized on the map as $\underline{\mid 5}$, where the horizontal line represents the direction and orientation of the dip. The number, 5, adjacent to the symbol represents the dip angle in degrees. In this example the dip is equal to 5°.

Figure 4-11 Strike and dip nomenclature and symbol. (Courtesy of USGS.)

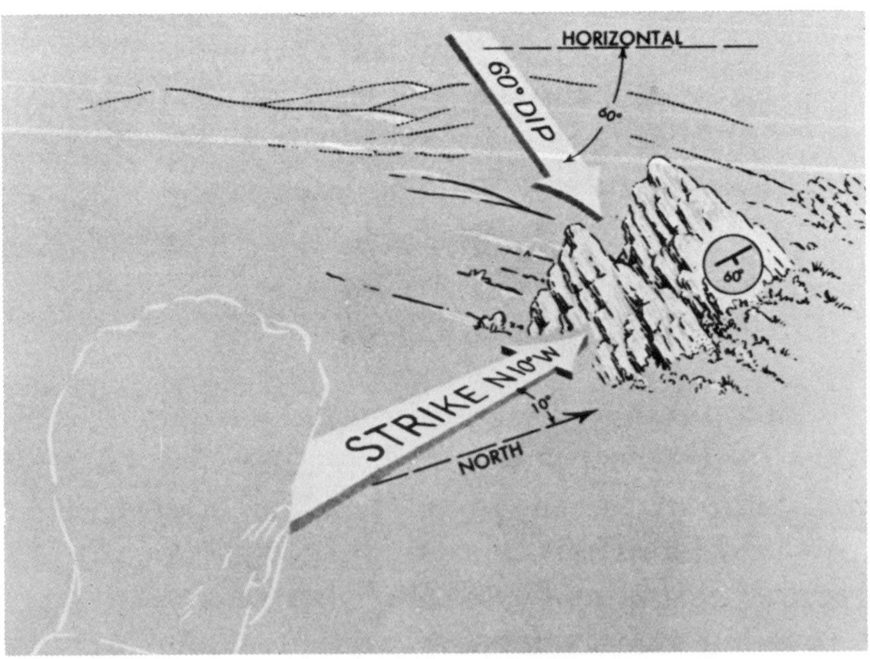

4.4.2 Interpreting Contour Maps

The configuration or pattern of contour lines is indicative of the shape or "lay" of the land. Some of the more prevalent land profiles and corresponding contour lines are shown in Fig. 4-12. A uniform slope is represented by evenly spaced contour lines. A steep slope or grade is indicated by lines closer together. A concave slope would have a greater number of contour lines at the top of the slope and fewer at the bottom; a convex slope would have the reverse condition. Obviously, the slope of the surface varies inversely to the distance between contour lines. Valleys are indicated by contour lines forming in a characteristic "V" which "points" uphill, while a ridge line, which forms a less distinct "V," points downhill. Cliffs or other extreme slopes are indicated by contour lines that appear to converge upon one another. Contour lines that close within the boundaries of the map indicate either a depression or a summit. You should be able to identify these topographic traits as they appear on Figs. 1-13 and 1-20.

The severity of relief indicated by the contour line plays a direct role in determining the location of entries, the direction and extent of headings, the type of mining, the methods of haulage (and other transportation), the method of drainage control, and other critical activities. For example, mainline haulage using mine locomotives is generally limited to grades less than $4\frac{1}{2}$%. The optimum grade for general haulage while still maintaining adequate drainage is closer to $1\frac{1}{2}$%. Thus the mine operator must have an

Figure 4-12 Typical relief and contour representation.

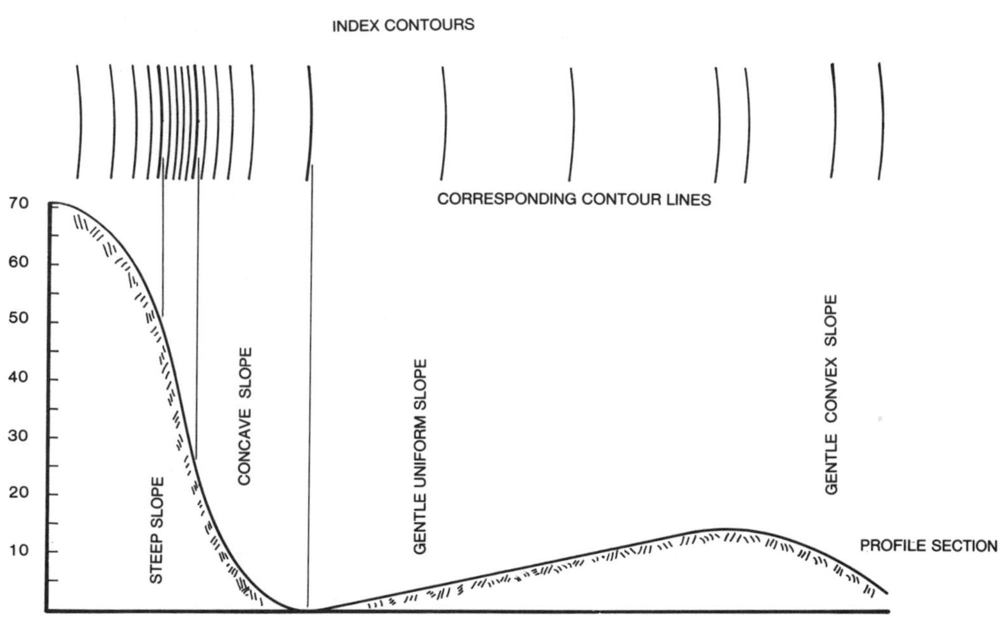

accurate record showing the variations in seam strike and dip. This record is conveniently provided by the contour map. A topographical map is basically a contour map which also depicts rivers, streams, lakes, vegetation, and human-made or cultural features, such as roads, bridges, towns, and utility services.

4.4.3 Survey Systems

To establish the location of a contour line on the paper, the mapper must be able to establish the horizontal and vertical position of each traverse point. These data are produced in one of several different forms. One form commonly used to report these data is the *grid layout.* With this system the region to be surveyed and mapped would have been divided into a system of rectangular grid points such as those shown in Fig. 4-13. The elevation of each point is determined through field leveling, and the whole-number contours that lie between the adjacent points are "interpolated" using linear graphical methods (described in Section 4.4.5).

Another form the field data might assume is the *contour survey.* In this system the survey team follows various site contours, recording the needed horizontal control data at differing points, as shown in Fig. 4-14. Although this method simplifies the mapping process, it significantly hinders the speed and efficiency of the survey team and is therefore rarely employed.

A third method, often used due to the speed it offers the survey team, is the *controlling point survey.* With this method those points of physical significance are located by the survey team. The contours are then approximated by the mapper, again using linear techniques of interpolation. This method is used when the exact contours are not required or when the relief of the region is fairly uniform. This method is shown in Fig. 4-15.

Figure 4-13 Grid system of leveling.

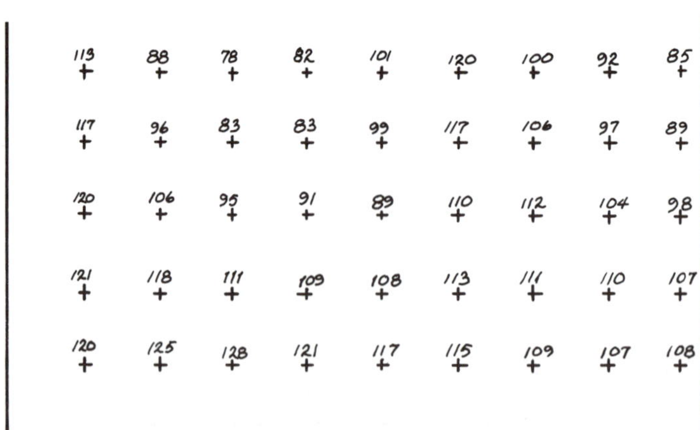

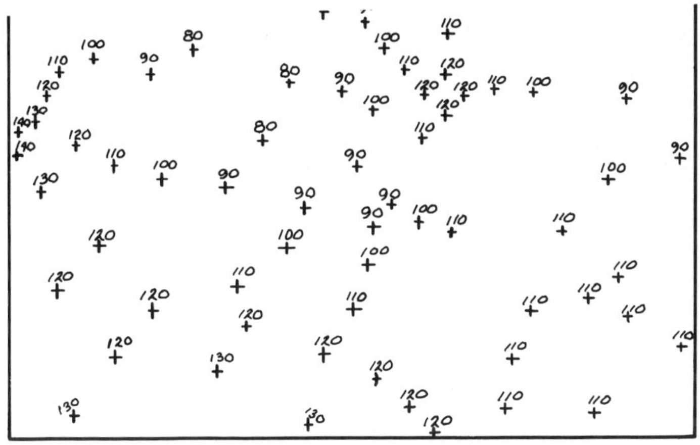

Figure 4-14 Contour system of leveling.

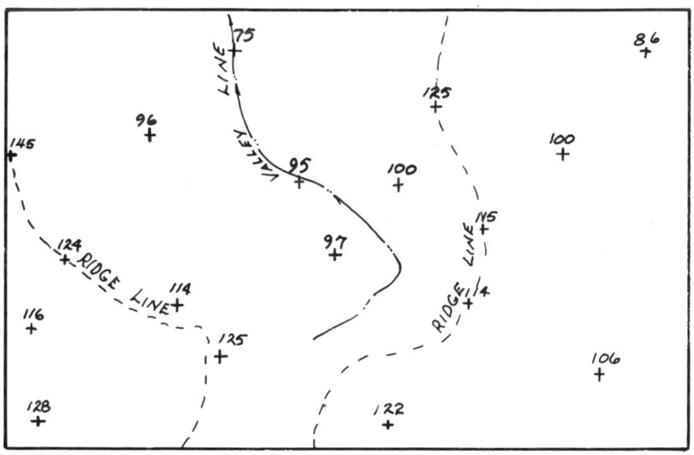

Figure 4-15 Controlling point system of leveling.

A fourth method or form that the field data may assume is the *random point system*. With this system the survey team records the horizontal and vertical control data points located randomly throughout the region. Like the previous system, this system usually increases the frequency of observations in the vicinity of significant features, such as summits, depressions, ridges, and valleys. Also like previous systems, the cartographer uses linear techniques to locate contour lines.

A final system, one used commonly in mines, is the *traverse leveling survey*, which records elevations along the path of the transit traverse. Since each entry and open crosscut is surveyed, this system, shown in Fig. 4-16, provides data in a form somewhat similar to the grid layout system. Also, like the grid layout system, the position of the whole-number contour lines must be determined from linear techniques of interpolation.

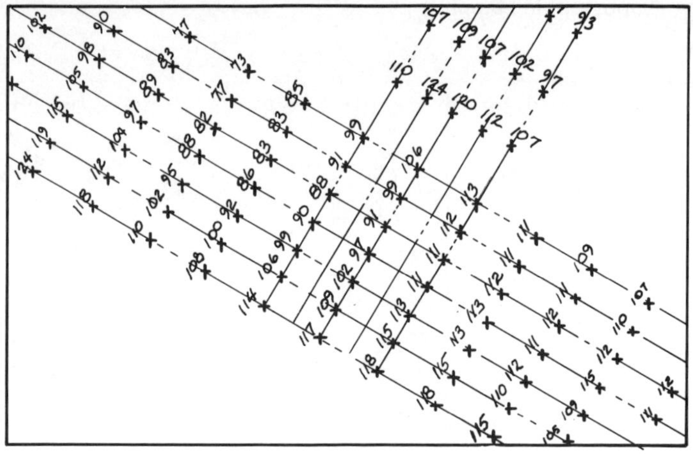

Figure 4-16 Traverse leveling.

4.4.4 Selecting the Contour Interval

Before the actual process of interpolating the contour lines, the cartographer must determine the most appropriate contour interval. The contour interval, you will recall, is the vertical distance between adjacent contour lines. The smaller the contour interval, the more detailed the topographic relief.

The selection of the most appropriate contour interval is not a purely arbitrary decision but is governed by four factors: the accuracy of the survey, the accuracy of the plotting, the topography of the area, and the ultimate use of the map. The four factors are not totally independent of one another but will be discussed below as if they were.

1. *Survey accuracy* is not a constant but depends on the equipment used, the purpose of the survey, the type of survey (geodetic versus plane), the number of survey stations, and so forth. If the accuracy of the survey is such that a variation equal to ±2ft in contour position resulted, a contour interval of 1 ft would be overly rigorous and therefore unjustifiable. Most modern survey equipment is quite accurate; therefore, equipment accuracy is of relatively less consideration than the other factors.
2. *Plot accuracy*, like survey accuracy, serves to limit the practical range of the contour interval. If a scale of 1 in. = 500 ft were used, the average accuracy of plot is ± 0.01 in. or ±5 ft at the site. A contour interval of less than 5 ft would be difficult to justify.
3. The *topographic relief* of the site most influences the range of contour interval selection. If the region is relatively flat a large contour interval would not show any topographical relief. Conversely, if a region were extremely hilly, too small a contour interval would result in numerous run-together lines which might interfere with the reading of the map. No hard and fast guidelines exist, however, aesthetically, maps requir-

ing more than 30 contour lines per inch or fewer than 2 contour lines per inch could be represented more effectively by changing contour interval.

4. The *end use* of the map also affects the topographic detail needed. For example, maps used by hikers need only display the general topography of a given area, whereas maps used for engineering work, such as drainage construction, must be quite accurate and detailed. The end use of a map also determines the scale of the map. In most cases the horizontal and vertical scales are the same. However, in rare cases the vertical scale may be made different from the horizontal when the cartographer needs or wishes to depict the vertical scale in a distorted fashion. For these cases a different contour interval will also be needed.

Most mining operations require accurate seam topography (C.I. = 1 to 5 ft) and less accurate surface topography (C.I. = 5 to 20 ft). The accuracy of a smaller contour interval depicting the working level is justified for coal mine mapping since (1) many closely spaced spad locations are used, thus the accuracy of survey is high; (2) map scale ratios are generally large, thus plotting accuracy is high; (3) the frequency of change in seam height of bedded deposits such as coal is usually small and almost always less than that of the surface; and (4) the maps will be used for the detailed engineering and layout of haulageways and drainage control.

4.4.5 Graphical Interpolation: Contour Construction

Contour lines are plotted only through points representing whole-number contours. Most survey systems determine the elevation either at points convenient for the survey team to locate, or at points considered critical to the mapping process; therefore, only rarely will the recorded elevation be equal to a whole number. Several methods exist for determining the position of whole-number contours located between two known elevations. Most of these methods assume that the profile between any two known points of elevation can be approximated by a straight line. Linear interpolation, as these methods are called, can be calculated analytically using ratio multipliers or can be graphically estimated. (Both methods are therefore based on geometric similitude.) The graphical method is more often used by mappers because it provides a very fast and accurate determination of the whole-number contours.

For purposes of example, assume that the contour interval is 1 ft and that a grid layout form was used to report the horizontal and vertical control data, as plotted in Fig. 4-17(a). Three whole-number contour lines, 57, 58, and 59, occur between the points 56.6 and 59.5 [see the rectangular box in Fig. 4-17(a)]. To find these whole-number contours, the draftsman must be able to lay off distances to the nearest 0.1 of an inch. This requires the following graphical/analytical procedures:

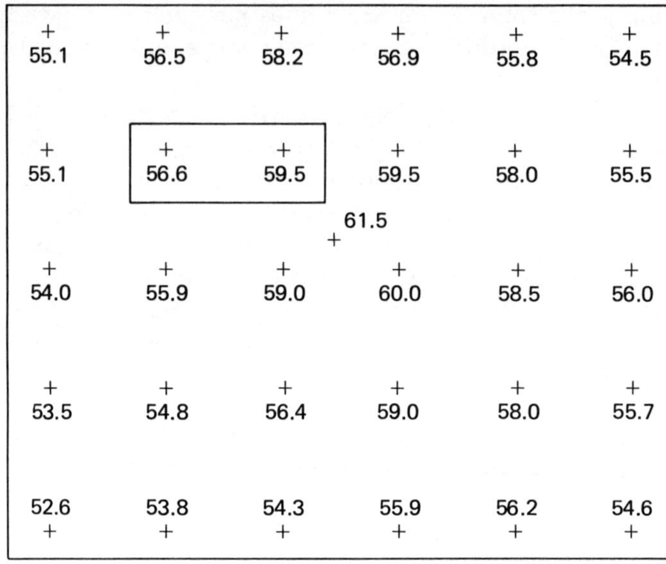

(a)

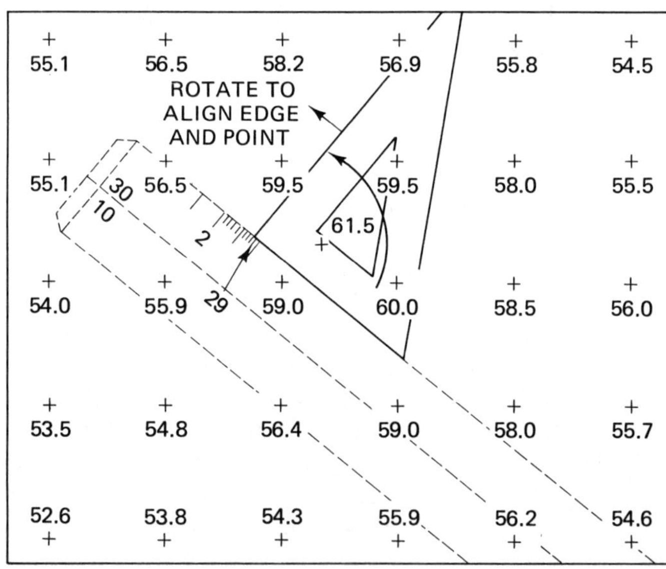

(b)

Figure 4-17 (a) Grid survey: graphical interpolation (C.I. = 1 ft). (b) Arrangement of scale and triangle. (c) Alignment of points, scale, and triangle (C.I. = 1 ft). (d) Movement of triangle and identification of whole-number contours.

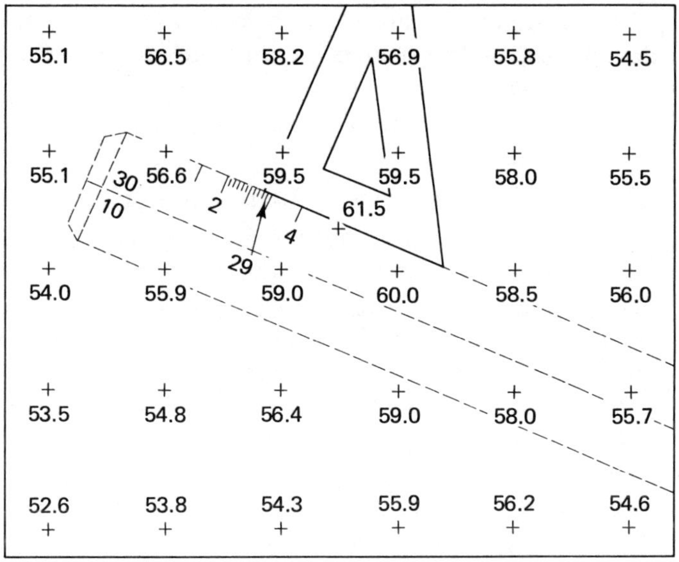

(c)

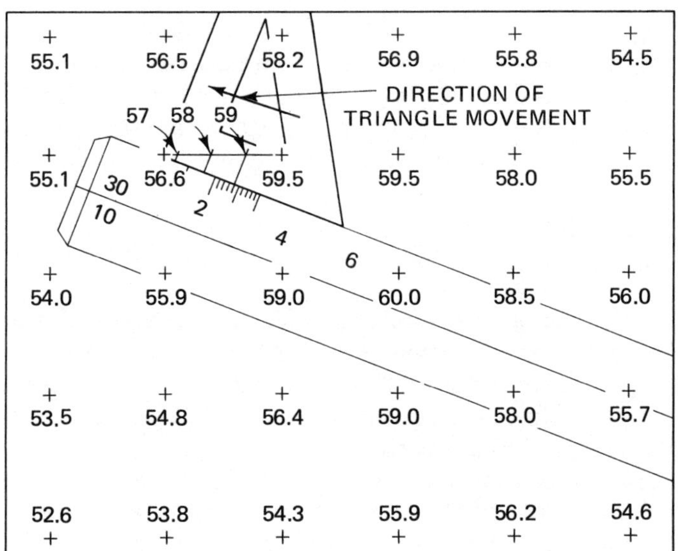

(d)

Figure 4-17 (continued)

1. The difference in elevation between the two known points must be determined and then divided into the number of 1-in. divisions needed. For this example:

$$\begin{array}{r} 59.5 \\ -56.6 \\ \hline 2.9 \end{array}$$ or 29 divisions of 1 in. each

   The distance on the paper between the two points (1 in.) must now be divided into this number of equal segments. To do this:

2. Select the side of an engineer's scale that has the 1 in. divided into either an equal or a greater number of divisions. For this example, the 30-divisions-per-inch scale is sufficient. Place the zero point of the scale at one of the points. (Which one is unimportant.) This is shown in Fig. 4-17(b).
3. Align one leg of a right triangle perpendicular to the side of the scale with the division on the scale that corresponds to the equal number of divisions needed between the two points. For this example, this point is 29. See Fig. 4-17(b).
4. Rotate both the scale and triangle (maintaining the 90° relationship) about the first point until the leg of the right triangle intersects the second point, 59.5. [See Fig. 4-17(b) and (c).]
5. The 1-in. distance between the two points can now be divided into the equal number of divisions needed by progressively moving the triangle toward the zero, stopping at each scale division or whole-number contour point to lay off a light mark between the two points. See Fig. 4-17(d).
6. Identify each contour line as needed. Then proceed to the next set of adjacent points, for instance, 55.1 and 56.6.

Although at first cumbersome, the method will eventually become "second nature" and in time you will find that the process can be completed quickly and accurately across any distance or elevation. (Like most drafting techniques, this requires practice. To facilitate this, a mapping notebook is available from the author.)

The final step needed to complete the contour map is the actual drawing of the contour lines according to the rules previously stated. Most contour lines are carefully drawn in using freehand sketching techniques; however, you may at first prefer to use a french curve, ship curves, or an adjustable curve until you reach a greater level of proficiency and confidence. Figure 4-18 shows the completed topographical map for the region.

The contour lines, called *intermediate* contour lines, should first be lightly (6H) approximated before final demarcation (2H to 4H). Every fifth contour line, the *index* contour line, is darker (H to 2H) and heavier to permit ease of reading. For engineering, supplemental and proposed contour lines may be desirable. *Supplemental* contour lines represent $\frac{1}{2}$-contour

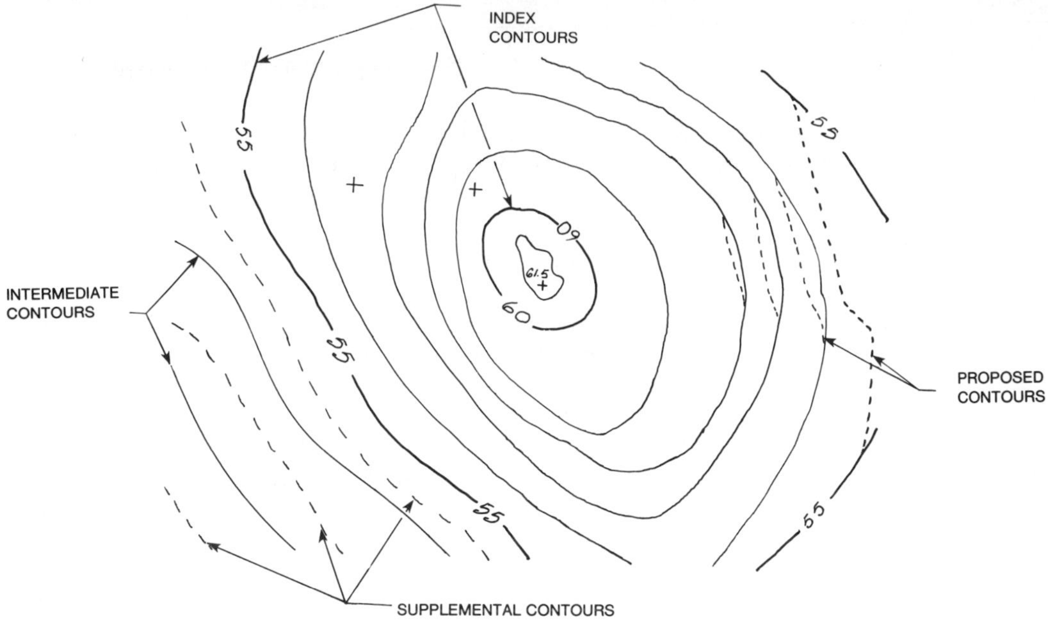

Figure 4-18 Completed contour map.

intervals and are used to show detailed relief, specifically for relatively level sites that show little vertical relief. *Proposed* contour lines are used to show proposed excavation changes to the topography of the site. Both supplemental and proposed contours are represented by dashed lines of medium weight (4H). The differences between intermediate, index, supplemental, and proposed contour lines are shown in Fig. 3-26.

4.4.6 Topographic Profile Construction

After the contour lines have been constructed, the draftsman or designer will often be requested to show a profile cross section of a given portion of the contour map. A topographic profile is a cross section of the earth's surface tilted so that the reader can better visualize the horizontal and vertical changes that are being displayed on the plan view. A profile, if you will recall, is a "side view" of some portion of the earth, constructed to scale from the data represented on the contour map. Topographic profiles serve a variety of purposes:

1. Display the apparent dip of the earth's surface along a predetermined "cut line" or route.
2. Show the amount of material that must be excavated or cut from the high spots and embanked or filled into the low spots to provide a given uniform slope along some predetermined route.
3. Show the thickness and stratigraphic positions of the sedimentary beds existing below the surface together with other geological conditions.
4. Aid the layman's interpretation of the vertical relieve that occurs along a predetermined route.

A typical profile is shown in Fig. 4-19. Although a profile construction is not a commonly observed trait of most topographic maps, its value for establishing drainage control, haulageway layout, geological layout, and route construction is practically essential. Therefore, the techniques and procedures required for profile construction will be presented at this time.

Procedures. A topographic profile can be taken anywhere on the earth and across any distance spanned by the map. Profiles can even be taken along curved paths and angular paths as may be needed for highway or railroad projects. Several construction methods can be used for directly projecting the needed profile. One method, shown in Fig. 4-19, uses a separate paper pinned or taped to the map. The individual contour planes are drawn as a series of lines parallel to the proposed cut line *AB*. The points formed by the intersection of the contour lines and the cut line *AB* are projected onto the appropriate contour planes in the profile. By connecting these projected points, the profile outline is generated. (This procedure is covered in more detail in the mapping notebook.)

Another construction method involves the construction of the profile directly on the contour map. This method is illustrated in Figs. 4-20(a)–(c) and described in detail on the following pages.

1. First, determine the region where a profile view is needed. Often this will be a proposed roadway, track heading, or region of suspected poor

Figure 4-19 Profile construction.

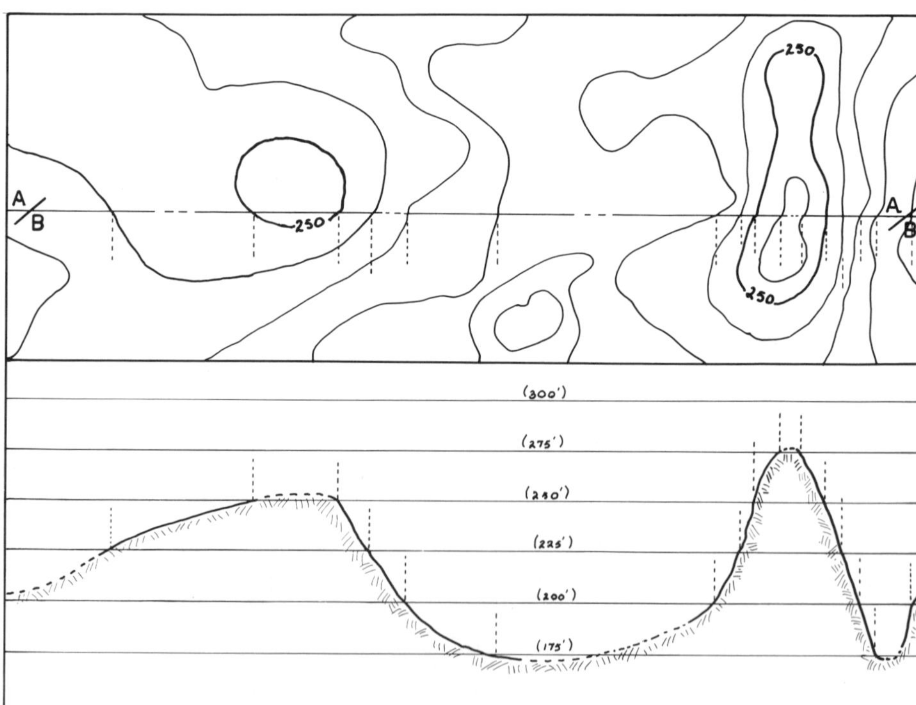

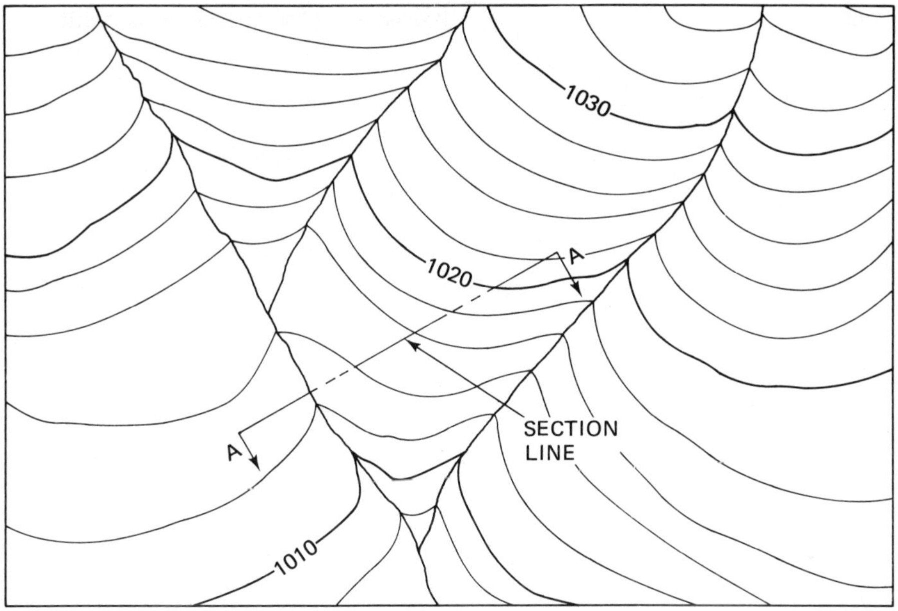

(a)

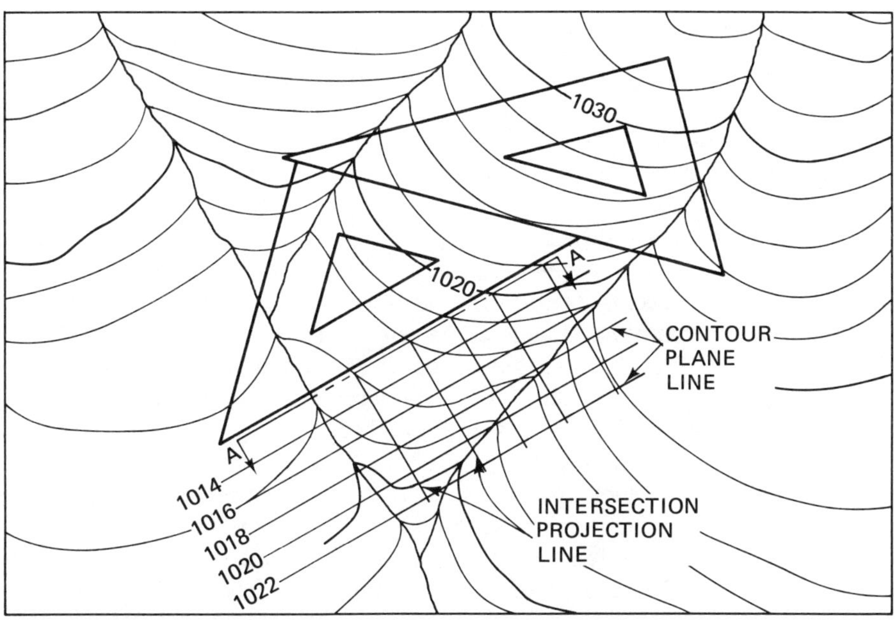

(b)

Figure 4-20 (a) Contour map and the desired profile section A–A. (b) Constructing contour planes and projecting points. (c) Completing the profile at section A–A.

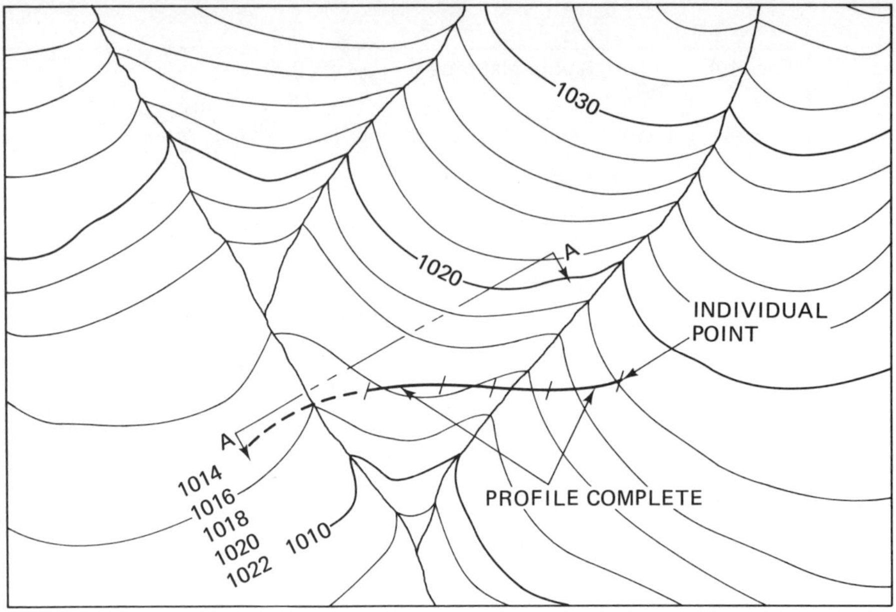

(c)

Figure 4-20 (continued)

top. The region to be profiled is defined by a cutting plane line (see Chapter 3).

2. Second, a series of equally spaced lines are drawn *parallel* to the cutting plane line. The lines, called *contour plane lines*, are uniformly spaced according to the *scaled* length of the contour interval. The number of lines should equal the number of contour lines cut by the cutting plane line. For best results, the contour plane line closest to the cutting plane line should be labeled with the elevation value of the lowest contour. Then successive lines are labeled with their corresponding contour elevations. Figure 4-20(b) shows the simplest way to construct these parallel lines using two triangles (when an indexing head is not available). Place the longer leg of your 30°-60° right triangle along the cutting plane line, then place the hypotenuse of your 45° right triangle along the hypotenuse or other leg of the first triangle. By moving the 30°-60° triangle along the 45° triangle, a series of parallel lines can be quickly constructed.

3. Next, draw extension lines (6H) from each contour line/cutting plane line intersection to the corresponding contour plane line. The extension lines must be perpendicular to the cutting plane line.

4. Connect the points on the profile that have been laid off, as shown in Fig. 4-20(c). Most draftsmen prefer using freehand sketching to complete the profile. Regardless of the drawing method used, the profile should be smooth and free of sudden changes unless called for by the field notes.

5. If the cutting plane line intersects the center of a depression or summit, the distance from the last contour plane to that known elevation can be scaled and laid off and the profile bottom or crest drawn as indicated. More often than not, however, the cutting plane cuts across a portion of the

summit or depression where the exact elevation between the extreme point and the last contour plane is not known. When this occurs, or for any estimated profile shape, the profile outline should be shown as a dashed line, similar in construction to a hidden line.

6. The final, although not always necessary step is to cross-section the region that has been sectioned. This demarcation of the surface is done using either general section lines, as shown in Fig. 4-20(c), or using strata coded sectioning, which is discussed next.

Lithological cross sectioning. The use of strata coded sectioning of profiles forms an important class of topographic profiling called *lithological cross sectioning* or "fense" diagramming. The relative position, number variations, and thickness of the various beds comprising the subsurface geology are characteristics of the lithology of the region. These characteristics significantly influence the type and success of the mining operations planned.

Lithological cross sections are constructed primarily from core logs, supported by additional information from surface outcropping notes and other geological data derived from different geophysical techniques. (These techniques are discussed in greater detail in Chapter 6.) The core logs provide local or point data, including the type, position, and apparent thickness of the subsurface structure. Using linear interpolation between the individual hole locations, an approximation of the lithological structure is derived. This approximation can later be modified using other available sources of data to effect a refined cross-sectional view of the subsurface. The mapping techniques needed to reduce the geophysical data to usable information, such as strike, dip, seam thickness, and so forth, are discussed in Chapter 6. (Additional techniques and problems are presented in the mine mapping notebook.)

Route mapping. The amount of excavation or cut and the amount of embankment or fill needed to provide a uniform slope along any given route can be determined using the contour map and corresponding profile. Figure 4-21 shows a proposed roadway route *AB* across a hilly terrain, the uniform profile (15% grade) of the centerline of that highway, and the original land form. The points of intersection between the roadway profile and the original contours are neutral points defining regions where material must be moved. The areas above the uniform roadway profile represent excavation areas and those below the roadway determine embankment regions.

The total amount of material that must be hauled depends on the size of the road and the desired side slopes. This requires the plotting of new or proposed contours to do this for the route illustrated, using the following procedures:

1. Determine the needed width of the road and the side slope ratio, usually measured as some horizontal distance to a 1-ft vertical distance

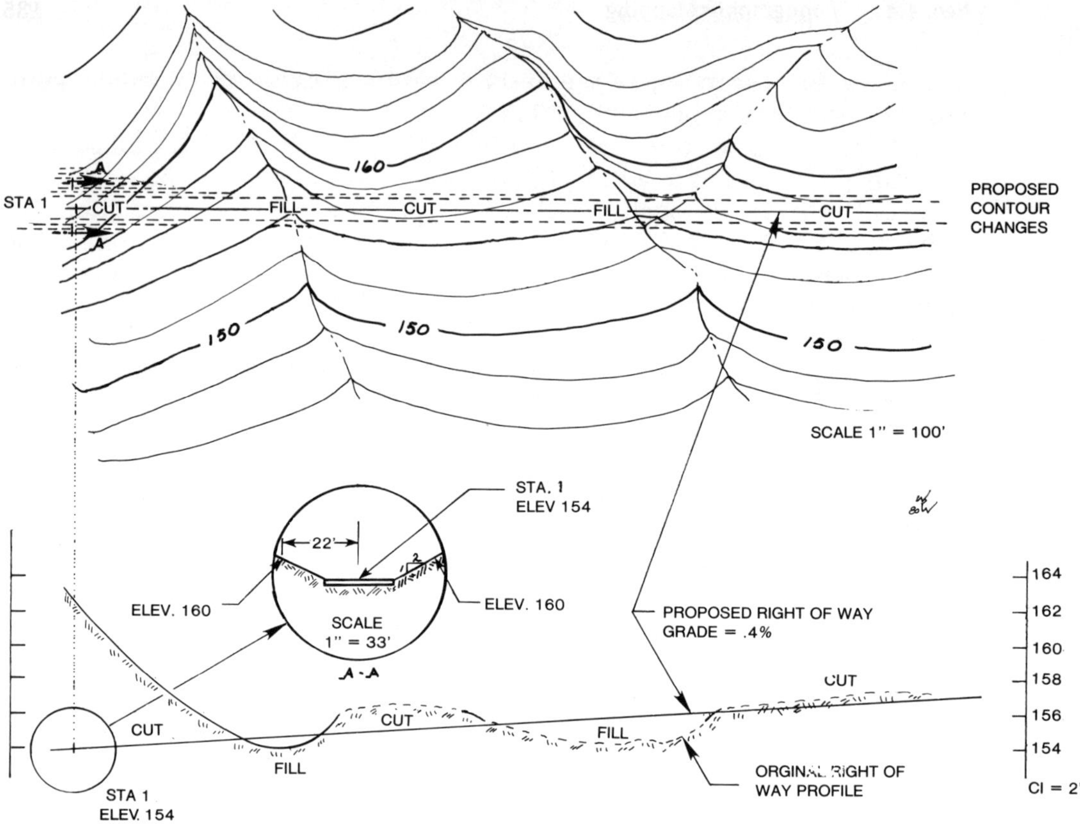

Figure 4-21 Cut-and-fill diagram for proposed roadway.

(e.g., 2 to 1). Let us assume a road width of 20 ft. Using a 2:1 ratio, the side slope will begin at a distance of 10 ft to either side of the roadway center. These should be drawn in as dotted lines since they represent proposed contours.

2. To locate points along contours to either side of a point on the roadway, you must subtract the elevation of the roadway at that point from different contour elevations, then multiply that amount times the side slope ratio, and finally scale that product and lay it off on the plan view. For example, at station 1 the elevation of the roadway is 154 ft. To find the position of the 160-ft contour, make the following computation:

$$(160 - 154)2 = 12 \text{ ft}$$

Therefore at station 1 the 160-ft contour is located 12 ft to either side of the roadway or 22 ft to either side of the centerline.

3. Repeat the process for various points and differing contours along the route, connecting the new contours as they will appear.
4. The boundaries of cut and fill as shown by the plan view can be de-

fined by connecting the points of intersection occurring between the old contours and the new contours.

4.4.7 Photogrammetry

The techniques discussed thus far are basic to your understanding of the drafting and use of contour maps. Before concluding this section of topographic mapping, we should review the powerful process of photogrammetry. Photogrammetry, as noted in Chapter 2, is a technique of measuremement using iconic images or photographs. Any surface that can be photographed can be measured using this technique. The greatest application to date is in the detailed construction of topographic maps. A detailed review of photogrammetry is beyond the scope of this text; however, several books have been written describing the procedures in detail.[7,8] Briefly, the process of generating a topographic map using photogrammetry is illustrated in Fig. 4-22 and proceeds as follows:

1. The region to be mapped is studied to define the best pattern and location of control points. These points will be used to establish horizontal and vertical control.
2. Field surveys are used to locate the control points physically.
3. The region is flown and photographed along straight predetermined paths. The number of paths and photographs taken within each path is

Figure 4-22 Photogrammetric procedures for topographic mapping. (Courtesy of USGS.)

determined by the height of the flight, the size of the region, the degree of detail required, and the inclusion of factors adversely affecting accuracy.

4. Once the photographs have been taken, they are assembled much like a mosaic, using only the central portion of the photograph. (As the edge is approached, the amount of distortion increases geometrically.)
5. A portion of the photograph is viewed through a stereoscope. By separating the photograph a given distance and viewing through the stereoscope, the viewer can "observe" the region three dimensionally.
6. By "following" a surface contour, the individual contour lines can be plotted with a stylus, which in turn is linked to an automatic plotter. (Several makes of stereoscopic plotters permit the viewer to select and follow surface contours.)

4.4.8 Completing the Topographical Map

After the contour lines have been completed, the remaining features of the topographical map can be laid off. These features include roads and highways, political and property boundaries, buildings, structures, and so forth. the symbols corresponding to the appropriate features and the drafting guidelines were presented in Chapter 3.

The following summarizes the procedures needed to construct a topographical map:

1. Determine the scale ratio.
2. Determine the necessary paper size.
3. Determine the placement of origin and baseline.
4. Plot the points defined by the horizontal control.
5. Label elevations corresponding to each point.
6. Establish an appropriate contour interval.
7. Find the whole-number elevations.
8. Draw the contour lines.
9. Locate boundaries, roads, waterways, and so forth.
10. Label and side-note as needed

4.4.9 Available Topographic Maps

In 1879, Congress established the USGS as a consolidation of four earlier organizations which had been engaged in topographic and geologic mapping throughout the continental United States. Shortly thereafter, in 1882, a plan for generating topographic maps of the United States was developed as the USGS National Mapping Program.[9]

A series of maps, known as the Topographic Map Series, are published by the USGS. The size and scale characteristics of the series are shown in Table 1-2. Other maps published through the National Mapping Program are (1) orthophoto maps, which are color-enhanced maps depicting the nature (marsh, sand, forested) and relief of an area; (2) county maps, which are 1:50,000 or 1:100,000 scale, showing political boundaries, road networks, topographical relief, and other cultural features; (3) state base maps (see Chapter 2); and (4) the National Park Series, which are shaded perspective maps depicting national lands, monuments, and historic sites. Information about the maps, charts, aerial and space photographs, geodetic control, and other cartographic details can be furnished by the National Cartographic Information Center, U.S. Geological Survey, 507 National Center, Reston, Virginia 22092.

4.5 CONSTRUCTING THE MINE MAP

The construction of the basic mine map, like that of topographic maps, follows from the data provided by the field survey. The basic mine map shows the controlling horizontal traverse of the mine property, the outline of existing entries and headings, location of gas and oil wells, existing mine structures and buildings, hydrographic data, utilities, roadways, and other data as mandated by federal and state regulations (see Section 4.2). This section concentrates on developing the techniques needed for projecting the outlines of existing mine excavations needed to produce a basic mine map. From this general mine map, various functional maps will be produced, including the ventilation layout (Chapter 7), the haulage layout (Chapter 8), the drainage layout (Chapter 9), and the electrical layout (Chapter 10).

As with all maps, the preparation of the general mine maps begins with an analysis of the survey data and the intended purpose of the map to determine the scale of the map and paper size needed. Mine maps showing excavation details, namely position and location of entries and pillars, should be represented by large-scale factors. Most mining firms require these maps to be drawn to a scale of 1 in = 100 ft. Maps depicting general operations or the relationships between operations are represented with scale factors of 1 in = 500 ft or smaller. Remember that as the scale of the map becomes smaller, the amount of detail shown on the map decreases and the extent of symbol generalization increases. Thus entries that are 20 ft apart in the real world are only 0.2 in. apart when drawn to a scale of 1 in. = 100 ft, and at a scale of 1 in. = 2000 ft, the rib lines merge to form a single line.

For certain types of layout (e.g., planning layouts) the mine map may be laid out either as an overlay or directly on an existing $7\frac{1}{2}$-minute topographic map. This is done when alternative portal locations, main entries, and facilities are being compared. In these cases, of course, the scale of the mine map must equal the scale of the topographic map (usually 1:24,000).

Following the selection of a map scale, the next step is to select the size of paper needed to represent the information. For limited data and smaller drawings, as contained within the examples of this text, the process is one of multiplying the extent of data by the scale ratio and selecting the most suitable paper size for the resulting dimensions. For most commercial mine mapping the data are so extensive and the scale factors so large that a single standard size of paper would not adequately represent the information. Instead, 36- or 48-in. rolls of drafting paper are used and the individual drawings assembled as a mosaic when a complete mine map is required.

The preparation of the control points on the paper for most mine mapping first involves demarcation of the rectangular coordinate system. Ideally, this system is directly tied into the state plane coordinate system. Care should be taken to ensure that the center of the paper and the center of the data coincide (see Section 4.3). The second procedure in establishing dimensional control is to define the surface control points, baseline, and bench marks. Next, the underground survey data can be represented.

4.5.1 Plotting of Underground Survey Notes

Two types of mine maps exist: (1) those showing the intended type and direction of future entries and crosscuts, and (2) those depicting existing entries and crosscuts. The former describe *projections*, which show only the centerline of the proposed entry. The procedures of designing and plotting projections are discussed in Chapter 6. The plotting of the latter is the result of careful deciphering of the underground survey notes.

If the underground survey notes have not previously been reduced to rectangular coordinates, the cartographer should first do this to speed the layout process. The first data the cartographer lays off are the spad locations and numbers. The spad identification code should be lightly noted since pillar outlines and other features remain to be shown. Each spad is ideally located along the centerline of the entry. If the mining process has been properly controlled, the spad positions should form a straight line corresponding to the ideal centerlines of the projections. To aid the subsequent operation of locating and denoting the outline of pillars, you may wish to draw in lightly (6H) the centerline of the entries and crosscuts.

The location of the rib line is determined from the side notes that accompany the survey notes. Several methods for describing the rib line were presented in Chapter 2, the simplest being the offset method. With the offset method, the location of the rib to the right and left of the entry centerline was denoted by first noting the position from one spad to a point along the centerline and then noting the distance of the left and right ribs from the centerline. One way in which these data can be symbolically reported is illustrated in Fig. 2-23. Another way is by showing the respective right and left rib distances on a field sketch, as shown in Fig. 2-28. Regardless of the method of side noting or the format used to report the rib location, points along the rib should be positioned carefully. As these points are laid off, caution should be taken by the draftsman not to lose perspective of the ribline by laying off too many points at one time. Rather, he or she should lay off one or two pillars (100 to 200 ft) and then lightly connect the rib line points. Most pillar outlines are drawn freehand. This is usually faster and some believe it to represent more accurately the inconsistent nature of the ribs within the mine. However, in those mines where good control of heading direction and size are observed, straight edges are often employed to represent rib lines. Either method is acceptable, and the conventions or standards existing within each company should govern your method of treatment.

The survey team will record the location of the last open crosscut or occasionally the face. These points are noted on the mine map by placing the data across the entry. The date should be numerically represented according to the month-day-year. The use of these dates for future planning, cost accounting, and operations research aimed at improving mining efficiency is of utmost importance.

As the mine is developed, pillar extraction will begin. To differentiate existing pillars from those partially or totally removed pillars, a system of crosshatching and dating is used. Figure 4-23 illustrates the designation of pillared sections.

Figure 4-23 Pillaring and caved symbols. (Courtesy of Consolidation Coal Corp.)

4.6 SUMMARY

This chapter has presented a review of mining practices, a discussion of the laws and regulations governing the preparation of mine maps, a presentation of topographic mapping techniques, and a presentation of the procedures needed to construct a general mine map. Other techniques of mapping needed for mining operations are presented in Chapters 6 through 10.

REFERENCES

1. C. F. Jackson and E. Gardner, *Stoping Methods and Costs*, U.S. Bureau of Mines Bulletin 390, U.S. Dept. of the Interior, U.S. Government Printing Office, Washington, D.C., 1936.
2. Paul W. Thrush, ed., *A Dictionary of Mining Mineral and Related Terms*, Bureau of Mines, U.S. Dept. of the Interior, Washington, D.C., 1968.
3. W. W. Staley, *Introduction to Mine Surveying*, 2nd ed., Stanford University Press, Stanford, Calif., 1964.
4. Charles G. Frush, "Mine Surveying," in Arthur B. Cummins and Ivan A. Given, eds., *SME Mining Engineering Handbook*, Vol. 2, Society of Mining Engineering, AIME, New York, 1973.
5. Federal Mine Safety and Health Act of 1977, Public Law 91-173 as amended by Public Law 95-164, U.S. Government Printing Office, Washington, D.C., 1977.
6. Roscoe C. Sloane and John M. Montz, *Elements of Topographic Drawing*, 2nd ed., McGraw-Hill Book Company, New York, 1943.
7. Frances H. Moffitt, *Photogrammetry*, 2nd ed., International Textbook Co., Scranton, Pa., 1967.
8. Paul R. Wolf, *Elements of Photogrammetry*, McGraw-Hill Book Company, New York, 1974.
9. Morris M. Thompson, *Maps for America*, U.S. Dept. of the Interior, U.S. Geological Survey, U.S. Government Printing Office, Washington, D.C., 1979.

PROBLEMS

1. Obtain a copy of the Federal Mine Safety and Health Law (Title 30, ref. 5) and the mine mapping regulations in your state. Make a detailed checklist of each content requirement to guide your future mapping.
2. Prepare a contour map from the data provided (on opposite page; use graphical interpolation.
3. Prepare a contour map from the data and construct profile *AB* as an overlay (on opposite page)
4. Using an overlay, construct a section line perpendicular to *AB* in Problem 3 which exactly bisects *AB*, and complete the profile.

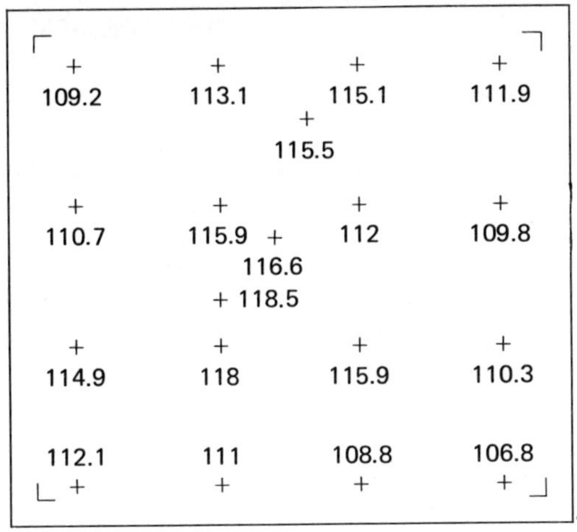

Problem 2

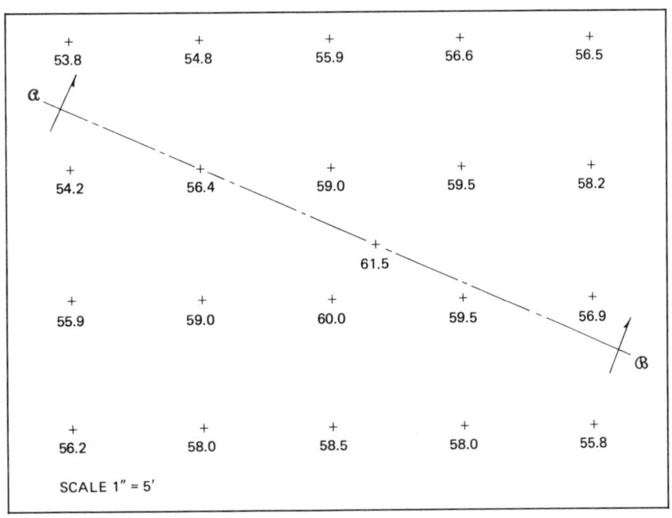

Problem 3

5. Using the topographical map below, show the cut and fill needed for a 25-ft roadway with a 2:1 side slope.
6. Given the following information and the layout (on page 143), plot the mine map: (Scale 1″ = 50′).

1-2	135°		1° 15′	245	H = 240
1-2	+ 6	R 5	L 6	Cors	Rt & Lt
	+20	R 6	L 8		
	+64	R 7	L 8		
	+74	+	88	X C	Lt. & Rt.
	+74	R 7	L 8	Cors	Lt. & Rt.
	+88	R 6	L 8	Cors	Lt. & Rt.
	+124	R 9	L 5		

	+146	R 8	L 6		
	+154	+	170	X C	Lt. & Rt.
	+154	R 8	L 8	Cors	Rt. & Lt.
	+170	R 8	L 7	Cors	Rt. & Lt.
	+200	R 8	L 7		
	+230	+	246	X C	Lt. & Rt.
	+230	R 7	L 7	Cors	Rt. & Lt.
	+246	R 6	L 7	Cors	Rt. & Lt.
1-7	225°0′		0°5′	80	H = 80
	+ 6	R–	L 6	Cor	Lt.
	+20	R 5	L 7		
	+56	R 7	L 8		
	+64	+	82	X C	Lt.
	+64	R 7	L 8	Cor	Lt.
	+82	R 7	L 8	Cor	Lt.
7-1	45°0′				
7-8	135°0′		1° 0′	160.2	
	+20	R 6	L 8		
	+79	+	88	X C	Lt.
	+74	R 5	L 10	Cor	Lt.
	+88	R 6	L 9	Cor	Lt.
	+140	R 6	L 9		
	+154	+	169	X C	Lt.
	+154	R 5	L 9	Cor	Lt.
	+169	R 6	L 9	Cor	Lt.
	+178	R 7	L 8	Face	
8-7	315°				
8-0	45°				
	+25	R 5	L 7		
	+35	R 6	L 9		
	+66	R 6	L 8		
	+92	R 5	L 10		
	+130	R 6	L 9		
	+154	R 6	L 9	Cors	Lt. & Rt.
1-5	45°	0°5′		80	H = 80
	+ 6	R 6	L 5	Cor	Rt.
	+50	R 7	L 6		
	+72	R 7	L 6	Cor	Rt.
	+84	R 7	L 6	Cor	Rt.
5-1	225				
5-6	135°		1° 05′	80.2	H = 80
	+10	R 8	L 7		
	+50	R 8	L 7		
	+72	R 8	L 7	Cor	Rt.
5-6	+87	R 7	L 7	Cor	Rt.
	+100	R 6	L 9		
	+125	R 6	L 9		
	+150	R 6	L 8	Face	

6-5	315°		
6-0	225°		
	+25	R 7	L 7
	+50	R 6	L 8
	+100	R 5	L 9
	+125	R 6	L 8
	+150	R 6	L 8

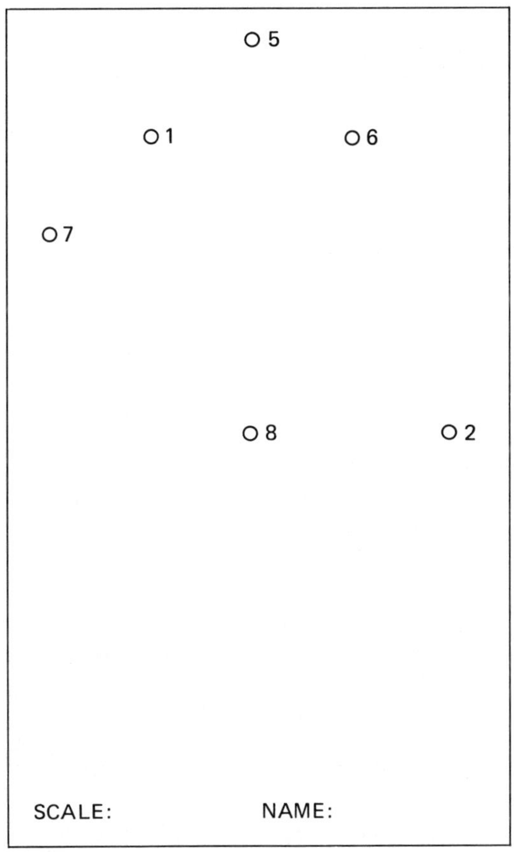

5

Functional Performance

5.0 INTRODUCTION

Functional layouts, also called special design or operational layouts, are maps that show the structures, equipment, and operations of one or more of the functional activities required to operate a mine. The methods of layout vitally affect the profitability, safety, and ease of mining. The major functional activities include premining operations, ventilation techniques, transportation operations, drainage operations, and electrical services.

Functional layouts serve a variety of purposes throughout each phase of mine design, development, and operation. Figure 5-1 depicts the six sequential phases typical of most mining operations.

Phase 1: Prospecting and Exploration. Prospecting is the systematic search for mineral reserves. *Exploration* is the systematic investigation of known reserves for the purpose of generating data about the reserve and the surrounding region. Since many of the techniques associated with these two activities are similar, they have been combined as a single phase. Data obtained within this phase of mining are manipulated to obtain maximum information, and this information is then used in subsequent phases.

Phase 2: Feasibility and Planning. *Feasibility* is defined as the evaluation of exploratory information to determine the economical, technological, and sociological capabilities of any given organization to extract the mineral resource successfully. *Planning* is that group of activities needed to schedule, organize, staff, and generally prepare for the design and operation of a mining operation.

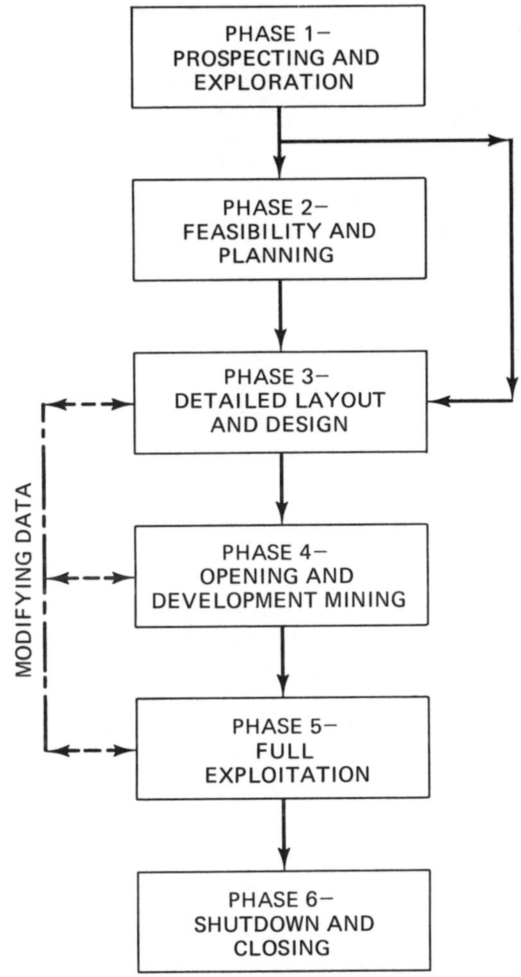

Figure 5-1 Sequential phases of mining.

Phase 3: Detailed Layout and Design. Phase 3 uses the technical information developed in phase 1 together with the management guidelines and philosophies from phase 2 to develop needed flowsheets, projections, layouts, and specifications for mining.

Phase 4: Opening and Development Mining. Phase 4 is the implementation of the plans generated in phase 3. Once mining begins, more information about the reserve will be made available. This information is used to update and thereby modify the initial designs. Thus a constant interchange of information and design revisions continues between the third and subsequent phases.

Phase 5: Full Exploitation. Exploitation is the extraction of mineral deposits. Although depicted as a separate sequential phase, exploitation often occurs simultaneously with development mining and with the shutdown of mined out areas.

Phase 6: Shutdown and Closing. The last phase of mining is the shutdown of operations, the removal of the salvageable equipment and materials, and the sealing or restoration of the excavations. This phase is just as much a part of mining as any other phase and is quickly becoming as important as other phases to the economic solvency of the mining company.

Layout purposes. Functional layouts are critical to each of the phases described above. Within phase 1, premining layouts serve to store accumulated data and information. Premining layouts are principally maps depicting topographical, geological, statistical, and any other information deemed valuable regarding the reserve or the region. The premining layouts also serve as an indication of where more data investigation may be needed, such as securing additional core samples within a region of high geological activity (faulting).

Within phase 2, function layouts serve to simplify existing complex elements and relationships existing in a given region and thereby facilitate planning activities. Functional layouts serve to summarize the qualitative and quantitative types of data in a region, for instance, showing the locations (qualitative) and numbers (quantitative) of qualified personnel within the region of the mine site. Functional layouts are also used to present alternative designs for accomplishing the same goals. Finally, preliminary layouts of major functions serve to guide planners in preparing purchase specifications for long-lead-time items. Waiting to order such equipment until final detailed design and layouts are completed may result in costly delays in subsequent phases of the mining operation.

Functional layouts are used in phase 3 to coordinate the relationships between the various functions. For instance, electrical power distribution and haulage normally coexist within the intake air course of the mine and therefore must be coordinated with the ventilation projections. Layouts also facilitate the design of installations, such as pump houses, fan placements, switching stations, and so forth, by detailing critical dimensional requirements, minimum equipment specifications, and interfacing constraints and limitations. Functional layouts are further used as legal documents needed to gain permits or variances for mining (see Section 4.2).

Functional layouts are used in phase 4 to guide the initial excavations, to plan entry development, and to install major equipment. Functional layouts of existing entries record the progress of development. Such layouts are of vital necessity in the event of major roof falls, ground failures, mine fires, flooding, or other significant occurrences.

In phase 5, functional layouts are used to optimize the extraction process and general cost effectiveness of the mining operation. Techniques of operations research are applied to different layout concepts to reduce operating costs while maintaining the same production level. As with phase 4, the layouts provide a legal record for communicating the state of excavations and projections.

Finally, functional layouts are used to show the techniques and methods needed to effectively close a mine and reestablish an acceptable environment. For surface mines this usually entails repositioning the topography and revegetating the land. For underground mines this involves sealing the mine to contain mine water contamination, reclaiming the entry sites, and accounting for future ground subsidence.

5.1 FUNCTIONAL ANALYSIS: COMMON CHARACTERISTICS

Although each functional system differs in the techniques, principles, and equipment used to achieve success, all functional operations require accomplishing a common set of goals. From the complexity of the haulage system to the relative simplicity of the drainage system, the common goals achieved through the design of each functional operation are:

1. To provide the function
2. To ensure safety
3. To maximize reliability
4. To improve maintainability
5. To optimize value

Primary and secondary objectives. Before analyzing in detail each of the objectives listed above, we need to develop the concept of primary and secondary objectives. The primary objective of a function is that objective which if not successfully completed results in a total failure of the system. Secondary objectives are any other purposes that, if not accomplished, may inconvenience the operation of the system but may not necessarily cause a total failure. Although many different purposes can be identified for different systems, all systems will have one and only one primary objective. Other objectives are secondary.

For purposes of example, consider a watch, clock, or other timepiece. The primary objective of a watch is to indicate the correct time. Certain secondary objectives include providing continuity of service (reliability), permitting maintenance or adjustment (maintainability), preventing injury to the user (safety), and so forth. Each of these secondary objectives are important; however, if the design of the clock fails to provide the user with one of the secondary objectives, the clock still operates—perhaps not as we would like or as long as we would like, but nevertheless the clock provides some indication of time. Now assume that the primary objective fails; that is, the clock does not indicate the correct time or we are unable to read it.

Regardless of how safe or reliable the clock may be, for all practical purposes it is useless to us.

So it is with the functional operations of a mine. Unless the system accomplishes the primary goal of the functional system, regardless of how maintainable, safe, or inexpensive it may be, it is useless. The primary objective of every functional operation is to provide a function when and where needed. Safety, reliability, maintainability, and cost are all secondary objectives.

This is not to say that the designer should subjugate his or her concern for safety to that of performance. It is, however, an attempt to bring the *supportive* role of safety and other secondary objectives into proper perspective. A functional system that does not do what is needed when it is needed is never safe, reliable, or cost effective. Without accomplishing its primary function, any system is useless. The remainder of the chapter is an overview of the objectives shared by functional systems.

5.1.1 Provide the Function

The primary objective of any functional operation within a mine is to accomplish successfully the task needed. In the case of ventilation, the primary objective is to deliver the required quality and quantity of fresh air when and where needed. In the case of power distribution, the primary objective is to deliver electrical power when and where needed. Because of the diversity of principles and methods used to accomplish the primary function in each case, this objective cannot be treated as an overview but will be specifically detailed in subsequent chapters.

5.1.2 Ensure Safety

The most important secondary objective existing within any functional operation is safety. Safety implies first, ensuring the health and well-being of any person interacting with the specific functional system and second, ensuring the protection and integrity of the equipment or property comprising that system.

Safety should not be partitioned into different categories of safety, such as electrical safety, haulage safety, explosive safety, and so forth. To do so implies that each category has a fundamentally different approach to ensuring safety. This is not the case. The techniques of ensuring both personnel and property safety are the same regardless of the specific function being treated.

System analysis. One method for ensuring safety falls under the category of systems analysis. A system is two or more objects that operate within an environment and interact with one another.[1] The mining system depicted in Fig. 5-2 consists of *objects* (e.g., workers, equipment, and sup-

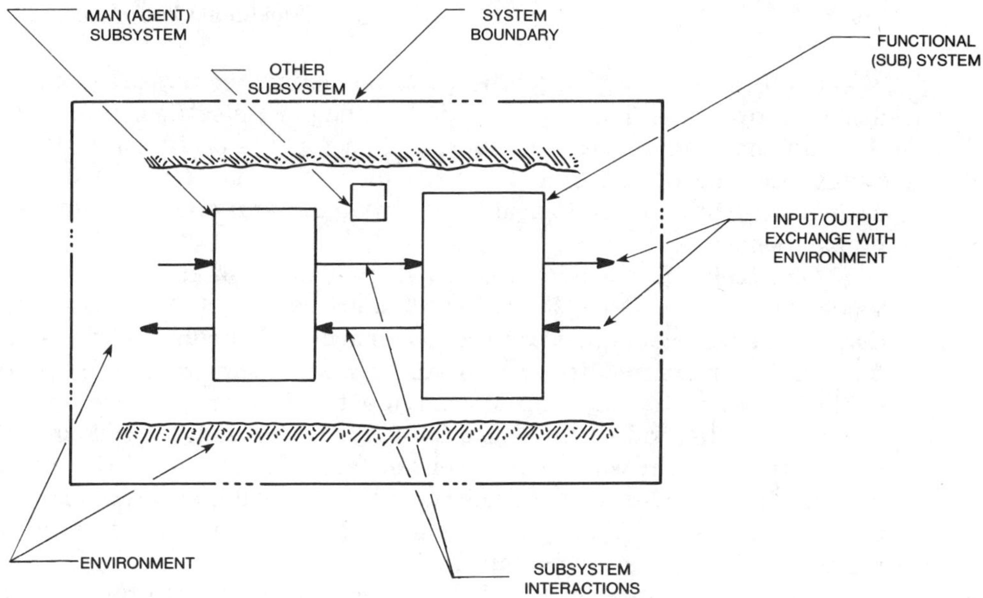

Figure 5-2 Simplified mine system scheme.

plies), and an *environment* surrounding the objects (e.g., geologic structure, air). The environment supplies certain *inputs* which support the objects (e.g., air to workers and equipment). It also provides *outputs* for wastes, which are also needed to support the operation (e.g., the discharge of carbon dioxide from workers and equipment). The environment provides a common link between the objects (e.g., the ground supports both workers and electrical equipment).

To ensure safety we must ensure that three actions are taken: (1) ensure that the inputs and outputs supplied by the environment are controlled, (2) ensure that the indirect (environmental) links between the objects are controlled, and (3) ensure that the direct interactions between the different objects are controlled. The first action, input and output control, is exemplified by ensuring that workers receive the proper quality and quantity of air, ensuring that electrical equipment can discharge heat or controlling the movement of ground relative to the objects. The second action, controlling the indirect links between objects, is exemplified by controlling the presence of water, which may form a conductive path between workers and electrical equipment, or controlling the presence of methane gas, which may be exploded by sparks from the electrical equipment. The third action, controlling the direct interfacing of objects, is exemplified by controlling the way in which a worker can operate a machine, or controlling the effect of a continuous miner cutting through a power cable.

To maximize safety, all three actions must be taken. Control, whether of the environment, of the indirect environmental links, or of the direct object links is predicated upon (1) our knowing or understanding the factors affecting those three cases, and (2) our ability to manipulate the factors to effect improved conditions. The physical phenomena with which we must deal in both cases include geological principles, electrical principles, fluid flow principles, chemical reactions, and so forth. Additionally, since human beings are one of the objects with which we must deal, we need to also understand the physiological and psychological phenomena governing human behavior.

One technique that is used to help organize and plan the control of systems is fault tree analysis (FTA).[2] Fault tree analysis is simply a relational diagram of the events that can lead to an accident. Figure 5-3 shows a partial fault tree constructed to account for a miner being struck by a roof fall (note that this fits into the first class of actions that must be taken to ensure safety). The method of forming a fault tree is based on the systems principle of partitioning,[1] in which each event or factor is considered to be composed of a smaller set of events or factors, and each smaller event, in turn, thought to be composed of even smaller parts. The partitioning continues until all significant factors have been outlined.

Symbols are used in Fig. 5-3 to represent different processes: the *lines* represent direct influences; rectangular *blocks* (▭) describe a dependent event, fact, or activity; a *diamond* (◇) represents an event about which little is known or for which factors cannot be identified; *circles* (○) represent independent events which affect other events; *window-shaped* elements (⊓ , ∩) are called logic gates. The first logic gate (⊓), the "and" gate, represents the requirement that all the factors heading into the path must be present prior to forming the next higher event. The second logic gate (∩), the "or" gate, represents a situation in which if any one of the possible inputs is present, the next-higher event will occur.

With these concepts in mind, review Fig. 5-3 to confirm the validity of the logic gates, elements, and factors that are presented. Questions you might ask include: Are all the factors that contributed to the accident shown? Are the relationships of a direct nature?

After you are reasonably satisfied with the factor relationships depicted by the diagram, the next stage of safety analysis is to evaluate your ability to control the factors and thereby eliminate hazards. Obviously, some factors are simple to manipulate, whereas others are more complex.

To manipulate the factors and thereby gain control of the system, you must:

1. Develop specialized controls, devices, and equipment to break one or more of the causal links.
2. Implement special inspections and audits of equipment.
3. Properly train, organize, and manage the workers interfacing with underground systems.
4. Control the environment.
5. Lay out the system.

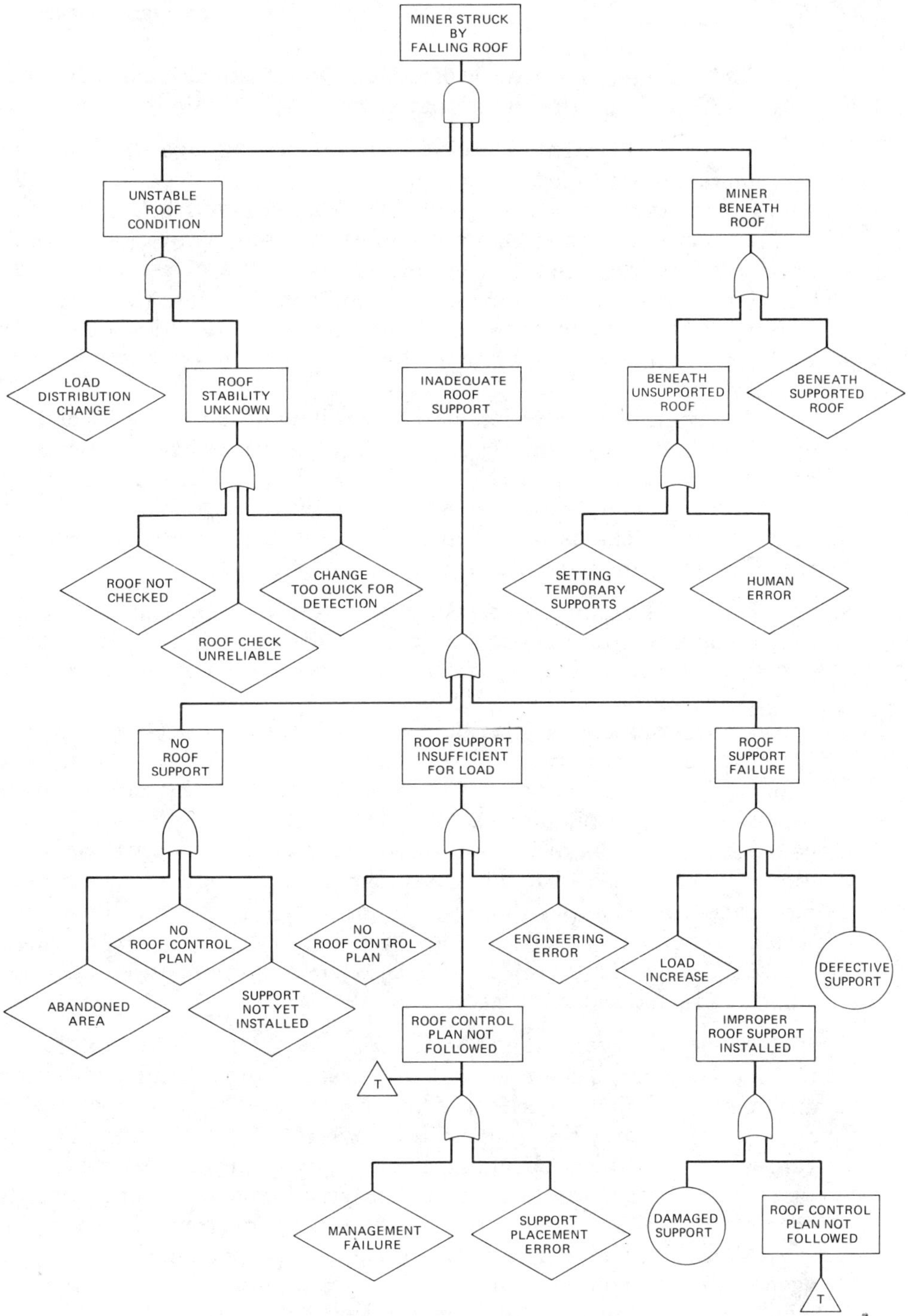

Figure 5-3 Partitioning of accident causes: fault tree analysis. (Courtesy of National Mine Health and Safety Academy.)

Each of these five areas is directly interlocked with every other area and is further limited by the traditional practices used underground.

Safety Equipment Development. One way to protect personnel, equipment, and property is to develop and use safety equipment. Safety equipment prevents vulnerable objects from contacting the hazardous parts of the functional system. To develop safety equipment, three types of hardware must be designed: (1) equipment that creates a physical barrier between personnel and hazardous system components, (2) equipment that protects the system components and environment, and (3) equipment that shuts down the system when malfunctions occur (termed "active" safety equipment).

An example of physical barrier equipment is protective clothing worn by certain personnel. An example of equipment designed to protect components and environment would be permissible enclosures surrounding certain types of equipment. An example of active safety equipment is methane monitor interlocks used on continuous miners to interrupt electrical power when the methane percentage of the atmosphere reaches 1%.

Periodic Equipment Inspection. Periodic inspection and safety audits of the equipment are needed to identify minor problems prior to their causing major failures. Through periodic inspection, failures, which may cause hazardous situations, can be avoided.

Inspections and safety audits are of two types: (1) observational and (2) instrument supported.[3] Perceptual inspections are those conducted using sight, sound, smell, and touch to indicate component problems. Instrument-supported inspections are usually performed to evaluate those component parameters that cannot be monitored easily using perceptual inspection techniques. Although instrument measurement is considered by some to be the most dependable type, the vast majority of minor failures will first be detected using perceptual methods, with instrument support being used to establish the severity. (More will be said in subsequent sections about the value of inspection for all secondary functions.)

Safety Management. The third method for ensuring safety entails the method by which personnel and the measures described above are organized and managed to ensure safety. Organization involves the way in which available personnel may be deployed to facilitate safety. An example of organization is the method of checks and balances that formally exists to ensure safe task performance. Normally, the checks and balances operate between management personnel, union representatives, and government inspectors.

Management involves two related phases, personnel training and management of acceptable behavior. Training requires that the workers understand the potential hazards contributing to accidents and that they know how to avoid these hazards. Training is easily accomplished using job orientation sessions, well-planned operation manuals, and periodic retraining sessions. Managing effective behavior is more important in producing an effective safety program but is correspondingly more difficult to achieve. Better techniques of behavioral engineering based upon principles of psychology are needed to achieve complete success in safety management.[4]

Environmental Control. The environment surrounds both the worker and equipment subsystems. As discussed earlier, the environment supplies inputs and outputs which support life. It also contains hazardous conditions, such as dammed water, unstable geologic forces, dusts, and methane. It further provides an indirect link between workers and equipment. This link is usually harmless, but when the environment is not controlled, it may become quite dangerous. Several methods of environmental control are discussed in detail in subsequent chapters, including geologic control, ventilation control, and drainage control.

Layout of Safety. Several different schemes of component layout can be used to achieve safety: (1) isolation schemes, (2) controlled link schemes, (3) protected agent schemes, (4) protected equipment schemes, (5) deenergized contact schemes, (6) fail-safe schemes, (7) alerted contact schemes, (8) environmental exclusion schemes, and (9) combinations of the above. These schemes are discussed below and are illustrated in Fig. 5-4(a)-(h). (For simplicity the functional system is abbreviated as FS and the other object, which may be a worker or equipment, is designated as the agent, A.)

Isolation Scheme. The strategy of the isolation scheme is to avoid contact between A and FS by maintaining a physical space between the two

Figure 5-4 Various schemes designed to ensure safety.

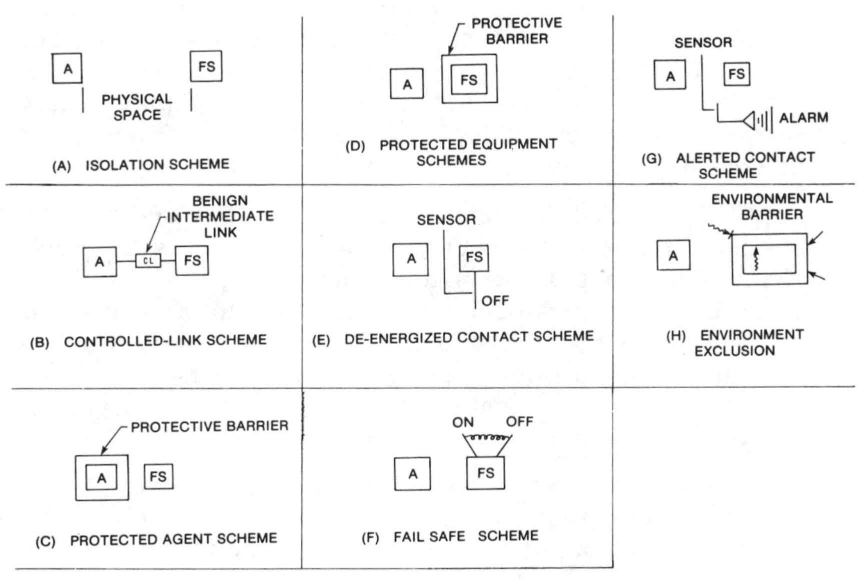

components. In most cases the isolation scheme is achieved by locating FS away from all potential agents. Isolation schemes are practical strategies for highly reliable equipment requiring little maintenance or continual monitoring.

Controlled Link Scheme. The controlled link scheme prevents an unsafe direct contact between FS and A by inserting a benign link between the two. Using this strategy, FS can be manipulated by A without damage to either object. The controlled link scheme is used when continuous contact is needed (e.g., controlling the electrical distribution system).

Protected Agent Scheme. The protected agent scheme uses firefighting clothing, auxiliary breathing devices, insulated gloves, or other artificial barriers to maintain the safety of the agent. Protected agent schemes are often used to protect or isolate A from the environment as well as from FS, particularly when the environment fails to provide the necessary life-support inputs.

Protected Equipment Scheme. The protected equipment scheme attempts to isolate the equipment from environmental or agent attack. Additionally, the protected equipment scheme protects the environment and agent from the effects of malfunction within the equipment. For instance, if methane gas is ignited within a shuttle car circuit breaker box, the function of the surrounding protective enclosure is to prevent the effects of that explosion from propagating to the environment. This type of scheme is exemplified by the permissibility requirements of face equipment.

Deenergized Contact Scheme. The strategy of the deenergized contact scheme is to monitor the proximity of A to FS. When contact becomes potentially hazardous, the functional system is deenergized, thereby preventing catastrophic contact from occurring. Examples of deenergized contact are methane monitors mounted on face equipment, grounded shielding surrounding power cables, and ground fault interrupters on electrical equipment.

Fail-Safe Scheme. The fail-safe strategy is one that causes the entire system to deenergize when any component of the system fails. For example, if the power fails, all mobile equipment stops. When this occurs, the first instinct of the operator is usually to dismount and check the power cable. If the power is restored while he is attempting to do this, he may be thrown from his vehicle. However, using a fail-safe scheme, any power outage causes all equipment to deactivate in an "off" mode, thus preventing the potential hazard described above.

Alerted Contact Scheme. Like the deenergized contact scheme, the alerted contact scheme monitors the A–FS interface and alerts A (the worker) to the potential danger of the situation using a warning device. However, it is not intended to prevent contact, nor is it intended to deenergize FS. Such a system may be needed for maintenance or special operation modes.

Environmental Exclusion Schemes. Environmental exclusion schemes are methods of controlling the environment and preventing harmful effects of the environment from affecting either A or FS. Examples include ground control measures, drainage control, and dust control.

Combined Schemes. The most pragmatic strategy is combining all of the various layout schemes described above into a single safety policy. All of the layout strategies are needed to effect a safe mine.

To summarize, we must be able to control (1) the direct effects of the environment, (2) the direct contacts between A and FS, and (3) the indirect environmental link between A and FS. Control implies that we are cognizant of factors that contribute to hazardous conditions and we are able to manipulate these factors to effect a safe working place. The technique presented for identifying the factors, fault tree analysis, is simply a relational diagram constructed for a hazard scenario. The techniques presented for manipulating the factors include developing safety equipment, using periodic inspections, using management and organization theory, controlling the environment, and layout objects comprising the system.

5.1.3 Maximize Reliability

Reliability is the probability that a system will perform its intended function over a predetermined period of time. An unreliable system is one that frequently breaks down or is otherwise out of service. Unreliable systems are, among other things, unsafe and uneconomical. The unsafe conditions result when workers bypass existing safety controls in order to improve operating continuity. Economic burdens associated with unreliable systems result from downtime[3,5] costs:

1. Lost revenue due to lost production
2. Direct and indirect labor costs needed to repair equipment
3. Component and material costs needed to repair equipment
4. Costs associated with the downtime of other equipment which depends upon the disabled equipment
5. Transportation costs
6. Consultation costs
7. Costs of injuries
8. General overhead costs
9. Subjective costs (morale, dissatisfaction, interruption of schedules, etc.)

Types of failures affecting reliability. The types of failures that can interrupt the performance of a functional system are varied and depend on the type of system involved. For those systems composed primarily of mechanical or electrical hardware, eight distinct modes of failures can be listed.[3,6]

1. *Defect failures*—failures that result from improper design, inferior material selection, faulty manufacturing, and poor quality control.
2. *Aging failure*—failures linked to the action of the environment acting on the system components over time (e.g., corrosion).
3. *Degradation failure*—failures resulting from use of the system (e.g., wear and erosion).
4. *Improper use or maintenance*—failures that occur due to incorrect installation, improper operation, abuse of the system, and incorrect maintenance procedures.
5. *Accumulative microscopic failures*—failures that link the use of equipment, the time of use, and the level of performance to effect changes in the microscopic material lattices (e.g., fatigue, creep).
6. *Accumulative macroscopic failures*—failures linking use, time, and level of operation to effect measurable dimensional changes in the system (e.g., overload).
7. *Spillover failures*—failures that occur as a result of the physical closeness or direct (flow) relationships that exist between components. (For instance, a malfunctioning circuit breaker may cause the armature insulation of a motor to burn out. Thus the failure of the motor is attributed to its relative proximity to the failed circuit breaker.)
8. *Simultaneous modes of failure*—failures that result from two or more modes of failure occurring simultaneously. This failure mode is the most common.

Enhancing reliability. The reliability of any given functional system can be substantially increased using the following principles of equipment layout:[3,5,6]

1. Keep the system simple.
2. Use reliable components.
3. Use redundancy when and where needed.
4. Use equipment derating.
5. Control the operational environment and installation.
6. Use human factors engineering (HFE).
7. Provide proper organization and management.
8. Inspect and maintain the system.

Obviously, many of the factors noted as affecting safety also affect reliability. This is not a coincidence, since a safe system is usually a reliable system and a reliable system is usually a safe system.

Keep the System Simple. Simplicity as used here implies that the number of component parts needed to operate the system have been kept to a minimum. As the number of independent components in a system increases, the reliability decreases. Independent components may be as obvious as a transformer unit or as inconspicuous as a series of splices within the power cable. Each component has its own reliability index and the more components, the less will be the reliability of the system.

Consider the conceptual system shown in Fig. 5-5, where three components are needed to accomplish a desired output. For such a series system, the reliability of the entire system, R_{sys}, is the product of the reliability of each of the components. Such a relationship is summarized in Eq. 5-1.

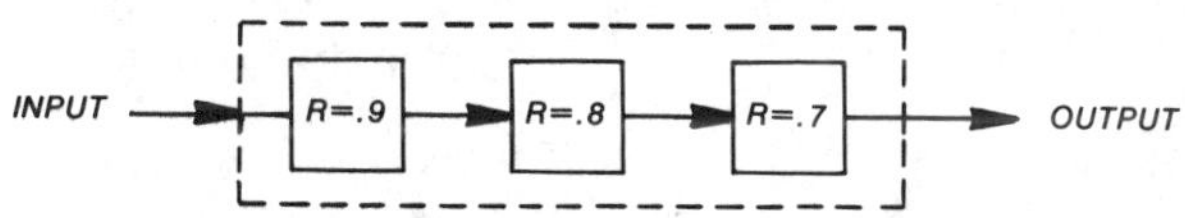

Figure 5-5 Three-component series system.

$$R_{sys} = (R_A)(R_B)(R_C) \ldots (R_N) \ldots \tag{5-1}$$

where R_{sys} = reliability of the entire system
R_i = reliability of each independent component within the system (expressed as a decimal equivalent)

Based on such a relationship,[1] the reliability of any series system must always be less than the reliability of the least reliable component in the system, and the greater the number of component parts in series, the lower will be the total reliability. These effects are made clearer with some examples.

Example 1

Assume that for Fig. 5-5 the reliability of component A is 90%, that of B is 80%, and that of C is 70%. Determine the reliability of this system.

Solution

$$\begin{aligned} R_{sys} &= (R_A)(R_B)(R_C) \\ &= (0.9)(0.8)(0.7) \\ &= 0.504 \end{aligned}$$

The probability of the system functioning over its defined time span is only 50.4%. R_{sys} is less than the reliability of any individual component part!

Example 2

Assume that instead of the three components, two systems (Fig. 5-6) were considered for adoption. Alternative 1 consists of four components, each having a 90%

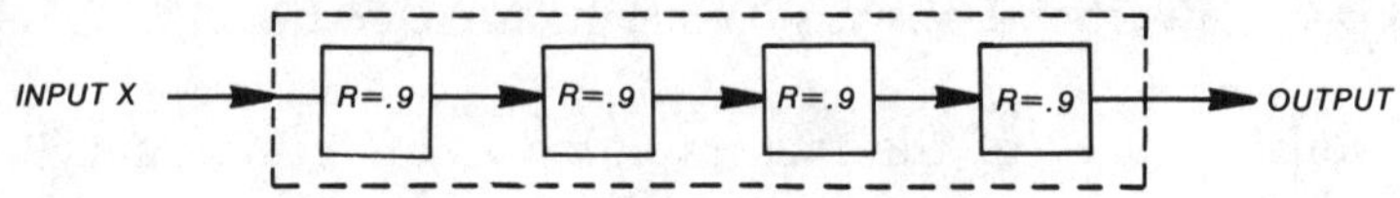

Alternative 1: Four Series Component

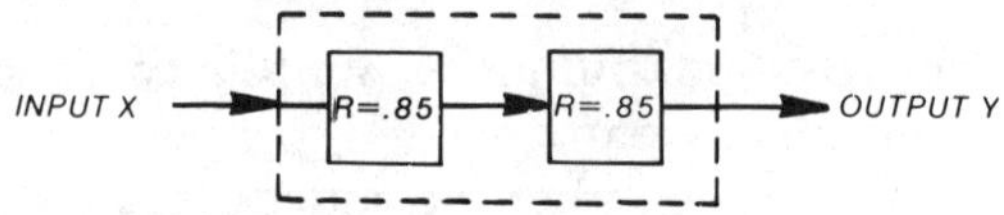

Alternative 2: Two Series Component

Figure 5-6 Contrast of two systems with identical inputs and outputs.

reliability value. The second alternative system is composed of two components, but each component had a corresponding reliability of only 85%. Which system is more reliable?

Solution

$$\begin{aligned} R_{\text{sys}1} &= (R_A)(R_B)(R_C)(R_D) \\ &= (0.9)(0.9)(0.9)(0.9) \\ &= 0.6561 \\ R_{\text{sys}2} &= (R_A 1)(R_B 1) \\ &= (0.85)(0.85) \\ &= 0.7225 \end{aligned}$$

Thus the reliability of the two-component system is higher than that of the four-component system, even though the reliability of each individual component in the second system is less than the individual reliabilities of the components comprising the first system.

Use Reliable Components. In a series system the reliability of the system is always less than the reliability of the least reliable component comprising the system. Thus to design a reliable system, the components making up the system must be reliable.

The selection of reliable components is not as simple as it may at first appear, since most equipment manufacturers and suppliers do not know the reliability indexes of their equipment or, when they do, they often find supplying such information to be counterproductive to their company goals. Thus to select the most reliable components the designer must understand those attributes shared by reliable designs. These attributes are related to the design, manufacturing, and quality control operations of each manufacturer, and include:

1. Simple designs (few components)
2. Redundant components where critical
3. Few "critical" areas
4. Designed protection from the environment
5. Overdesigned components

6. Obvious use of human factors engineering
7. Ease of maintenance procedures
8. The use of proven designs and components
9. Adequate maintenance and operation manuals

In addition to these nine direct attributes, component reliability can be correlated with characteristics of the manufacturer's organization, including:

1. An independent quality control group
2. An available training package for the equipment
3. Readily accessible service personnel or troubleshooters
4. Generous warranty periods
5. Easily obtainable parts and materials

Use Redundancy When and Where Needed. A *series system* is one in which the output from one component is directed as input to another. A *redundant system* (Fig. 5-7) is one in which the output from one component has several possible paths, leading to an equal number of successive components. The components that form these parallel paths perform the same functions and are thus redundant in their operation. For a system configuration to be considered truly redundant, each parallel path must be (1) capable of functioning independently of the other path(s), and (2) capable of functioning under the entire input of the circuit without damage. Redundancy allows the operator to switch the load from one component to the other without interrupting the functioning of the system. Although not critical to the definition, some redundant circuits can be operated simultaneously by dividing the input equally, thereby placing a lower load through each component. This mode of operation is discussed below under "derating." Redundant circuits are needed to meet several operating conditions:

1. When the component(s) have low reliability.
2. When the function(s) performed are absolutely critical to the system operation.
3. When failure of the component(s) leads to unwanted after-effects.
4. When maintenance of the component must be performed under "no load," but when the function must be continuously provided.

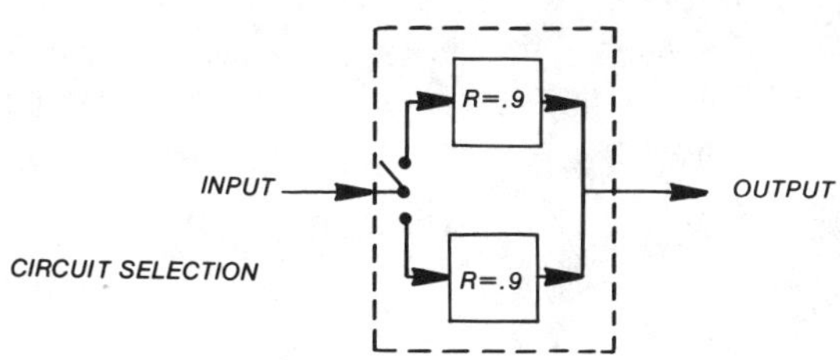

Figure 5-7 Two-component redundant system.

5. When component derating (the operation of the component well below its rated operation level) greatly extends the useful life of the components.

The reliability of a two-path redundant circuit can be computed using Eq. 5-2.

$$R_{sys} = R_A + R_B - R_A R_B \tag{5-2}$$

where R_{sys} = reliability of the redundant system
R_A = reliability of system A
R_B = reliability of system B

The following examples illustrate the advantages of redundancy.

Example 3

If the reliability of component A in a two-component system is 90% and the reliability of the second component in the system is also 90%, determine the reliability of the system.

Solution

$$\begin{aligned} R_{sys} &= R_A + R_B - R_A R_B \\ &= 0.9 + 0.9 - (0.9)(0.9) \\ &= 0.99 \end{aligned}$$

Thus the reliability (99%) of a redundant system is significantly greater than when a single component (R = 90%) is used to provide the same processing.

Example 4

Given a series system with the reliabilities shown in Fig. 5-8, if any one of the three components could be replaced with a redundant subsystem accomplishing the same function, which component replacement would most improve the total system reliability? The three alternatives, (a), (b), and (c), for such a proposition can be calculated as follows:

For the original series system,

$$\begin{aligned} R_{sys} &= (0.6)(0.8)(0.9) \\ &= 0.432 \end{aligned}$$

For system (a):

$$\begin{aligned} R_A &= 0.6 + 0.6 - (0.6)(0.6) \\ &= 0.84 \end{aligned}$$

The new R_{sys} is:

$$\begin{aligned} R_{sys\text{-}a} &= (0.84)(0.8)(0.9) \\ &= 0.6048 \end{aligned}$$

For system (b):

$$\begin{aligned} R_B &= 0.8 + 0.8 - (0.8)(0.8) \\ &= 0.96 \end{aligned}$$

The new R_{sys} is:

$$\begin{aligned} R_{sys\text{-}b} &= (0.6)(0.96)(0.9) \\ &= 0.5184 \end{aligned}$$

For system (c):

$$\begin{aligned} R_C &= 0.9 + 0.9 - (0.9)(0.9) \\ &= 0.99 \end{aligned}$$

The new R_{sys} is:

$$\begin{aligned} R_{sys\text{-}c} &= (0.6)(0.8)(0.9) \\ &= 0.4752 \end{aligned}$$

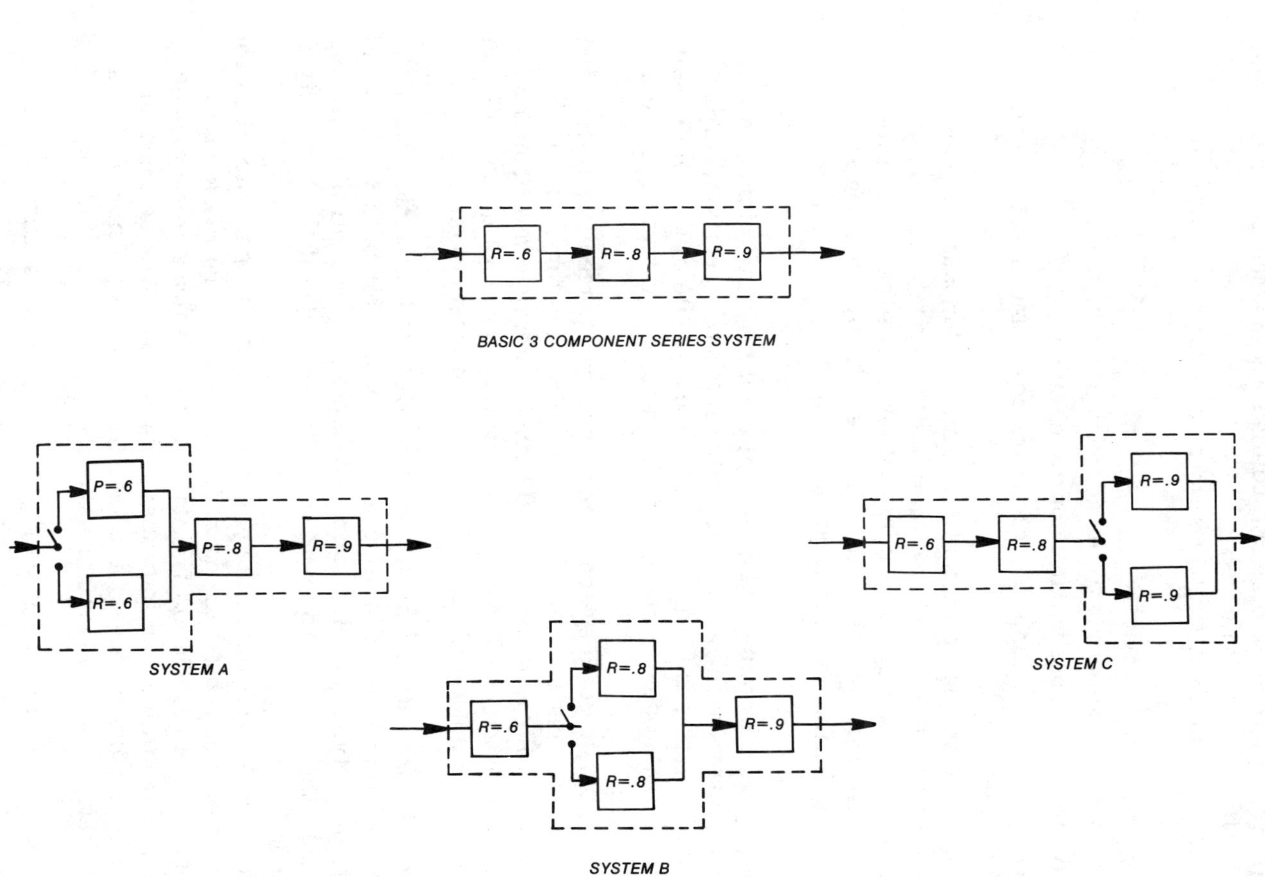

Figure 5-8 Use of redundant strategies to increase system reliability.

Thus for any given series system, the least reliable component should be replaced with an equivalent redundant configuration when the greatest increase in system reliability is desired.

The use of redundant strategies does carry with it certain penalties, including:

1. Added cost, corresponding to the added components.
2. Added complexity, particularly control circuitry.
3. Added maintenance time and therefore cost (since more components need to be serviced).
4. Some questionable improvement; since most redundant components are located within the same area, vibration, temperature, humidity, and other factors that may cause failure of one leg of the configuration will just as likely cause failure in the other leg, thereby removing any advantage offered by the use of redundancy.

Use Equipment Derating. Equipment *derating* is the intentional operation of components well below their rated levels of operation. Normally, the strategy of derating is used to extend the useful life of the components within a particular system, although it also acts to improve the reliability of the component.

The derated value of a system is measured as a percentage of the rated operation level. For instance, an electrical system rated for 25,000 volts and operated at 2500 volts is derated at a 10% level. Derating is most common at levels ranging between 10% and 30%; however, derating up to 50% can successfully extend the useful life and reliability of most systems.

Control the Operational Environment. As noted in Section 5.1.2, the environment of underground mining is quite hostile for both workers and equipment. Aging types of failures are accelerated by uncontrolled environmental attack; thus the environment must be controlled if the reliability is to be extended.

In addition to controlling the environment, the designer must control the specific procedures used to install the component hardware within the environment. Proper installation is usually quite significant to the reliability of a system. For instance, an improperly aligned conveyor belt installation may decrease the useful life of the belt to one-tenth its normal life and therefore increase maintenance time and costs.

Use Human Factors Engineering. Human factors engineering is a design and layout strategy of considering the physiological and psychological limitations of those persons who might interface with the system. The role of HFE in extending or ensuring reliability is to:

1. Prevent accidental contact and inhibit intentional contact between personnel and potentially hazardous components of the system.
2. Prevent accidental or purposeful abuse and misuse of the equipment by personnel.
3. Facilitate proper use, operation, and maintenance by altering the layout and design of the system components.

Some of the methods used to accomplish this include:

1. Providing "limited-access" designs or keys so that only those persons who are qualified can contact the system.
2. Use of special configurations for components to facilitate proper replacement following maintenance.
3. Inclusion of automatic limit switches to control range of motion.
4. Use of standardized layouts (components and systems) to facilitate "standard" operation.
5. Use of standardized procedures and instructions to aid operators and maintenance personnel.

HFE will extend service reliability when individuals are familiar with the layout, procedures of operation, and practices, and cannot significantly deviate from these acceptable standards.

Provide Proper Organization and Management. Organization and management in ensuring system reliability is identical to the use of that strategy in ensuring safety. Principally, organization and management attempts to:

1. Organize the personnel skills needed to accomplish the tasks required.
2. Supply training and education concerning the function of the systems, limitations of the system, and proper operation and maintenance of the system.
3. Control the behavior of the individuals interfacing with the system in order to maximize reliability.

Inspect and Maintain the System. All other factors being nearly equal, a system that is periodically inspected and maintained is always more reliable than one that is cared for only haphazardly. This is true regardless of the type of system.

Many surveys of automobile owners who have accumulated over 100,000 miles have shown that the only maintenance or operation practice these owners have in common is that they regularly perform routine inspections and maintenance. Few controlled studies have been performed in actual operating mines, but all equipment manufacturers agree that periodic inspection and maintenance are critical to high service reliability.

In summary, it should be noted that reliability is only a probability of maintaining service continuity. This probability is mainly contingent upon (1) the reliability of the components, (2) the method by which the components are laid out (e.g., the use of redundancy when and where needed), (3) control of the environment and other systems (e.g, workers) that interact with it, and (4) the nature of inspection and maintenance the system experiences over its lifetime.

5.1.4 Ensure Maintainability

In its strictest sense maintainability is the probability that a failed system can be restored to service in a given period of time at a given expenditure of resources (e.g., workers, money, and time).[5,7] However, for the purposes of layout, we will extend the definition to include the relative ease of modifying, inspecting, and maintaining the functional systems.

Modifications. Underground, the function system must advance, retreat, and change according to the extraction progress. Factors affecting the ease of system modifications will be (1) the rate of advancement, retreat, and alteration; (2) the selection of equipment for the various modification stages contemplated (principally the level of derating, overload capability of equipment, cost factors associated with underutilized capacity, etc.), and (3) the nature of the layout (e.g., is the layout of operations principally "straight line" like an assembly line, or is it highly branched?). These factors are contingent upon the specific functional system and are discussed in subsequent chapters.

Inspection. System inspection entails the periodic survey of the condition of the system relative to a set of operating norms. The further the state of the system deviates from normal values, the greater the likelihood of malfunction.

When first presented, inspection was divided into observational surveys and instrument-supported measurements. The greatest value is derived from observational surveys. Observational surveys involve (1) looking for anomalies such as leaks, cracks, discoloration, or missing parts; (2) listening for unusual operation sounds, such as squeaks, squealing, tapping, or differing frequencies; (3) feeling excessive qualities, such as vibration, temperature, tightness, or looseness; and (4) smelling to detect ozone, burning, or other usual odors caused by unacceptable conditions.

Observation, although not quantitative in nature, is still quite accurate and is the most important method of equipment inspection. The major disadvantage of observational surveying is that it requires an inspector to have significant, and usually recent, equipment familiarity and operational experience.

Instrument-supported measurements are used when the operational parameters that define the condition of the system range beyond human sensory limits. For instance, many motors operate at temperatures 40 to 50° C above the ambient temperature (e.g., 20° C). Although the motor may feel overheated to the touch, only by using a thermocouple or other suitable temperature-monitoring device can an inspector properly establish the operational state of the motor.

Depth and Frequency of Inspection. Additional factors involved in system inspection are[5] the depth of inspection and the frequency of inspection. The depth of inspection is the extent of detail in which the various operational parameters of the system are evaluated. Like the inspection of individual pieces of equipment, the depth of system inspection can be ranked according to five levels. Level 1 is a general observational (sometimes called walking) examination of the external hardware of the system. Level 5 is a highly detailed instrument-assisted examination of the system in both the assembled operating state and in a disassembled static state. Obviously, the cost and time required of a level 5 inspection is much greater than that of a level 1 inspection, but the increased information gain is often needed.

The frequency of inspection is the number of inspections per unit time. Inspection frequency depends on the reliability of the equipment and the relative value of the equipment to normal operations. Low-reliability equipment requires higher inspection frequencies, as does essential equipment. In general, as the frequency of inspection increases, the cost due to downtime decreases.[7] This decrease is, of course, offset by the added cost of the inspection; thus the benefit (reduction of repairs and downtime) of higher inspection frequency needs to be weighed against the total cost to the mine (cost/benefit analysis).

Maintenance. Maintenance is commonly divided into several categories:

1. Failure-cued maintenance
2. Preventive maintenance
   (a) Routine maintenance
   (b) Preventive replacement

Failure-Cued Maintenance. *Failure-cued maintenance* is the strategy of correcting a failed system component or subsystem. The procedures leading to the implementation of failure-cued maintenance are shown in Fig. 5-9. The decision to repair the failed component is not always a positive one. Several factors need to be considered before failure-cued maintenance is implemented:

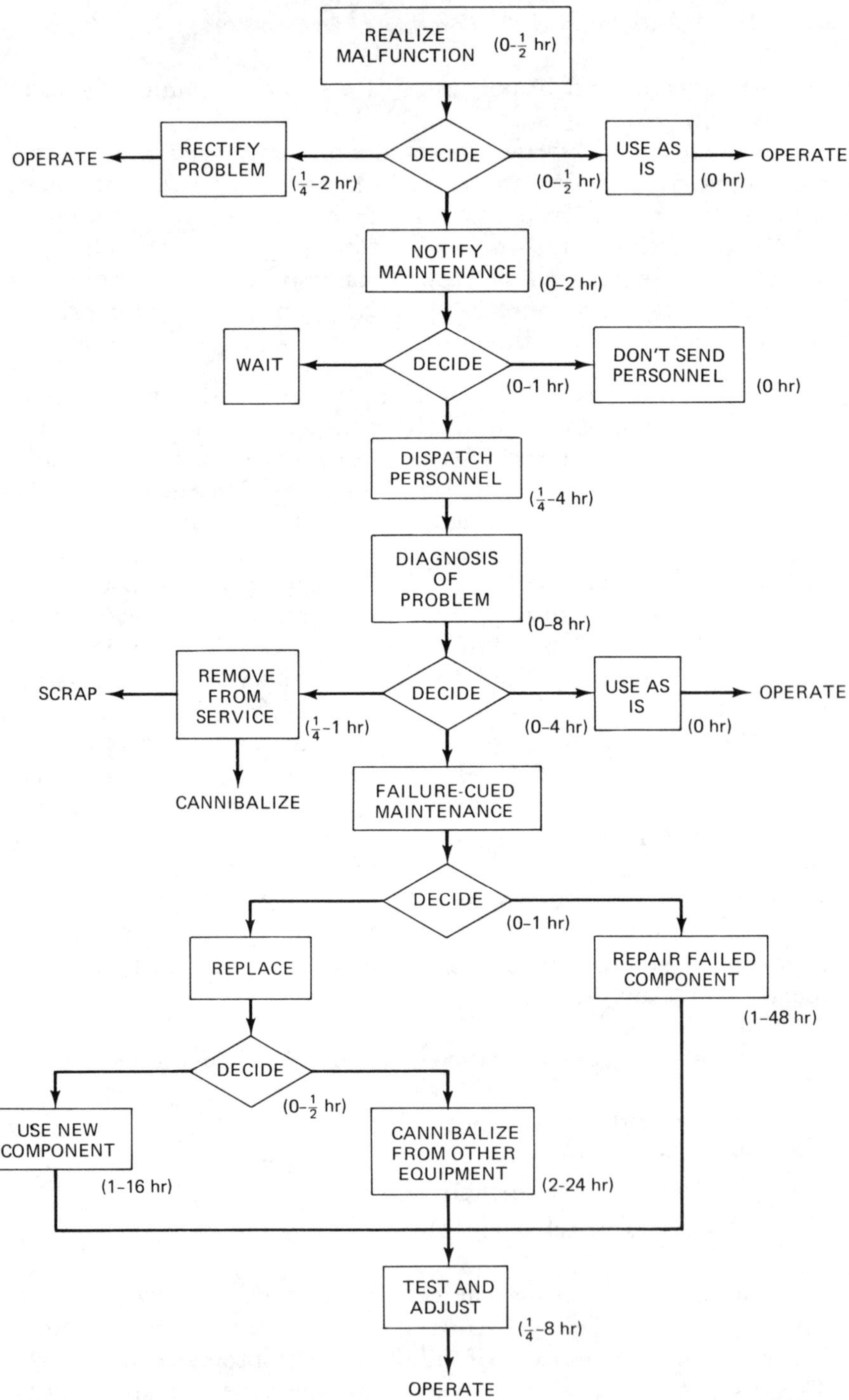

Figure 5-9 Procedures and times leading to failure-cued maintenance.

1. What is the cost of repair compared to the value of the system function?
2. What is the relative necessity of the component?
3. Can the component be bypassed without adversely affecting the func- the time period needed?
4. Can the component be bypassed without adversely effecting the functioning of the total system?
5. Can the component self-repair or otherwise adapt without failure-cued maintenance?

If failure-cued maintenance is used, one of several methods is available. The failed component can be replaced from inventory stores, purchased parts, or cannibalized components. The first two methods are common new or rebuilt part replacements. The latter, replacement from another cannibalized system, is interesting for two reasons. First, the cost of the part is usually much less than either of the two former methods, and second, cannibalized parts offer the "advantage" that they have been used. The replacement of defective parts with used parts is associated with a technique called component "burn-in,"[5] in which a new component is operated for a given period of time prior to its service installation. This is done to reveal defect failures which usually occur during the break-in period of component operation. By using burn-in techniques or used parts, this premature failure period is effectively avoided. Of course, part replacement from another cannibalized system can have the disadvantage that the component is worn out and may fail soon after replacement. The decision as to whether to use cannibalized components depends on the length of time the component was used.

Figure 5-10 shows graphically the general rate of failure of any component, subsystem, or system over time. If a candidate replacement component had been previously operating within phase 2, the constant hazard zone,

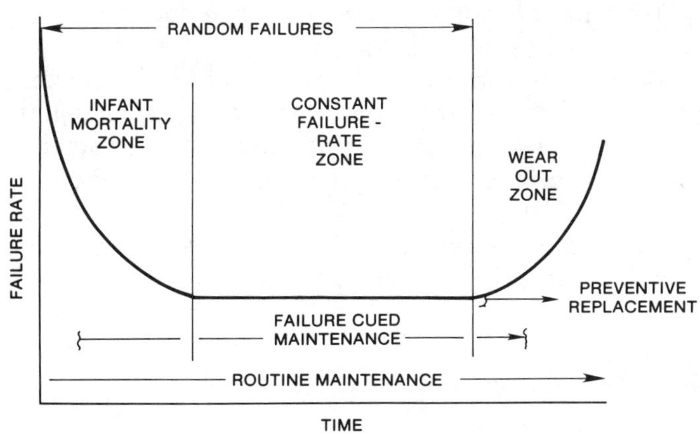

Figure 5-10 Generalized mortality curve.

replacement would probably be beneficial to the system. However, if the replacement component was within phase 3, the wearout zone, the use of the component would not be beneficial.

Preventive Maintenance. Preventive maintenance is a set of procedures initiated in an attempt to maintain the state of the system. Preventive maintenance assumes that if the level of performance can be maintained, the rate of downtime failures can be decreased.

Preventive maintenance is divided into routine and preventive replacement strategies. *Routine maintenance* involves the periodic and scheduled operations of cleaning, lubricating, tightening, making minor adjustments, maintaining storage levels (e.g., electrical charge, hydraulic fluid, etc.), and so forth. The purpose of routine maintenance is to control the specified operating ranges and tolerances of the equipment. If these tolerances are allowed to drift too far, a more rapid degradation of the state of the system would result. Routine maintenance is the only type of maintenance that begins from the time of installation and continues until the system is removed from active service (see Fig. 5-10).

Preventive replacement is the strategy of replacing a component or subsystem *before* it fails. Ideally, a component is not replaced until just before it actually fails. However, since we never really know the exact moment of a future event, we must rely on past experience to somehow predict the approximate failure time. To assist in our efforts we use whatever data we have and the mortality curve for this component. The only "predictable" failures are those that result from components wearing out. Therefore, the only time we can justifiably use preventive replacement is when a component or system is operating within phase 3 of the mortality curve. If we replace a component prematurely, that is, prior to phase 3, we lose the cost of that component (over the remainder of its actual life) and we may be replacing a perfectly good component with another that may fail due to infant mortality (defect failures).

In addition to requiring that preventive replacement be used only when operating within the wear-out zone, we must also be certain that the cost of replacement, that is, the downtime cost needed to perform the maintenance, is less than the cost incurred if we allow the component to fail and then use failure-cued maintenance. Obviously, if the cost of preventive replacement were equal to or more than the failure cost, the greatest value would dictate the use of only the failure-cued strategy.

The level of system maintainability (ease of modification, inspection, maintenance) is influenced by certain layout characteristics:

1. Keep the system simple.
2. Select reliable and maintainable components.
3. Use redundant layout when needed.
4. Control the environment and installation of equipment.

5. Use human factor engineering.
6. Specify built-in test equipment.
7. Provide proper organization and management.

These factors are essentially the same as those presented previously and therefore will be treated briefly.

System Simplicity. The fewer the components, the quicker and simpler will be the procedures of inspection and maintenance. In general, the simpler the system, the easier it will be to extend, retreat, and modify the system.

Reliable and Maintainable Component Selection. Reliable components are those that do not exhibit a high frequency of failure and therefore will not require a large degree of failure-cued maintenance time or cost. A maintainable component is one in which repairs can be implemented in a given period of time requiring a given set of resources. By knowing the time and resources needed to effect a restoration, the ability to maintain a system at the needed level of performance will be greatly enhanced.

Redundant Layout. Redundancy, the use of parallel networks in the system configuration, aids maintenance in that the system leg requiring repair or preventive maintenance can be removed from service, or simply deenergized, without interrupting the continuity of service close to the point of work.

Control of Environment and Installation. As noted earlier, proper environmental control and installation improve system reliability and safety. Environmental control also facilitates the inspection and maintenance function by providing freer access to equipment and less working hardship for personnel.

Human Factors Engineering. HFE plays a large role in inspection and maintenance by (1) designing for human access to the failed zone, using logical equipment placement and layout (those components with lowered reliability or those requiring more frequent inspections and routine maintenance should be designed with greater access than other equipment); (2) ensuring the use of standard layouts, components, and interchangeable parts, thus facilitating cannibalization if needed; and (3) using layout strategies that facilitate correct procedures and practices.

Built-in Test Equipment (BITE).[5] BITE is used to provide rapid and constant instrument-supported inspections and to decrease the diagnosis time in the event of failure. These advantages are offset by certain disadvantages, such as increased cost, weight, volume, and complexity. BITE must also be more reliable than the system it is monitoring; otherwise, the system may be unjustifiably repaired.

Organization and Management. Management of the maintenance functions requires that personnel be trained and that their behavior be controlled. In addition to these "typical duties," maintenance managers must consider the need for recording reliable information and historical data about existing equipment particularly pertinent to preventive replacement decisions.

5.1.5 Optimize Value

Several methods can be used to define value; however, the most useful is benefit received per dollar spent (value = benefit/cost). To improve the value of any system either the functional performance must be improved while maintaining the same costs, or the costs (required to deliver the function) must be lowered while maintaining the same performance. To optimize the value of a system, both function and cost must be evaluated simultaneously. We have previously discussed four major functions: operation, safety, reliability, and maintainability. Each of these can be partitioned into progressively smaller groupings until a fairly complex set of functions is produced. The cost analysis of these various functions is equally complex. Some value considerations are developed in subsequent chapters; however, a proper analysis of this involves economic systems analysis and modeling,[7,8] which are beyond the introductory scope of this text.

5.2 PROCEDURES FOR CONSTRUCTION OF LAYOUTS

Thus far you have seen that functional layouts are needed for guiding all phases of mining. Also, each functional system carries a single primary goal of performing certain operations. And you now know that each functional system has certain secondary goals which share common techniques to ensure safety, reliability, maintainability, and value.

The procedures needed to construct the individual design layout maps and drawings are also the same, regardless of mine size or functional activity. These procedures are depicted as successive activities in Fig. 5-11.

The first step, perform functional analysis, is a process of information gathering and assessing the needs of the mine. Qualitatively, you must ask what the system is to do and when it is to do it. These questions are asked of all five functions delineated in Section 5.1. Only the "needs" of the system are outlined; thus questions of how the system is to do what it does and what the circumstances of the operation are are reserved for later stages.

The second stage, preparing performance specification objectives (PSOs), is a method of defining in terms of *measurable* quantities what you want the functional system to do. It consists of a set of times, activities, quantities, and costs by which you can guide the design and measure the performance of the systems. For example, the haulage system cannot be designed without specifying the rate of throughput (tons/day). Similarly,

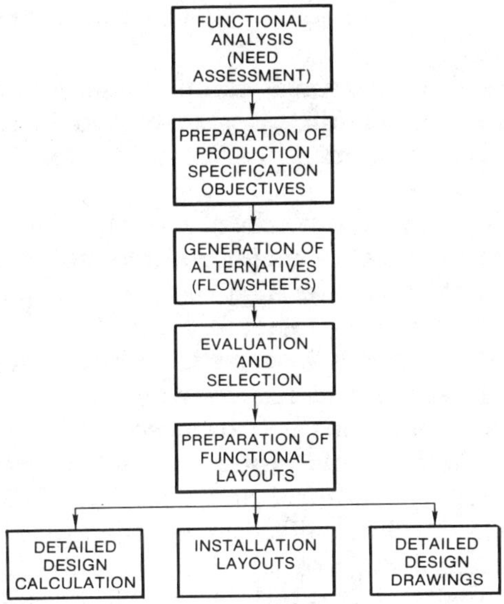

Figure 5-11 Procedures of layout and design.

the desired reliability of the haulage system should be declared, the level of maintainability known, the approximate cost estimated, and even the acceptable safety levels approximated. It is not a good practice to *assume* these. One design is never equal to another in reliability, safety, maintainability, and cost just because they perform the same operations.

Pragmatically, however, a designer cannot write a PSO which states that "the haulage system design should limit the lost-time accident rate to 4 worker-days per year." First, too many variables inhibit carrying out this goal accurately. Second, the designer might then be legally responsible for any injury incurred by workers who must operate the design. Third, and most important, the designer would be violating all ethical standards of engineering. To avoid these direct ethical conflicts, the common practice is to ignore the issue and accept whatever level of safety, reliability, and so forth the industry will bear. This is not to say that we do not try to improve each of the secondary performance levels. We do—but after the fact. What we do is design a system, allow it to operate, *then* say "let's improve the safety or reliability." (The ethical values of these two design approaches are left for individual and group thought.)

The third step is to generate alternative solutions for accomplishing the PSOs. These designs may be simple flowsheets depicting the major subsystems, specifying the approximate input and output requirements for each component, and showing the "flow" relationships that exist between components.

These alternatives are evaluated using the PSOs and knowledge of the standard practices of the mining industry. Various methods of evaluation can be used, including value engineering,[9] operations research,[10] and systems engineering.[11]

The next step is the preparation of preliminary functional layouts.

Functional layouts show the structures, components, and equipment positions in relation to the general mining operation and in relation to other functional systems. Usually, the general mine map is overlaid with the information needed to produce these layouts.

Following the preliminary layout constructions, installation layouts are produced to detail the installation of major equipment. Installation layouts contain detailed component and material specifications, specific installation procedures, installation tolerances, assembly specifications, and critical dimensional data needed to locate and position the components. Simultaneously detailed design calculations and drawings that are needed to support the installation of the equipment are produced.

The application of these phases produces a set of calculations, maps, and drawings that are used to control and guide the field layout of the various functional systems. Additionally, the various types of drawing serve to clarify each developmental stage of the design process. Flowcharts serve best to illustrate the functional subsystems and their relationships. The functional layout best depicts the physical or operational relationships that exist between the system and the rest of the mine. The installation layout shows the information necessary to locate and install the permanent subsystems physically. The detailed drawings provide the many smaller component designs that are needed to support installation of the system.

5.3 FUNCTIONAL (SYSTEM) LAYOUTS

A functional system is one in which activities occur in succession to produce a desired outcome. Based on such a definition, almost any organized mining activity can be considered a functional system. For the purposes of limiting the range of functional systems, we will constrain our future analysis to the ventilation system (Chapter 7), the haulage system (Chapter 8), the drainage system (Chapter 9), and the electrical service system (Chapter 10). Premining data, layouts, and techniques, topics that directly affect the functional layouts, are discussed in Chapter 6.

REFERENCES

1. William Randolph Williams, *The Philosophical Foundations of Educational Systems Analysis and Modeling*, Doctoral dissertation, West Virginia University, Morgantown, W.Va., March 1979.
2. J. E. Rankin and G. O. Tolley, *Safety Manual No. 8 Fault Tree Analysis*, U.S. Dept of the Interior National Mine Health and Safety Academy, U.S. Government Printing Office, Washington, D.C., 1978.

3. H. Dotson Cather and W. Randolph Williams, *Mine Equipment and Machinery*, Fairmont State College Printing, Fairmont, W.Va., 1979.
4. Frank J. Landy, "Behavior Approaches to Safety," in Nicholas P. Chironis, ed., *Coal Age Operative Handbook of Underground Mining*, Vol. 1, McGraw-Hill Book Company, New York, 1977.
5. David J. Smith and Alex H. Babb, *Maintainability Engineering*, John Wiley & Sons, Inc., New York, 1973.
6. Martin L. Shooman, *Probabilistic Reliability: An Engineering Approach*, McGraw-Hill Book Company, New York, 1968.
7. AKS Jardine, *Maintenance Replacement and Reliability*, John Wiley & Sons, Inc., New York, 1973.
8. W. J. Fabrycky and G. J. Thuesen, *Economic Decision Analysis*, 2nd ed., Prentice-Hall, Inc., Englewood Cliffs, N.J., 1980.
9. Arthur E. Mudge, *Value Engineering*, McGraw-Hill Book Company, New York, 1971.
10. George K. Chacko, *Applied Operations Research*, Vol. 1, North-Holland Publishing Co., Amsterdam, 1976.
11. Arthur D. Hall, *A Methodology for Systems Engineering*, D. Van Nostrand Company, Princeton, N.J., 1962.

PROBLEMS

1. Identify the primary and secondary functions of:
   (a) a pencil
   (b) an automobile
   (c) a mine
   (d) a ventilation system
2. Prepare a fault tree analysis diagram for an automobile accident. Can this be extended to a haulage system accident?
3. Use the eight layout schemes for achieving safety to prevent:
   (a) a child from drowning in a pool
   (b) a driver from being killed in an automobile accident
   (c) a miner from being electrocuted
4. Determine the reliability of a system comprised of six series elements, all having 95% reliability.
5. If an identical group of six more elements were arranged in parallel, what would the reliability be?
6. Figure 5-8 shows a system of three series components with reliabilities $R_a = 0.6$, $R_b = 0.8$, and $R_c = 0.9$. If the cost of these components is a = \$1.52, b - \$3.10, and c = 0.35, determine the configuration that will give the greatest reliability/cost ratio.
7. Prepare a set of PSOs for a haulage system.
8. As a group, discuss the advantages, disadvantages, and unknowns associated with PSOs about safety.

6

Premining Data and Layout Techniques

6.0 INTRODUCTION

Mining as an industry has grown increasingly more complex over the last several decades. Increasing health and safety regulations, environmental impact assessments, greater unionization, fluctuating markets, and rising costs have all contributed to the complexity of operating a mining venture. More than at any time before, the success of a mining operation depends on a relevant premining data base and on a management organization capable of using those data to their fullest potential. This chapter discusses the nature of premining data, its acquisition, and its treatment. The selection of capable staff to implement these data is left to the mining companies.

6.1 PREMINING DATA NEEDS

Chapter 5 introduced the various operational phases of mining. Exploration/prospecting, phase 1, was presented as an organized group of premining investigations and surveys needed to describe a proposed mining site. The goal of this first phase is to generate a data base upon which subsequent management and engineering decisions depend. The three broad categories of premining data forming the data base are technological, economical, and sociological. Technological premining data, the most traditional type, describe geologic, hydrologic, environmental, topographical, and engineering factors of the proposed site. Economical premining data define the market practices and trends, local competition, cash flow, cash markets, leasing factors, and financial options concerning the proposed operation. Sociological premining data describe legal requirements and constraints, political realities, demographic descriptions of the available working personnel, union strengths, and other cultural effects.

Figure 6-1 (page 176-77) shows a relational diagram of some of the types of premining data that form the initial data base. Note that many topics overlap between the subcategories. This is the nature of these data and also the nature of the decisions that will eventually depend upon the data base. You should examine this breakdown of data needs and know why each individual datum is required.

6.2 DATA ACQUISITION

Data can be obtained by three logical processes: (1) sense experience, (2) reasoning, and (3) the use of authority.[1] *Sense experience* is knowledge directly gained through the five sensory functions of the body. *Reasoning* is the process by which one can infer a conclusion based on a set of facts or premises. *Authority* is a method of data acquisition using the information supplied from another person. These methods of knowledge acquisition as they are applied to obtaining a premining data base are encompassed within a set of investigative procedures displayed in Fig. 6-2 (page 178).

The first stage of data acquisition, historical and literature survey, involves the systematic search for any and all information that exists pertinent to the site or region. A vast amount of applicable data can be obtained by conducting an extensive review of previous work. Usually, this method will provide the simplest, most cost-efficient means of obtaining large quantities of data. However, since the method depends on the accuracy and truthfulness of someone else, the validity of the information is *always* in question. The techniques, advantages, and limitations of historical and literature surveying are developed in more detail later.

Surface surveys are field methods of gaining information. The data sought are either not available from examining the literature or are sought to confirm the validity of the information reported. Surface surveying involves techniques that do not significantly disturb the surface of the site. Unlike the treated data obtained from the literature, surface techniques generate raw data. Raw data, you will recall, are data that result directly from readings of instruments or observations made in the field. Since raw data are not treated or otherwise mediated by the observer, their validity is not in dispute. (This does not imply that the data are accurate or that they were derived from the best possible instrumentation—only that what was recorded was what was measured or observed.)

Subsurface methods of data acquisition require searching below ground for information. The data that are produced from these techniques are raw and are often described as hard data. "Hard data" is a term used to describe

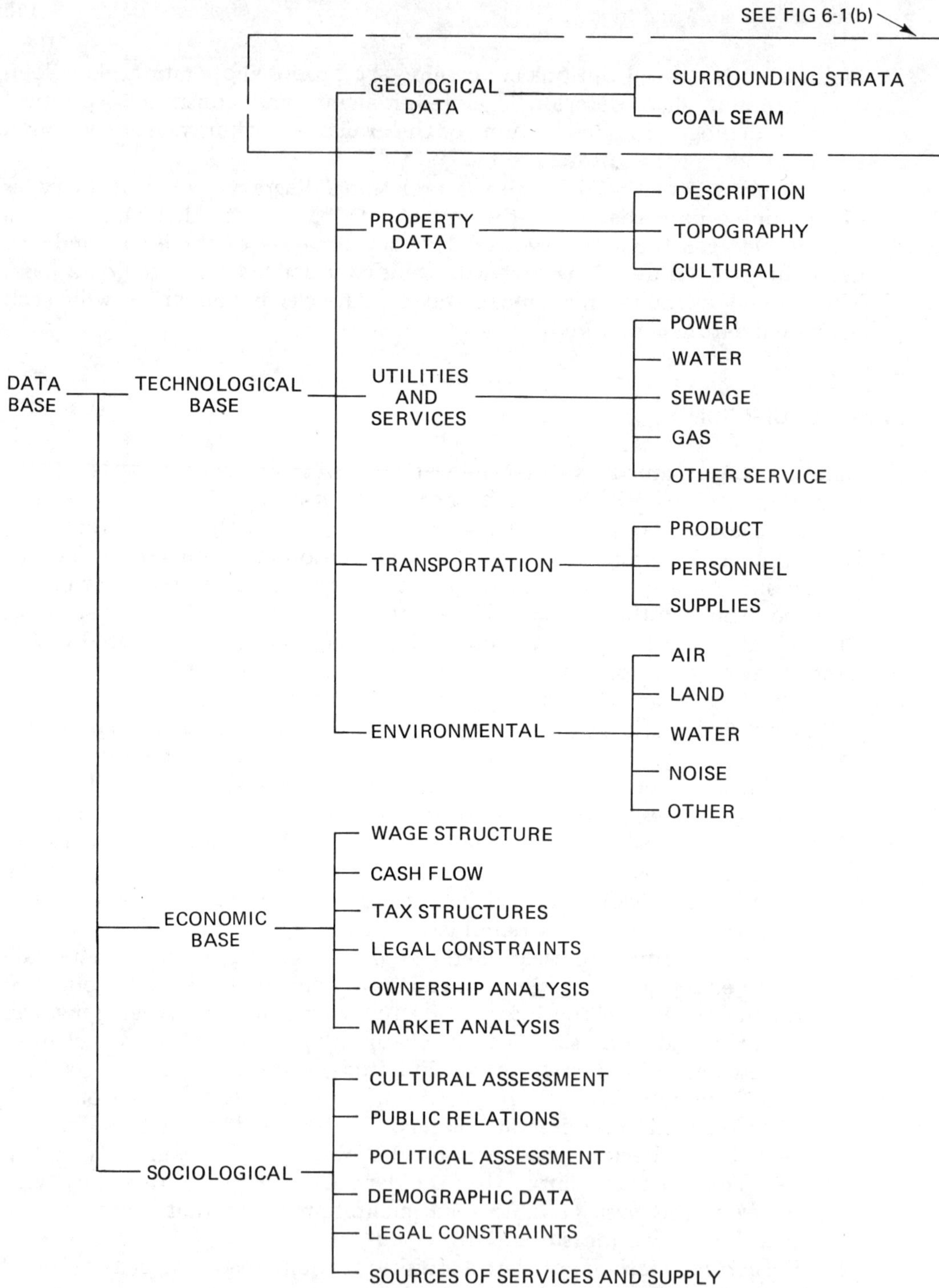
SEE FIG 6-1(b)
DATA BASE
TECHNOLOGICAL BASE
GEOLOGICAL DATA
SURROUNDING STRATA
COAL SEAM
PROPERTY DATA
DESCRIPTION
TOPOGRAPHY
CULTURAL
UTILITIES AND SERVICES
POWER
WATER
SEWAGE
GAS
OTHER SERVICE
TRANSPORTATION
PRODUCT
PERSONNEL
SUPPLIES
ENVIRONMENTAL
AIR
LAND
WATER
NOISE
OTHER
ECONOMIC BASE
WAGE STRUCTURE
CASH FLOW
TAX STRUCTURES
LEGAL CONSTRAINTS
OWNERSHIP ANALYSIS
MARKET ANALYSIS
SOCIOLOGICAL
CULTURAL ASSESSMENT
PUBLIC RELATIONS
POLITICAL ASSESSMENT
DEMOGRAPHIC DATA
LEGAL CONSTRAINTS
SOURCES OF SERVICES AND SUPPLY

(a)

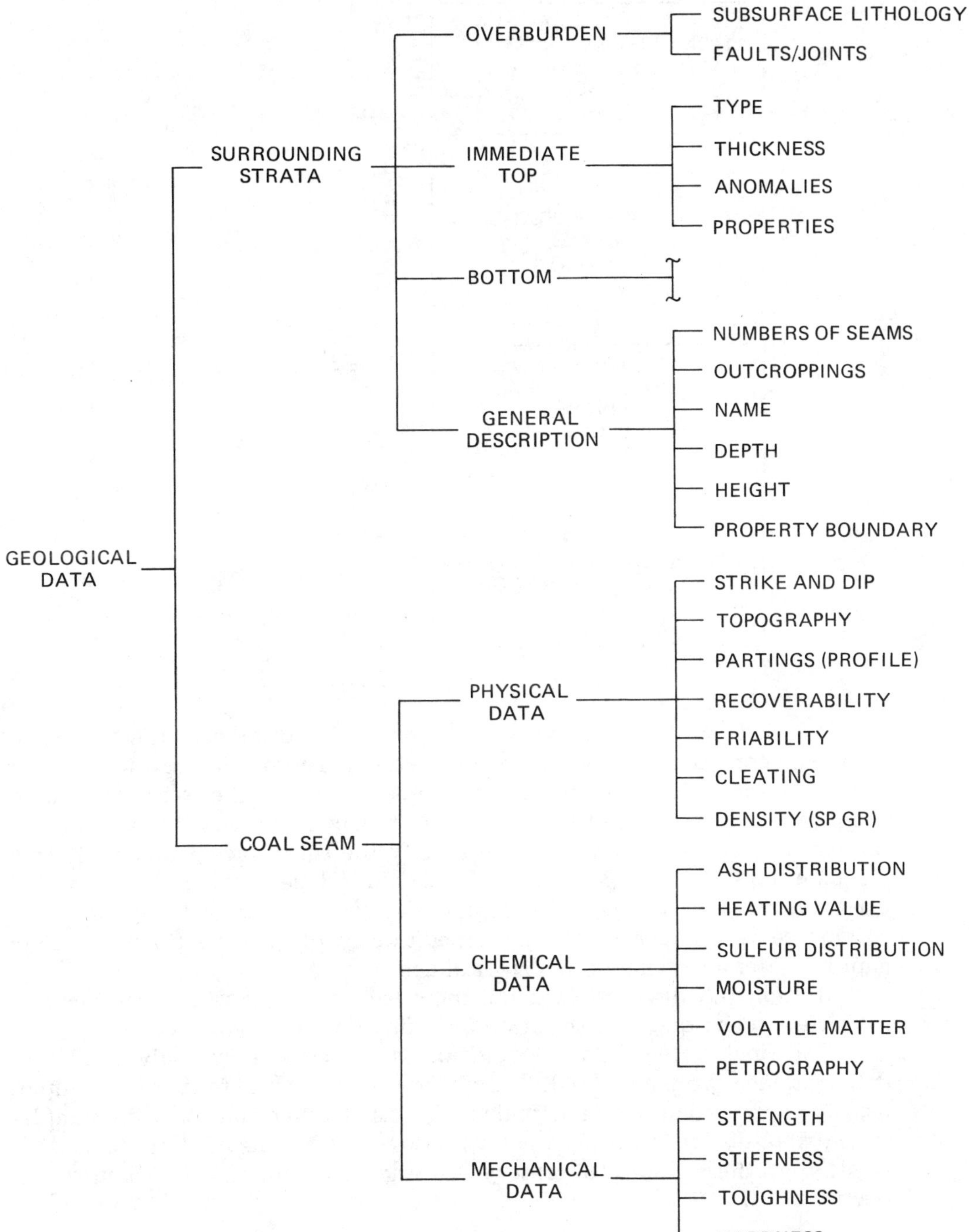

(b)

Figure 6-1 (a) Major categories forming the data base. (b) Partial breakdown of the needed geological data base.

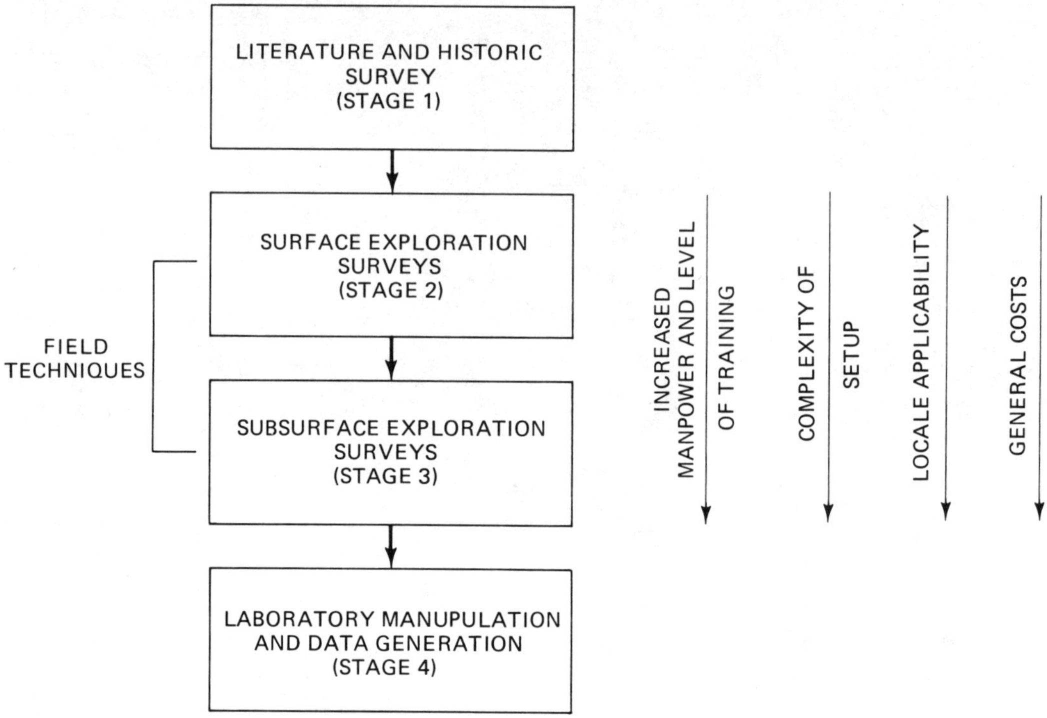

Figure 6-2 Data acquisition procedures.

physical evidence that can be seen, touched, and otherwise directly examined (e.g., core samples taken from a bore hole). These data provide the most specific information about a finite point. However, the extrapolation of these data to a more general population or region is limited. That is, one core sample cannot yield data that will specify the subsurface geology within a 5-square mile site. The generality of the data is dependent on the method(s) used and the consistency of the region. Because of their relatively high cost, subsurface techniques should be expeditiously planned to fill in the data gaps left after analyzing both historical and surface data.

Section 6.3 discusses in detail the selection of the various surface and subsurface techniques and the advantages and limitations of each.

The final stage of data acquisition is the laboratory analysis of hard data obtained from the field. Laboratory data derived from field samples usually cannot stand alone as "reliable" data. The very fact that the samples tested were no longer in place, or "in situ," at the time of testing alters the validity of these data. Laboratory acquisition techniques are discussed in Section 6.1.6.

6.2.1 Historical and Literature Acquisition Techniques

So much information is available from so many different sources that the procedures of obtaining this information and subsequently analyzing it can become overwhelming for the mine designer. The result is that the designer avoids a thorough examination of the available literature and begins site planning using surface and subsurface techniques. By so doing, he or she often duplicates data that are obtainable from the literature.

A complete survey of the literature and historical records is imperative. The advantages of the literature and historical surveys are:

Planning for subsequent surveying. An analysis of the available information gives the designer an overview of what is known and unknown about a given site. By organizing these data, the designer can reveal gaps in the data base. The gaps can later be filled by ordering appropriate field surveys.

Cost effectiveness. In many cases the published information that exists about a site is either free or available at minimum cost to the investigator. Examples include government-sponsored surveys, geological data, and maps. Personnel costs are inexpensive compared with those of other acquisition techniques.

Personnel effectiveness. Since the procedures required to obtain the information are nontechnical in nature, the level of personnel training and experience is quite moderate. Additionally, the number of people required is substantially lower than that of any other phase. The greatest personnel cost is associated with the organization and analysis of the data base after the information has been obtained. Of course, organization and analysis are activities that are needed for all acquisition phases.

Time effectiveness. Since the data have been previously treated to allow for publication, interpretation, and so on, the amount of time needed to treat and interpret these data is less than that of other phases.

Not site limited. The physical characteristics of the site (e.g., presence of water, buildings, roads, etc.) play a large role in limiting the field methods for data generation. Since the data available from historical sources were generated at a different time, with different constraints, they may represent data that are not obtainable today.

Lowest complexity. Literature and historical searches are simple, involve fewer activities, and require less equipment than any field technique.

Data completeness and contrast. The contrast of data is the availability of different types of data. (The three major classes are technological, economical, and sociological.) Using literature surveys the entire range of data shown in Fig. 6-1, from an economic analysis to specific stratigraphic positioning, can be made available. The completeness of data is measured by the amount of relevant data available from literature sources. Usually, more relevant data are available from the literature search than from any other source.

The only major disadvantage in using historical and literature derived data is that the validity of the data depends on the honesty of the reporter. The mere fact that something someone reports has been published does not imply that the information is valid. For instance, the reporter may have applied an invalid treatment to the raw data, thus unknowingly may have altered the outcome, or he or she may have falsified the data to make it support an opinion not otherwise tenable.

Regardless of the appearance of validity, the acceptance of data derived from literature sources depends on an affirmative answer to the following questions:

1. Is the person reporting the information an expert?
2. Does he or should he have access to these data?
3. Are these data consistent with what has been previously reported or consistent with commonly held theories?
4. Can you, given enough time and other resources, duplicate or confirm what was reported?

If for any reason you cannot respond positively to each of these questions, the validity of the information reported is in question and further confirmation is required.

Various types of information sources exist. Figure 6-3 depicts certain major sources of existing literature and data. An analysis of the types of premining data sought (Fig. 6-1) relative to these various sources is left as an exercise.

6.2.2 Field Acquisition Techniques

Several surface and subsurface techniques are employed to generate the required field data. Various methods are needed because of the wide variations in surface and subsurface conditions encountered. Each method has certain limitations which must be supplemented by additional evidence before an accurate and valid interpretation of the data can be made. The methods eventually employed are selected based on the following factors.

Geophysical diversity yielded. Geophysical information includes the structural, lithological, mechanical, and other physical properties concerning the surface and subsurface nature of the sites. Some methods reveal only small amounts of data about one data class, whereas other methods yield data about several classes at the same time. An example of the latter method is

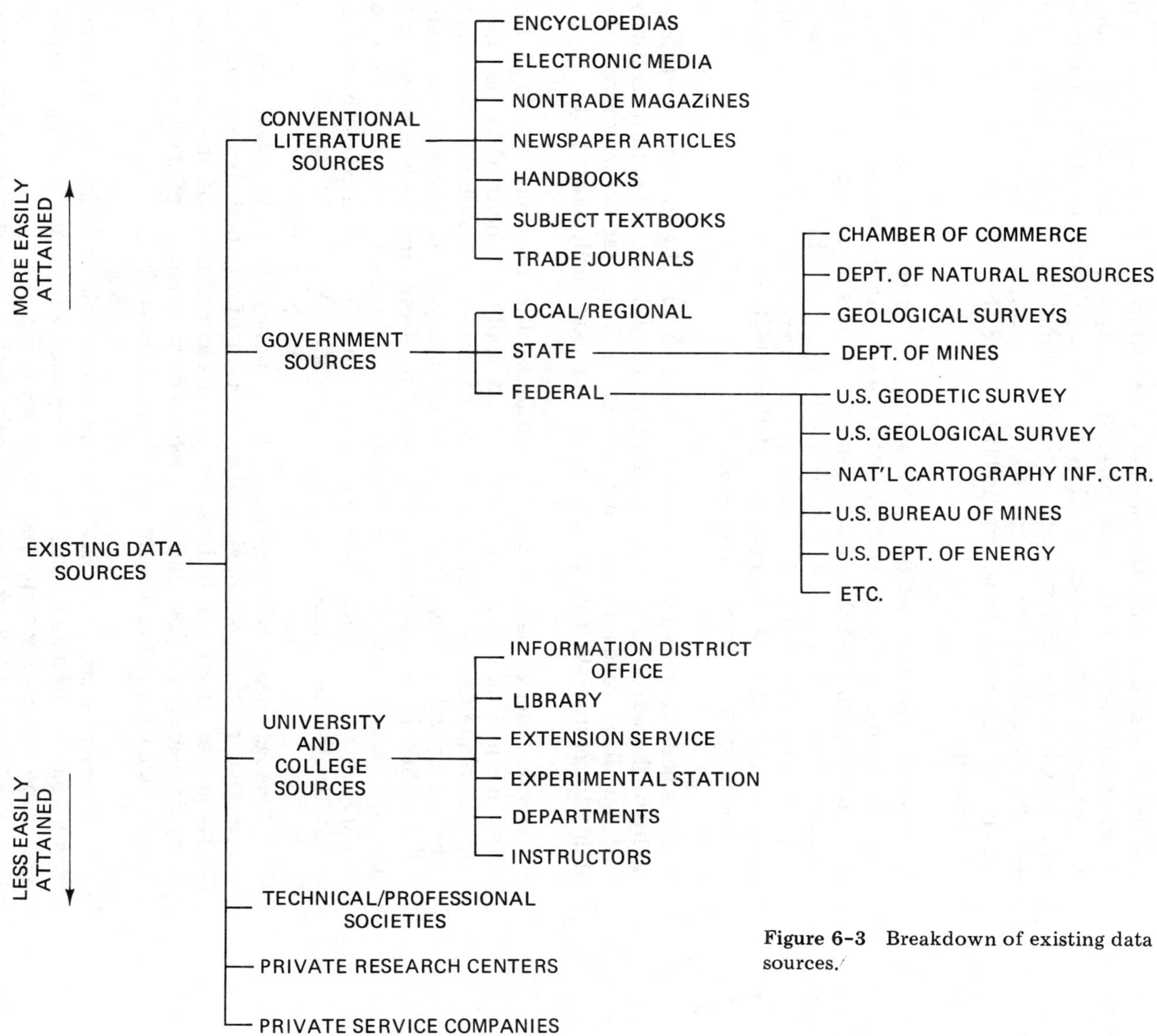

Figure 6-3 Breakdown of existing data sources.

core drilling. Examination of the core simultaneously reveals mechanical, physical, lithological, chemical, and some structural data. Methods that reveal many classes of data are called *diverse*. Often, the diversity yielded is in inverse relationship to the area covered. For example, although core sampling provides high diversity yield, it is limited to local specificity. Without other supplemental information, core data cannot be extrapolated to represent the properties of the larger region. As a rule and assuming the same amount of coverage, those methods providing the greatest diversity yield are selected.

Degree of raw data resolution. Resolution is the ease with which the data may be treated and the resistance of the raw data to treatment or convergence distortion. High data resolution corresponds to a lower degree of subjectivity. Those techniques that presently require either several treatment methods or a high level of subjectivity in interpretation are considered to have low resolution. Surveying has relatively high resolution, whereas satellite telemetry has relatively low resolution.

Applicability to the site. Factors relating to the site include the topography of the surface, the presence of water, the existence of cultural features (e.g., buildings, roads, towns), and vegetation. These factors may affect the ability to set up field instruments, explode charges, drill, and so forth.

Relative value of the coverage. To optimize the resources allocated for premining investigations, the planner must apply a benefit (data yield)-to-cost technique to select the most appropriate data acquisition method. Core sampling provides a very high diversity of data yielded; however, the mapper's ability to extrapolate the information from one bore hole is quite limited. Thus more holes are required at an average cost of $8,000 to $10,000 per hole. This data/cost ratio must be compared with similar ratios for other techniques.

Time and complexity of the set up. Like money, time is a valuable resource that must be weighed in the selection of the best data acquisition technique or group of techniques. As the complexity of the technique setup increases, so does the time required, the amount of staff required, the training level of the personnel, and the dollar cost.

These five areas all play a simultaneous role in evaluating the most appropriate set of data acquisition techniques. Of the many techniques available, those commonly employed are shown in Fig. 6-4.

The basic principles associated with each type of technique shown in Fig. 6-3 are discussed below. For a more detailed discussion of each, you should refer to the references at the end of the chapter.

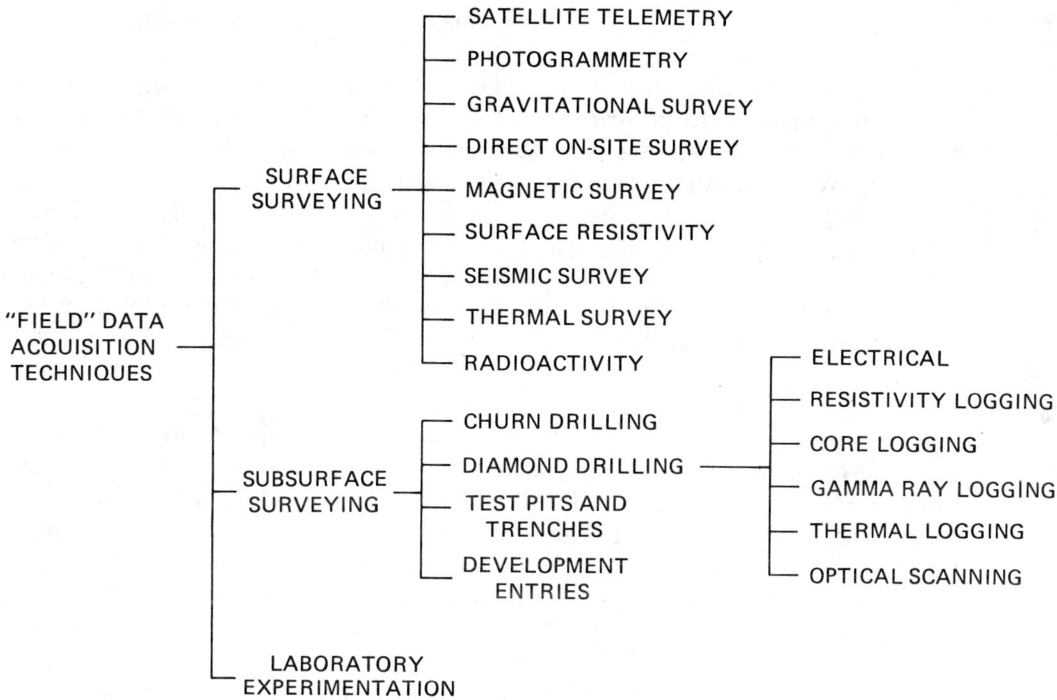

Figure 6-4 Relationships between "field" data acquisition techniques.

Satellite telemetry.[2,3] In the early 1970s NASA began the Earth Resources Technology Satellite (ERTS) program, the objective being the installation of several satellites to assist researchers in managing the earth's natural resources and protecting its environment. On July 23, 1972, the first of three ERTS satellites, Landsat 1, was launched into a circular orbit 570 miles above the earth's surface.

The earth's surface is photographed by Landsat using multispectral scanner imagery (MSS), which is so sensitive that 63 different shades of gray can be distinguished from the black-and-white pictures. Such photographic resolution allows analysts to delineate small changes in the moisture retention or the vegetation within a particular region. The presence of groundwater is greater within highly faulted or fractured strata. By identifying such regions, geoanalysts can identify possible fault zones or structural lineaments.

Faults, shear zones, and joint fractures usually occur along a linear path. Because of this, the term adopted for possible faulted regions identified on aerial photographs is *photolinear* or *linear*. Linears that can be specifically linked to a geologic disturbance (faults, joint set, etc.) are then called *photolineament* or *lineament*.

The actual procedures for treating and converting the data generated by Landsat to useful data are discussed later. The principal limitation of Landsat imagery as a data acquisition technique lies in the number of subjective interpretations and manipulations of the data necessary to derive useful information. In addition to faults and fracture zones, linears are correlated with aquifiers, slump features (e.g., paleochannels), sand channels, and clay veins. Analysts are not completely able to distinguish among the various types of linears. When a linear is known to have geological lines, it is not known what class of feature it represents or precisely what the effect may be to a mine. Specific interest is now being directed by MSHA personnel toward delineating the various types of linears observed and ranking the relative importance of each to roof control problems encountered within mines.

Photogrammetry.[4] The relative advantages and limitations of photogrammetry were discussed in Chapters 2 and 4. In terms of premining data, photogrammetry can be used for magnetic surveying, topographical mapping, and infrared scanning. The basic advantages are its relatively fast generation of data and its ability to provide a more continuous overview of a given site. For large reserves, photogrammetry may be 50% of the cost of a similar on-site survey.

Direct investigations. Still the most used methods for generating raw data about a site are those involving direct on-site land surveying and geologic feature identification. Geologic data, plotted directly on topographical maps, include the surface dip (read from transit, clinometer, or theodolite), surface faults, suspected subsurface faults, the type and extent of surface soils, and rock outcroppings. It should be noted that these rough geologic maps constitute treated data since much of what is placed on the map is inferred from the observer's impressions. These data, like all other data, must be mediated with other investigations before final interpretation and convergence of the data are possible.

Gravitational surveys.[5,6] The attraction one body has for another is directly proportional to the masses of the two bodies and inversely proportional to the square of the distance between them (Eq. 6-1).

$$F \alpha \frac{M_1 M_2}{r^2} \tag{6-1}$$

Gravitational surveys are performed to determine variations in the gravitational force within a given region. These differences are associated with the relative densities (masses), sizes, and shapes of the subsurface structure. *Anomalies*, variations in the gravitational index, can be correlated with structural anomalies. These variations may produce stress concentrations and result in poorer ground stability for a mining operation.

Magnetic surveys.[5,6] The earth's magnetic field is not constant but varies over time and from place to place. Local variations in the magnetic field are called magnetic anomalies and indicate the presence of subsurface size and shape as well as material having magnetic susceptibility. Magnetic surveys have been used extensively for gas field mapping and for locating abandoned shafts and wells.

Surface resistivity.[6] Electrical resistivity is used to determine absolute or relative resistance of subsurface structures to current flow. Certain materials, such as clay and metallic sulfide, tend to be better conductors than denser, more impermeable rock, such as tight sandstone or limestone. The relative differences between formations can be used effectively to map the apparent substructure and anomalies in the subsurface, such as faults, porous aquifers, and the like.

Seismic methods.[5,6] Rocks will support or transmit mechanical vibration (sound waves). Each type of rock and each type of structure will perform somewhat differently when subjected to a shock wave. When a small shock is created, as in a surface explosion, the wave that is created spreads through (down and out) the subsurface structure. The rate at which the wave propagates depends on the type of rock and the physical structure of the subsurface strata. Interface (or faults) between two differing rock types can be identified by the fact that part of the energy of the wave will be transmitted across the joint and part of the energy will be reflected. By measuring the magnitude, direction, and rate of the reflected wave, the interfaces can be identified.

Coring.[7–9] *Coring* is the physical extraction of a given diameter (usually 2 in. in diameter) and length (usually in 15-ft. sections) of subsurface structure. Coring is a traditional and much-favored procedure because it provides hard data; that is, the physical data can be seen, touched, and sent to the laboratory for further testing and experimentation. Since the evidence is physical, the data that are derived characteristically have a higher level of resolution. The data are also highly diverse, providing various geophysical and geochemical properties. Since the core is so small, the data are considered "point" data and apply only to the hole locality, which varies from 250 to 2500 ft around the hole. To make these data representative of the region, several cores are needed. The number and grid spacing of holes depends on the uniformity of the structure within the region. For a uniform seam and geology, spacing of 1500 to 3500 ft would be appropriate, whereas for an extremely variable structure, spacing of 500 to 700 ft would provide the best fit for data.

The first data derived from drilling are those in the well log. The well log contains such information as:

1. Location of the hole (vertical and horizontal positioning) and other relevant site details

2. Drilling parameters
   (a) Type of drill, drill bit
   (b) Drilling medium
   (c) Diameter of hole, diameter of core
   (d) Orientation of hole (vertical, horizontal, or angled)
3. Hole and core log
   (a) Date(s) of drilling
   (b) Depth of hole at each date
   (c) Depth of core—starting and stopping
   (d) Depth and size of casing
   (e) Depth and diameter of hole and core
   (f) Bit condition: wear, fractures, etc.
   (g) Time required to drill and core
   (h) Type and amount of drilling mud or fluid
   (i) Amount of water inflow
   (j) Field description of strata encountered
   (k) Percent of core recovered and loss accounting
   (l) Description of in situ sludge
   (m) Report of fragment or chip sampling
   (n) Notation of drilling delays
   (o) Analysis of in situ experimentation
   (p) Description of backfilling or grouting used

Other data generated from subsequent laboratory analysis of the core may include mechanical property analysis (e.g., shear, tensile and compressive strength, elasticity and toughness of impact test, etc.), an essay or petrographic analysis, washability, proximate and ultimate analysis, and physical properties (e.g., specific gravity, moisture, hardness or grindability, etc.).

Several cores analyzed together and mapped provide an analysis of the distribution of physical, mechanical, and mapped and geological properties over a given region. These techniques are discussed later.

Electrical resistivity logging.[5,8] Electrical resistivity logging is merely an extension of the resistivity method used for surface analysis except that various measurements of strata resistivity are recorded at different depths within the bore hole. In this way more accurate subsurface data can be generated relative to possible fault locations, discontinuities, major fractures, and so forth.

Gamma-ray logging.[8,10] Gamma-ray logging is the measurement of the natural radioactivity of the individual stratum at various depths within the bore hole. Sedimentary rock, such as coal, limestone, marine shales, and the like, can be distinguished from one another on the basis of their differing radioactivity.

Thermal conductivity logging.[5,8] Similar to electrical resistivity, the thermal conductivity of each rock type or strata varies sufficiently to allow differential measurement. Thermal conductivity logging is merely a measurement of the conductivity along various locations within the bore hole.

Optical scanning.[8] The development of fiber optics together with miniature closed-circuit television cameras has made visual inspection of the bore hole wall a reality. Optical scanning provides the investigator with the ability to observe "undisturbed" faults and fissures in situ. The present disadvantage of optical methods is its high cost.

Trenching and test pits.[6] Another subsurface method used for data acquisition is that of removing the unconsolidated topsoil to investigate outcropping migration and the extent of the bed. Because of the expense, such techniques are confined to quite shallow outcrops.

Development entries. The greatest amount of data regarding the immediate top, nature of fissures, cleating faults, and so forth can be generated during actual mining. These data must be analyzed relative to the premining data to allow for modifications of the mining projections.

6.2.3 Laboratory Acquisition Techniques

Field samples procured from surface and subsurface techniques are analyzed under controlled laboratory conditions and instrumentation. It should be noted that laboratory experimentation of rock samples, particularly analysis of engineering properties (e.g., strength toughness, stiffness, etc.), yields data of lower resolution than can be produced by similar methods testing the engineering structural materials (e.g., steel, brass, cast iron, concrete).

Several facts contribute to this disparity in data resolution: (1) rocks obtained through field methods are "disturbed" (i.e., they no longer represent the in situ material, due to the method of sample extraction); (2) geologic materials lack the homogeneity associated with engineering materials, thus a sample taken at point *A* may not represent an adjacent point *B;* (3) the failure theories and engineering properties depend on the assumption that homogeneity exists in the sample, but for most geologic samples it does not; and (4) consideration must be given to the performance difference between in situ confinement and the modeled confinement of the laboratory methods.

Each of the factors described above contributes to the lack of data resolution. Thus the laboratory data must be mediated in light of the other information obtained about the site. Various types of laboratory data[8] and laboratory procedures[11] are available to the geologist and mine designer. For an analysis of these methods, examine the references at the end of the chapter.

6.3 EVALUATING PREMINING DATA

The data that are acquired have different value to the planning process. Some data are general and give the designer an overview of the region; other data are quite specific and define in detail the nature of a point. The data also differ in accuracy; some will be quite accurate, whereas others may be quite erroneous. Data, whether acquired from the literature or from field techniques, have the following characteristics which determine their usefulness.[12]

Timeliness. The timeliness of information refers to the time the information is received relative to when it is needed. Information received too late is useless. For instance, knowing that a fault zone exists at a particular location after a roof fall injures a worker does not help. The information is too late (not timely) and therefore valueless. However, receiving the same information in advance of mining could be of great value in avoiding potentially hazardous roof conditions.

Accuracy. The accuracy of information is measured by its freedom from error. A map that shows an entry 90 ft from its actual location presents useless data when the information is needed to drop a bore hole into that entry.

Validity. The validity of information depends on the truthfulness of the reporter of the information. This was discussed in detail earlier in reference to the usefulness of literature sources.

Relevancy. The relevancy of information depends on the need the receiver has for the information and his or her ability to use it. Relevancy is a measure of the application of the information. For instance, telling a mine mechanic the precise location and size of a sandstone aquifier is of little pragmatic value, but providing the same information to someone developing the drainage system would have significant value and utility.

Completeness. The completeness of the information is the degree to which all relevant information is included in the message. To tell a mine foreman that the mine fan will eventually need to be shut down for maintenance is not as valuable as his knowing that this maintenance must occur between 5200 and 5300 hours of operations.

The feasibility decisions that must be made, the planning of the overall operation, and the specific engineering plans and decisions all depend on the timeliness, accuracy, validity, relevancy, and completeness of the data made available to the mine planners.

6.4 TREATMENT OF DATA

All premining data acquired from the various sources must be treated to derive the maximum degree of information. Various types of treatment procedures are needed, depending on the form and nature of the data. Data acquired from literature and historical surveys, although they represent treated data, probably need to be reorganized so that the gaps and unknown facts in the data base are revealed. Following this, the mine designers and developers select, schedule, and implement specific field methods needed to fill in the data gaps.

The treatment of the field or raw data follows the format presented in Chapter 1. To review, the notion of raw data treatment was divided into two classes, "active" and "passive" manipulation. Passive strategies for manipulating data do not effect the nature or dimensional quality of the raw data. Normally, three passive methods are used: (1) simplification of data, (2) enhancement of data, and (3) distortion of data. Simplification is the process by which extraneous information is eliminated from the raw data, thus clarifying the remaining data. Simplification may also entail categorizing, combining, or collapsing data. Regardless of the specific operations, the purpose of simplification is to make the important parts of the raw data easier to understand.

Enhancement is a data treatment in which additions are made to the raw data base to make the data appear more continuous or complete than they actually are. The development of contour lines from discrete survey points is one example of enhancement. The contour line represents points of equal elevation, however, only a very few points on that line can be traced back to the raw data from the survey. The other points and the line itself are are additions. Enhancement is a common treatment process that we often use unconsciously, but its effect is to *change* the appearance of the raw data from the discrete or discontinuous form to a continuous form.

The last passive data treatment is distortion. Distortion involves either the deletion of conflicting parts of the data to cause them to appear more uniform, or changing the data to show a trait more clearly. An example of the first type would be the removal of a datum point which deviates from the trend established by the rest of the data. An example of the second type is an equal-area projection of the earth. Although the map truly depicts equal areas, other characteristics, such as shapes, distances, and directions, are all distorted. Thus a technique used to obtain a one-to-one correspondence of one trait produces a distortion of the others.

Active techniques of data manipulation are those that result in substantial changes to the nature or dimensional quality of the data. These may be simple statistical manipulations producing new data, such as means, variances, and standard deviations, or complex mathematical or graphical treatments, such as multiple differentiation and integration, which result in differing data structures.

Each of the manipulative processes discussed above results in a change in the raw data. The validity of the change and the validity of the new data base depend on (1) the reliability and accuracy of the raw data, (2) the assumptions and intentions of the person manipulating the data, (3) the applicability and the number of methods used to treat the raw data, (4) the degree and nature of the change that results, and (5) the end-use requirements, reliability, and accuracy of the treated data. The next section describes some of the common graphical techniques and formats available to analyze and present premining data.

6.5 REPORTING FORMAT OF TREATED DATA

Treated data are reported in verbal, mathematical, and graphical formats. Of principal interest to those individuals involved in the functional layout of mining activities are the graphical formats, specifically maps. The various types of maps needed for mine layouts are regional maps, topographic maps, cadastral plats, geologic maps, hydrologic maps, mine projections, and functional layouts. The production of these maps depend on each other somewhat and follow the hierarchy shown in Fig. 6-5.

Regional maps are planimetric maps (commonly available from government sources) that depict major hydrographic data (e.g., rivers, bays, ports, etc.), various highways (e.g., interstates, principal highways, secondary roads, etc.), major landmarks, population centers, municipal property boundaries, and so forth. These maps are useful as "foundation maps" upon or over which can be placed other regional data, such as population distributions, economic distributions, and other statistical data. The resulting map is called a *thematic* or *statistical map*, the production of which is discussed in detail in Section 6.5.1.

Topographic maps are essentially regional maps which specifically detail the vertical relief of an area. Topographic maps are available from government sources or can be generated from field sources (spirit leveling, photogrammetry, etc.). The topographic map is used as the foundation map for practically all subsequent mapping.

Geologic maps are topographic maps upon which are plotted various types of geologic data. The type and detail of geologic data depicted on geologic maps depend on the use to which the maps will be placed. One example of a geologic map is the lithographic profile discussed briefly in Chapter 4. Geologic mapping is discussed in Sections 6.5.1 and 6.5.2.

Mine projections are maps that depict the principal entry points and the direction, number, and size of entries 6 months to 5 years in advance of mining. Other features often included are well locations, significant geologic conditions, and other mining and major cultural features. The location of the entry points, direction and number of entries, and other projection data are

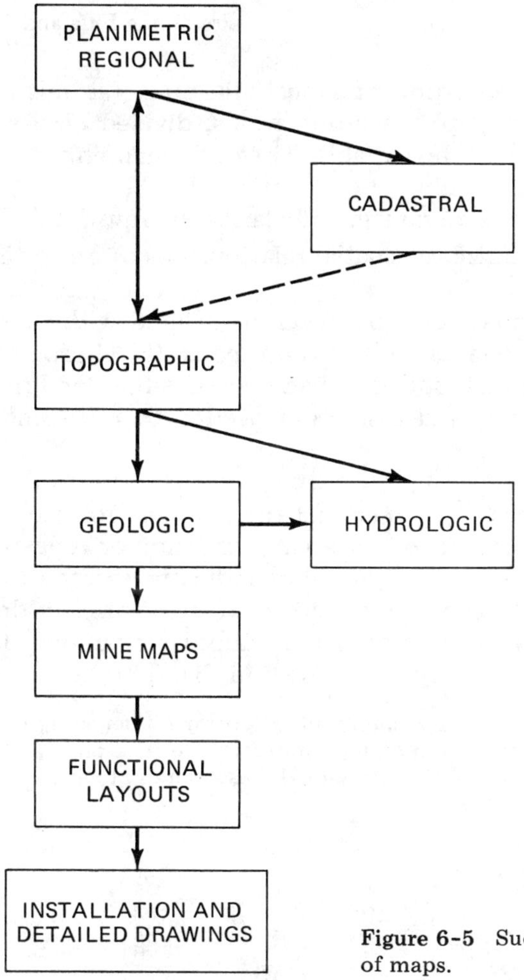

Figure 6-5 Successive production of maps.

based upon analysis of geology, haulage requirements, drainage needs, and other factors noted in Chapter 4. Techniques of descriptive geometry are particularly useful in solving mine projection problems. These techniques are described in Section 6.6.

Following or simultaneous to the production of mine projection maps, the specific functional layouts for the operational activities are prepared. Functional layouts denote major system components and the spatial relationships existing between this equipment and the other activities. Functional layouts of ventilation, haulage, drainage, and electrical systems are presented in Chapters 7 through 10.

6.5.1 Thematic (Statistical) Mapping

Thematic or statistical maps show the quantitative or qualitative aspects of one or more variables in relation to the region of interest. These maps are differentiated from statistical diagrams (as might be represented by line graphs, bar graphs, scatter graphs, divided circles) by depicting the spatial distribution of the variable. Thematic mapping serves to:

1. Simplify, organize, and clarify the raw data.
2. Emphasize the spatial relationships of one or more variables.

The methods most often associated with thematic mapping are (1) dot maps, (2) qualitative or Venn maps, (3) choropleth maps, and (4) isopleth maps. Each has specific characteristics that lend the degree of simplification, enhancement, and distortion needed to represent properly the variable of interest.

Dot maps.[13,14] Perhaps the simplest form of displaying qualitative and quantitative data of an area is the dot or repeated-symbol map. The position of the dot or symbol is indicative of the distribution (qualitative information). The relative size or density of each dot is representative of the quantitative characteristic. A familiar example illustrating this map is the population density map shown in Fig. 6-6(a).

Figure 6-6 (a) Example of a "dot map" depicting the population distribution across the eastern United States. (b) Correlation of linear intersections and mine problem areas. (Courtesy of U.S. Bureau of Mines.)

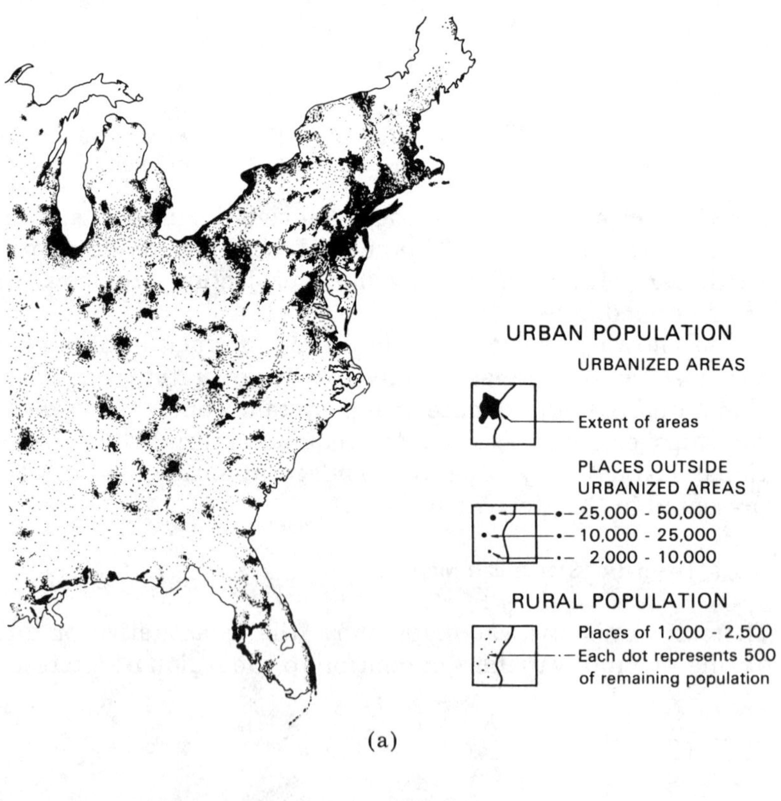

(a)

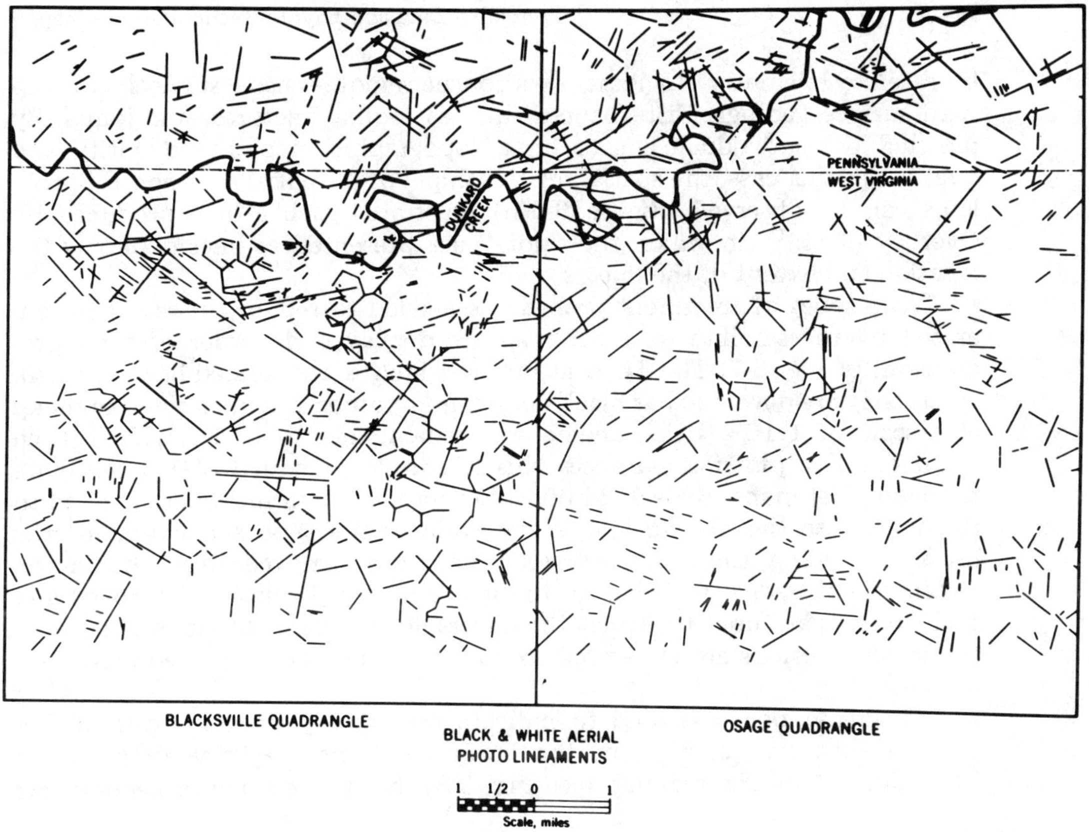

Figure 6-6 (continued) (b)

The dot method assumes discrete (point) values for the variable and may or may not be continuous over the area mapped. The mapper must determine the size of the dot that represents a given size quantity. This depends on the impression the mapper wishes to convey. A small dot representing a large density (number of items per area) may give the incorrect impression that very few items exist or that they are only sparsely distributed. Conversely, larger dots representing a small number of items might incorrectly portray the area as being inundated with the item, showing little if any distribution variability.

The mapper must also determine the nature of the variable distribution; that is, if the variable is uniformly distributed, it should be represented as such. However, if it occurs only discretely, the placement of dots within an area should indicate this. An even placement of dots is simplest to achieve and gives the impression that the variable is evenly distributed. An uneven distribution is more cumbersome for the mapper and reader alike, but an uneven distribution will always provide a better *qualitative* representation than an "averaged" or even distribution.

One application of the dot-mapping concept for mining purposes is the plotting of photolinears within a region. Recall that linears "photographed" by satellite telemetry (Landsat) or airborne remote-sensing surveys correlate with surface geologic disturbances such as faults, fractures, and joints. By plotting the individual linears on a topographic map, the quantity and quality (in this case, the spatial relationships) of potential roof control problems can be observed. Figure 6-6(b) illustrates such a map produced by Overbey et al.,[15] in which the "dots" are presented as lines depicting the location and extent of the linears.

The types of treatment procedures needed to reduce the raw data generated by the satellite to a dot map are displayed, in order of their progression, in Fig. 6-7. The Landsat infrared images are radioed back to earth, where they are developed as black-and-white photographs or as transparencies at a scale of $1{:}10^6$. Using analog enlargements a scale of 1:24,000 is then obtained. For plotting purposes this is correlated with USGS $7\frac{1}{2}$-minute topographical maps (scale 1:24,000). Several data treatments are applied to the images to aid the analyst in determining the type and extent of the linear, including simplification—scan lines are removed from the image, enhancement—edges and lines in the image are electronically improved and grid overlays are used to depict the linears more clearly, and manipulation—the image densities are converted to voltages which are displayed on a color monitor.[13]

A grease pencil is used to indicate preliminary linears. After further enhancement and correction, the linear locations are transferred to the $7\frac{1}{2}$ -minute topographic map (see Fig. 6-8). Next, the mine projections are

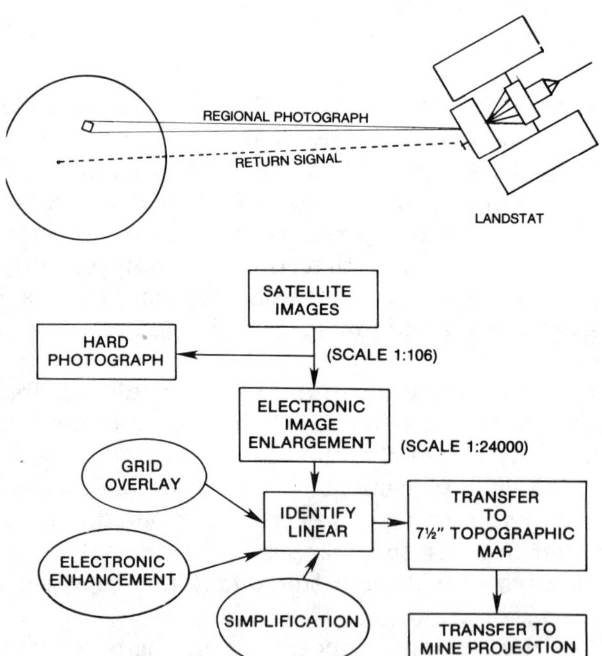

Figure 6-7 Reduction of infrared satellite images to a mine map.

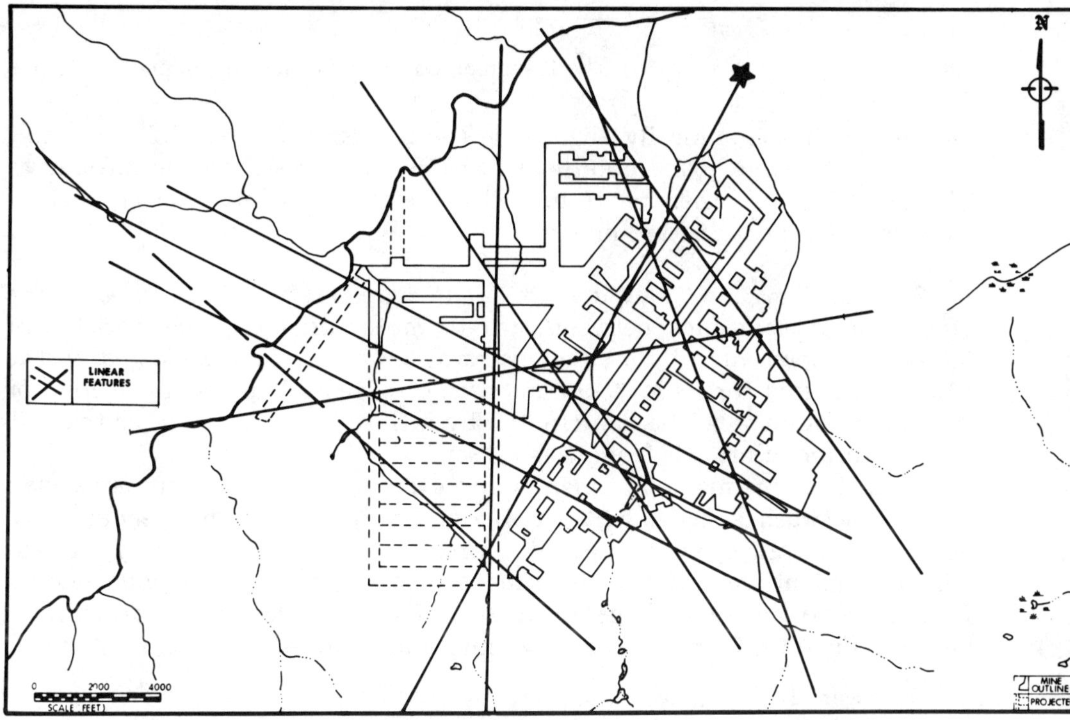

(a)

Figure 6-8 Linears from Landsat interpretations plotted upon a mine map. (Courtesy of U.S. Bureau of Mines.)

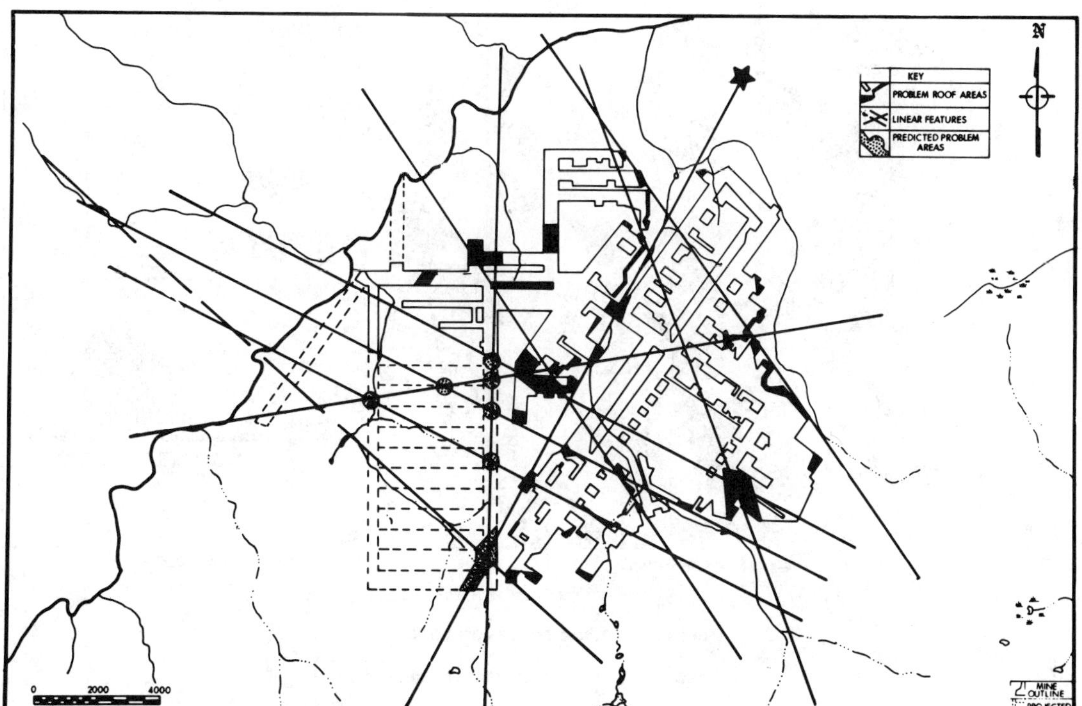

(b)

plotted or photographically aligned on the topographic map. (If the mining maps use the state plane coordinate system, this presents little difficulty.) The two plots can then be compared to identify possible intersections between mine projections and linears.

Qualitative maps. Qualitative or Venn maps are used to represent the qualitative relationships between two or more variables. Each variable of interest is represented by a color, pattern, or texture and is plotted on the map with regard for its location and area of extent. The area of overlap or "set intersection" can be represented by a third color, pattern, or texture, thereby differentiating it from the two sets.

This type of qualitative map is important when correlations are shown to exist between a dependent variable and two or more independent variables. For instance, Overby et al.[15] correlated the frequency of roof falls (the dependent variable) with three geological factors (the independent variables): stream unloading, unstable roof rock facies, and fracture trace density. Figure 6-9 shows the Venn map that resulted from this treatment

Figure 6-9 Venn map depicting the intersection of three hazardous roof conditions. (Courtesy of U.S. Bureau of Mines.)

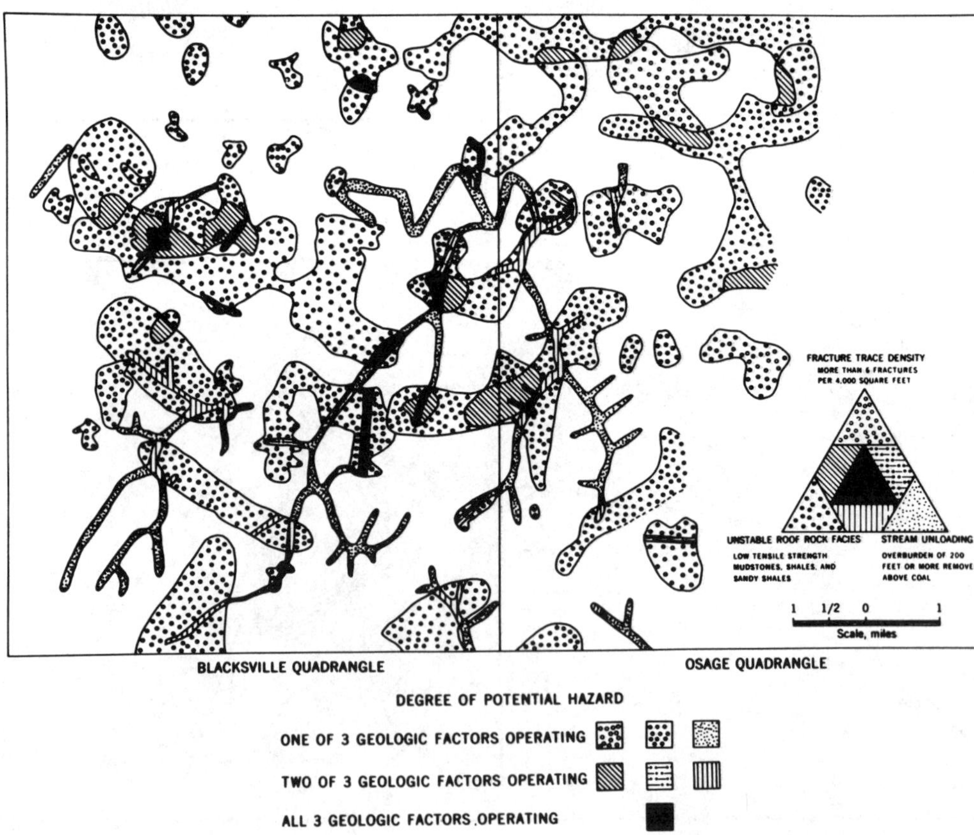

of the data. The value of qualitative information in this format is that those regions of multiple intersection can be avoided in the layout of permanent entries.

Choropleth maps.[13,14] Choropleth or shaded maps combine the qualitative and quantitative aspect of the data by averaging the density of distribution over finite areas and representing each averaged density with a shade color or pattern. With this type of data treatment the mapper assumes a *uniform* distribution of the variable or item throughout each area. The mapper must be careful to establish a numerical dimension that is not dependent on the size of the areas. If not, the quantitative value of the method will be destroyed. Common forms of density dimensioning include density per unit area, ratio, and percentages.

The numerical range of each grouping is very important from the point of view of the psychological impressions that can be conveyed to the reader. Divisions should be based on engineering principles, "threshold" values, or naturally occurring breaks in the data. The shading or pattern system chosen should be proportional to the mathematics of the group; that is, groups twice as dense should be twice as dark. Figure 6-10 shows the percentage distribution of sulfur in bituminous coals in West Virginia.[16]

Isopleth maps.[13,14] A fourth and final statistical map is the isopleth map (also called *isohyet*—for maps showing equal rainfall, *isopach*—for maps showing equal thickness, and *isobaric*—for maps depicting equal pressures and so forth). An isopleth is a line drawn on a map showing points of equal value or quantity. Therefore, an isopleth map depicts the relative distribution of a variable by showing the successive isopleths. Contour lines, lines of equal height elevation above some reference elevation, are isopleths of landform elevation.

The value of using isopleths methods as opposed to choropleths or dot techniques is that the continuous and uniform nature of the distribution is better conveyed to the reader. Any continuous item can be plotted using the isopleth technique. Some common types of exploratory data that lend themselves to isopleth mapping are percentage of sulfur content of coal (Fig. 6-11) and percentage of ash content of coal (Fig. 6-12).[16]

The techniques needed to construct isopleths are essentially the same as those described previously for constructing contour lines.

1. Lay off the horizontal control stations from which the raw data was initially derived (e.g., bore holes).
2. Establish the reference value or datum plane to which each isopleth will be referenced.
3. Select the most appropriate isopleth interval that will best describe the statistical relief.
4. Use either graphical or mathematical methods of interpolation to lay

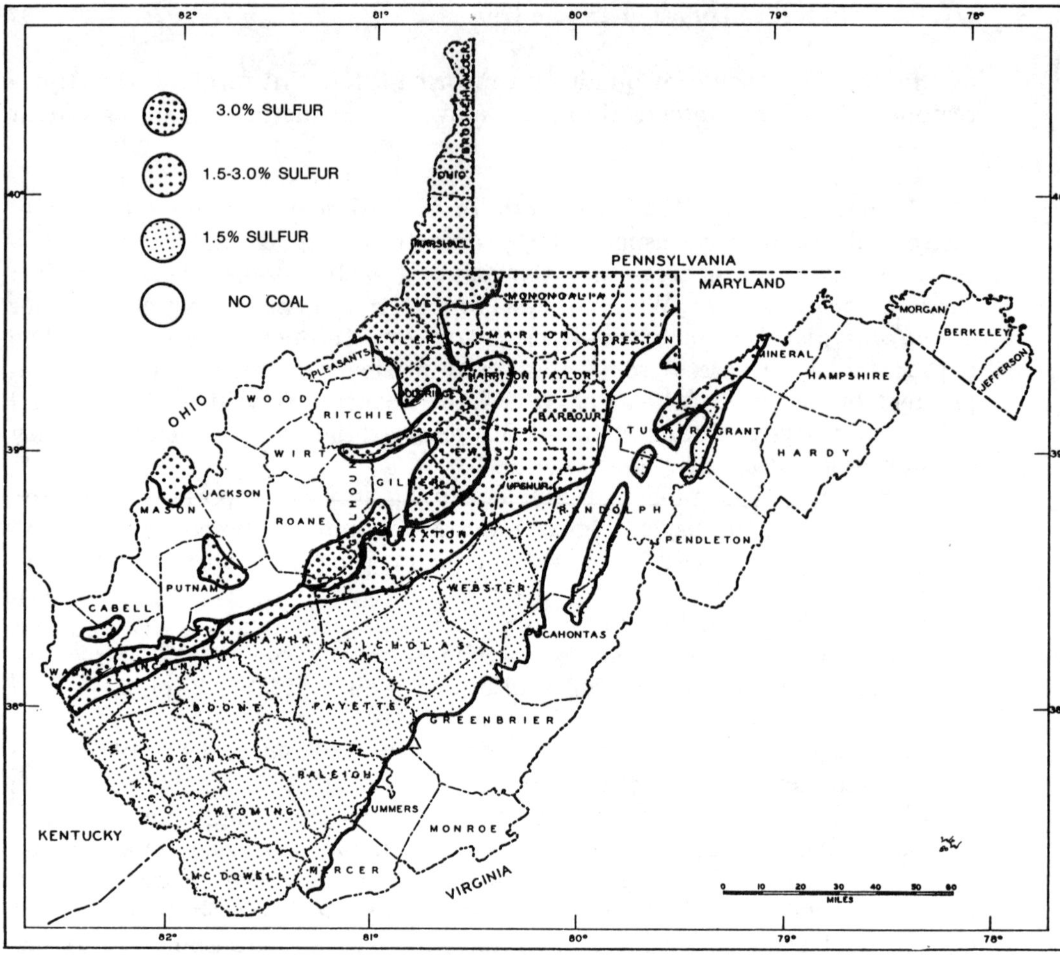

Figure 6–10 Choropleth map showing the generalized sulfur percentage of bituminous coals in West Virginia. (Data from the West Virginia Geological and Economic Survey.)

off those points between any two adjacent data stations that correspond to the needed isopleth intervals.

5. Connect those points of equal value, darkening every fifth isopleth.

The location of the data stations (bore holes) depends on several factors, including (1) the continuous nature expected subsurface of the area (inferred from literature searches, surface surveys, and other subsurface exploration), and (2) the purpose of the bore hole (the initial purpose of the bore hole will probably be data acquisition; however, a subsequent use may be to permit predrainage of methane, preleveling of water tables, or eventual electrical drops). For purposes of data collection and isopleth mapping, the most conventional pattern of data station location is the grid layout. Regardless of the type of station pattern, the methods of linear interpolation remain the same.

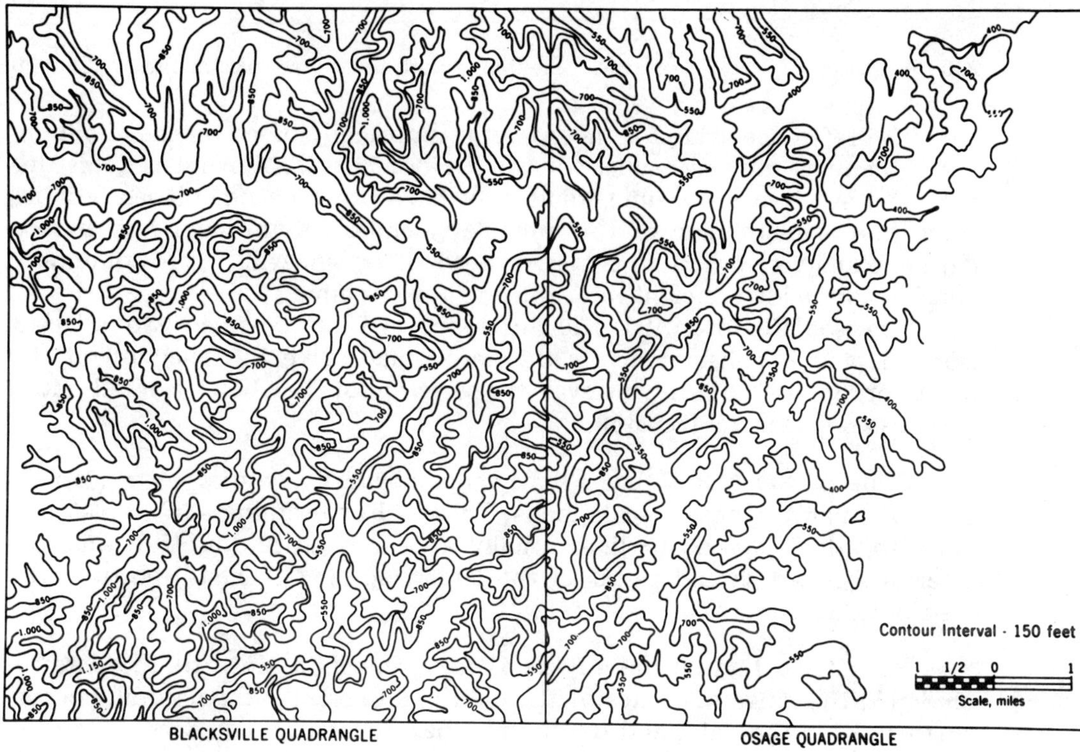

Figure 6-11 Isopleth (isopac) map showing overburden thickness within a region. (Courtesy U.S. Bureau of Mines.)

Figure 6-12 Combined isopleth and choropleth map depicting regional variations in the as-received volatile matter percentage of West Virginia coals. (Courtesy of the West Virginia Geological and Economic Survey.)

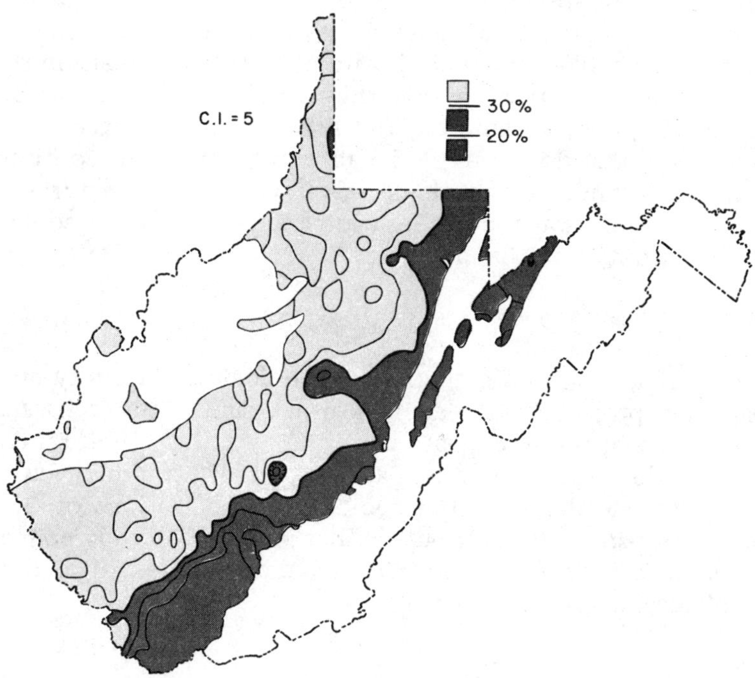

The reference value for the isopleths is normally based on the minimum (or maximum) limiting values for the item of interest. For most items this will be zero, however, some items may have other datum planes. For instance, an isopleth map of the minable thickness of the coal may have a datum plane referenced to 22 in. instead of 0, since 22 in. is commonly considered the lower height limit for profitable mining.

The isopleth interval is the difference in value between two adjacent isopleths and is constant across the entire map. Like the selection of the most appropriate contour interval, the selection of the "best" isopleth interval is based on several factors:

Scale of the Map. Large-scale maps obviously require larger isopleth intervals than do smaller-scale maps. The need or usefulness of isopleths as close together as contour lines usually appear is quite doubtful; thus the interval selected should result in the largest spacing that still portrays the distribution trend.

Percentage Limitations of the Data. The limitations of the data include (1) the range of values of the data in an area, (2) the naturally occurring breaks in the data, and (3) the number, reliability, and accuracy of the data. The range of the data is perhaps the most important consideration. If, for example, the data of sulfur content of coal in a region ranged from a low of 1.7% to a high of 8.2%, the contour interval should be such that the distribution changes can be represented. In general, the number of possible isopleths for a given map should number between 5 and 10. Thus, for this example, the isopleth interval should range between 1.3 and 0.65% (resulting from dividing the total range 6.5 by 5 and 10, respectively), with 1% providing the most likely interval. Naturally occurring breaks in the data also affect the contour interval, in that they present values of discontinuity. If these discontinuities are ignored, the mapper may erroneously lay off isopleths for values that do not exist. To account for these discontinuities, the mapper must determine the range of values within each naturally occurring group and then position the isopleth interval such that no isopleth falls within or on a discontinuous band.

6.5.2 Geologic Mapping

Most of the data affecting the technological feasibility and initial mine design and projections are geological in origin. Data such as the strike and dip of the seam, the immediate top lithology, the structural competency of the overburden, the orientation and extent of structural faults and joints, and so forth, are derived from geological investigations of the site. Since such data are regionally distributed, that is, geographically constant and oriented, it is not surprising to find that such data are most often presented as maps—geological maps.

A geological map is any map that describes the nature, structure, or orientation of geological features within an area. Major data and features contained on geological maps include:[6,17]

1. Stratigraphic and lithological progression data
2. Structural data (strike, dip, folds, faults, joints, stress anomalies, discontinuities, and bedding interrelationships)
3. Bedding planes and other boundary relationships
4. Geological history, or origins of region
5. Statistical data
6. Geohydrology

More often than not, a single map will depict only one geological feature, since too much information drawn on a single map usually limits the useful interpretation of the map. Thus various types of geological maps which best depict certain types of data have become customary. For mining, the major types of geological maps are described below.

Surface Outcrop Maps. Surface outcrop maps illustrate the surface topography in relation to the outcropped geological structure. With such information, data about surface ground stability can be inferred and details about subsurface exploration planned. Figure 6-13 shows a typical surface outcrop map of a region in northern West Virginia.[16]

Geologic (Lithological) Sections. Maps that show the nature of the stratigraphic or lithological order underlying a region are called geological models, sections, or fence diagrams. The advantages of such models are that (1) the various contacts or interfaces between differing rock types can be depicted, (2) changes in structure that might adversely affect the stability of a mine can be explored, and (3) faulting or other anomalies in the structure can often be inferred from the nature of the model. Figure 6-14 shows a geological section of "immediate top" in northern West Virginia and southeastern Pennsylvania.

Geolineament or Element Maps. Lineaments and elements are structural discontinuities resulting from the separation, fracture, or relative movement of rock masses. The importance of defining the orientation and nature of structural discontinuities is summarized by Roberts[8]: "Many engineering problems of strength and stability are dominated by the discontinuities in rock mass rather than by the material structure of the rock itself; it is important that joint, cleavage and fault fracture patterns be observed in detail and recorded." The photolineament maps shown in Figs. 6-6 and 6-8 are excellent examples of such maps.

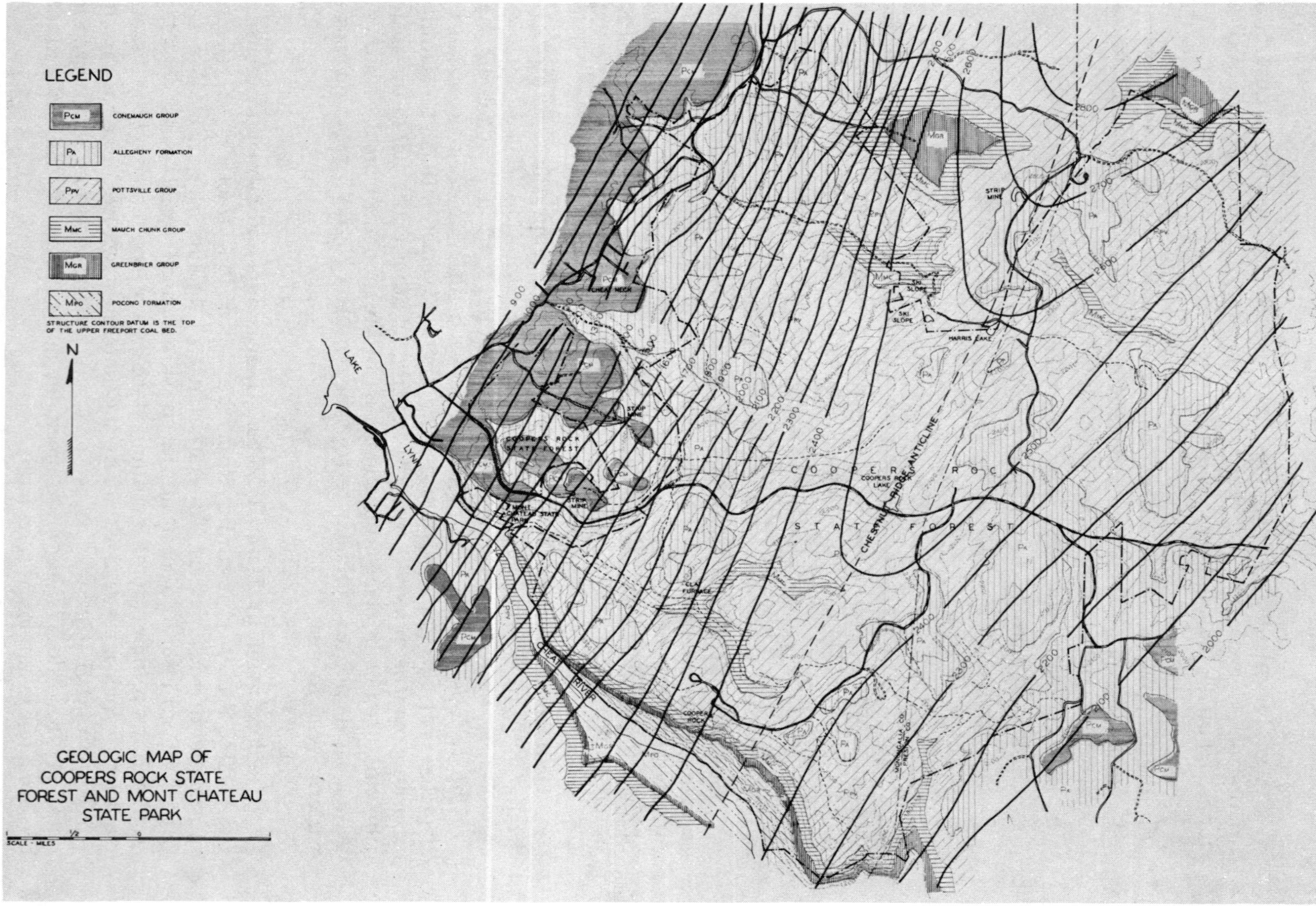

Figure 6-13 Geological map. (Courtesy of the West Virginia Geological and Economic Survey.)

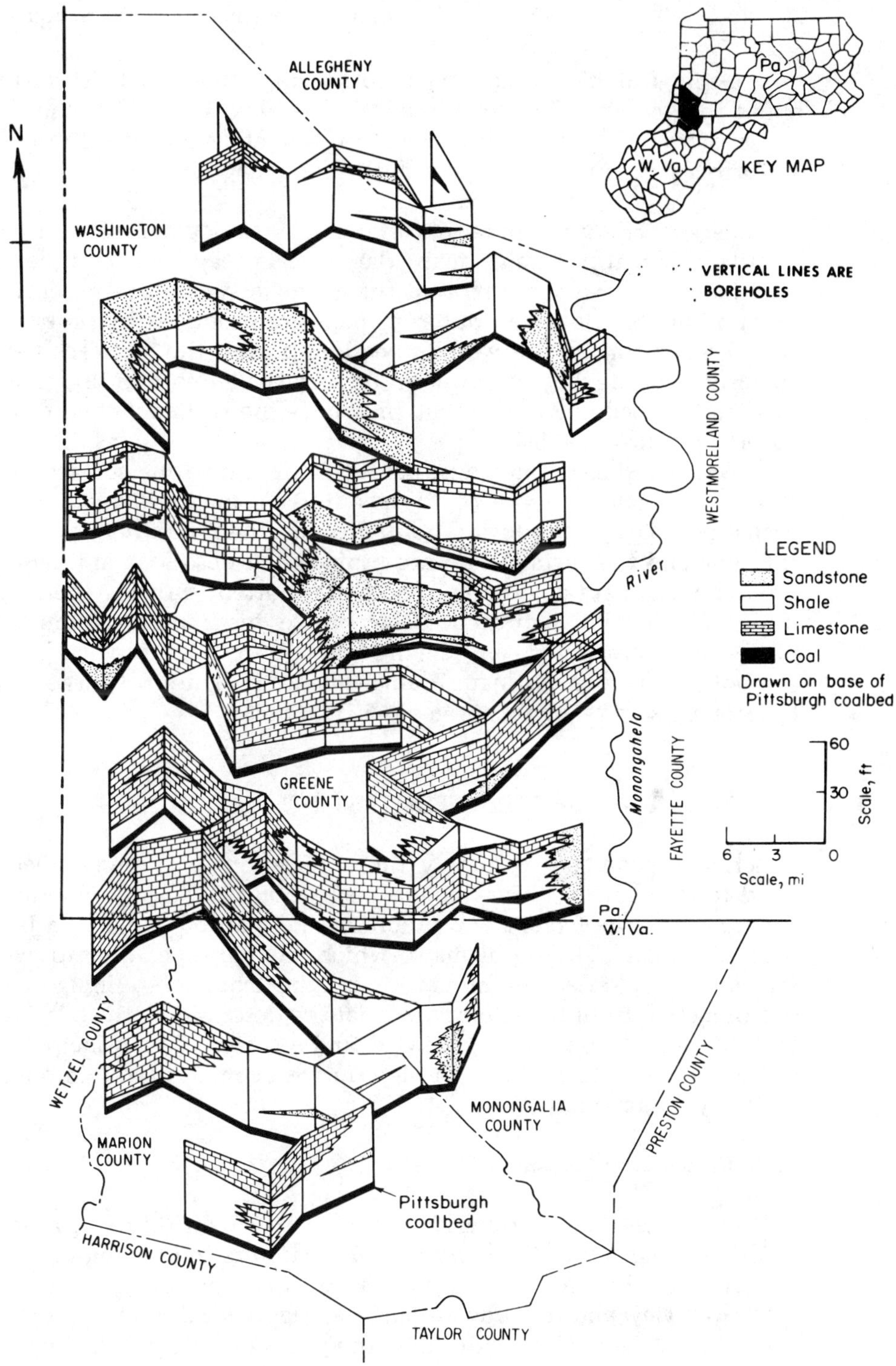

Figure 6-14 Combination of geological sections to form a fence diagram. (Courtesy of U.S. Bureau of Mines.)

Geological Thematic Maps. Any map that displays statistical data about the geology of an area is called a geological thematic maps. Maps such as isopach maps of the immediate top, isopleth maps of sulfur content, and so forth are examples of geological thematic mapping.

General construction of geological maps. A geologic map is usually a $7\frac{1}{2}$-minute USGS topographic map, (larger-scale maps can be used) called the base map, upon which is plotted the geologic symbols representing the features of interest. Surface outcrop maps are produced by plotting the horizontal migration of the various bedding planes (the interface between two adjacent beds) as they outcrop on the surface. Outcrop in this case is defined as a bed or bedding plane that protrudes the surface (whether or not it is covered by surface soils).

Geological sections are produced from correlating surface outcrops together with one or more of the geophysical methods detailing subsurface structure, usually core logs. The desired profile is first delineated with a section line and the surface data are projected as described in Chapter 5. Next, the structural data at each core hole are plotted, showing the thickness and dip of the individual stratum. Using simple linear interpolation the bedding planes are constructed, and finally the appropriate lithogical crosshatching symbol is drawn to distinguish adjacent beds. Figure 6-15(a) and (b) shows the step-by-step sequences of geologic mapping.

6.6 APPLICATIONS OF DESCRIPTIVE GEOMETRY[18,19]

Descriptive geometry is merely the application of graphical techniques applied toward solving three-dimensional problems. These techniques are used primarily to supplement the premining design and projection layout activities of mining. The problems to which techniques of descriptive geometry are applied are varied and numerous, ranging from determining the strike and dip of a seam from the exploratory data to determining the bearings of headings to provide optimum drainage and haulage. The principal advantages offered by the techniques of descriptive geometry are simplicity, speed, accuracy, and *visualization.*

6.6.1 Orthographic Projection

All techniques to be presented here are based on third-angle orthographic projection concepts. *Orthographic projection* is the technique of representing an object by projecting a single view of an object along lines perpendicular to the view and the drawing surface. Figure 6-16 shows a front or vertical projection of an object. Note that the projection lines, called *projectors*, are perpendicular to the surface on which the view appears as well as perpendicular to the view. A *third-angle projection* refers to the third quadrant positioning of the object and thus the relative positions of the *principal views* on the drawing. Principal views are the front, top, bottom, right side, and back views. Figure 6-16 shows the relative positioning of six views of the object as projected in a third-angle orthographic projection.

Figure 6-15 (a) Procedure for geological map construction. (b) Steps in geological mapping.

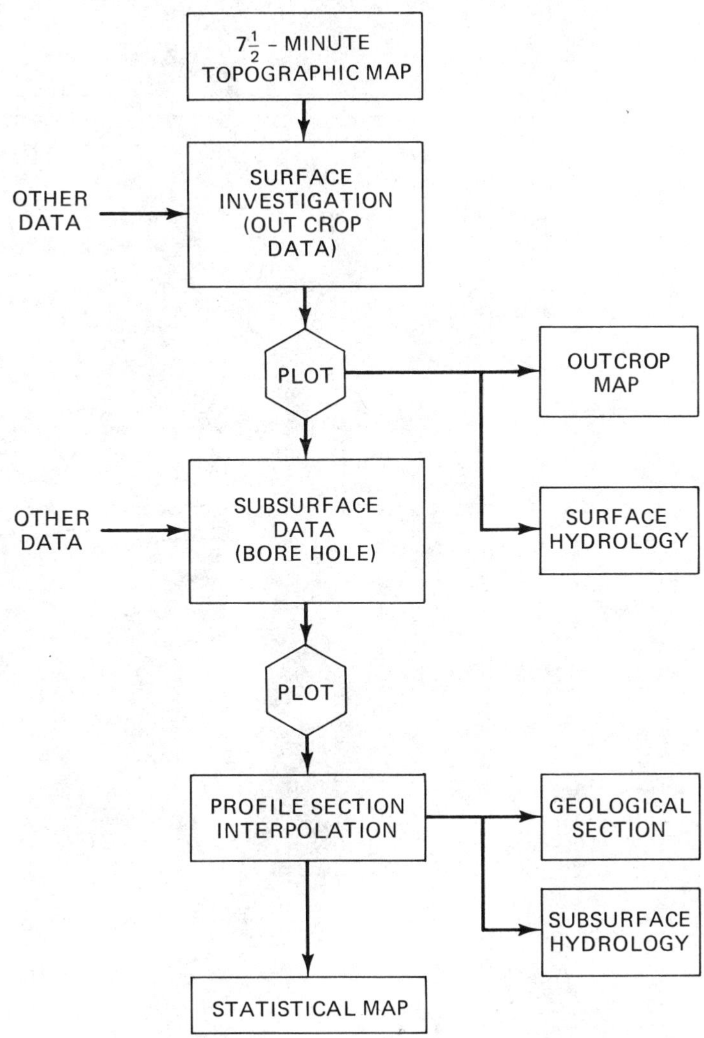

(a)

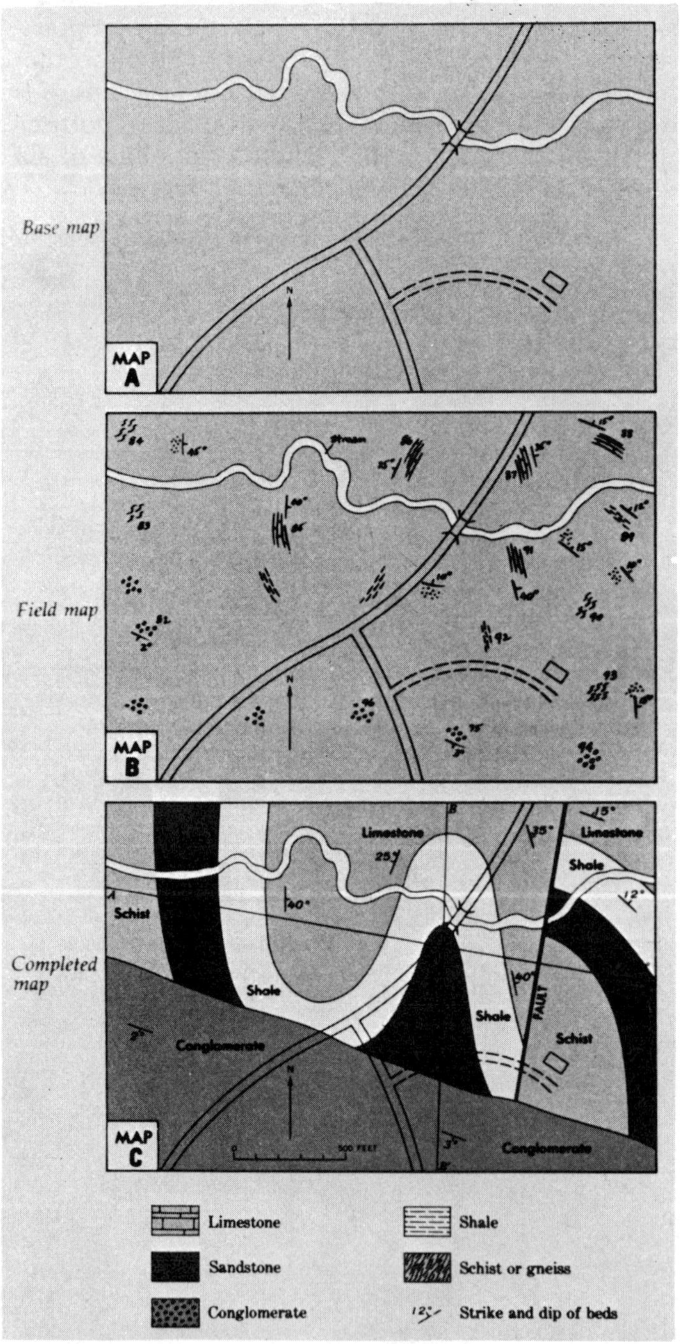

Figure 6-15 (continued) (b)

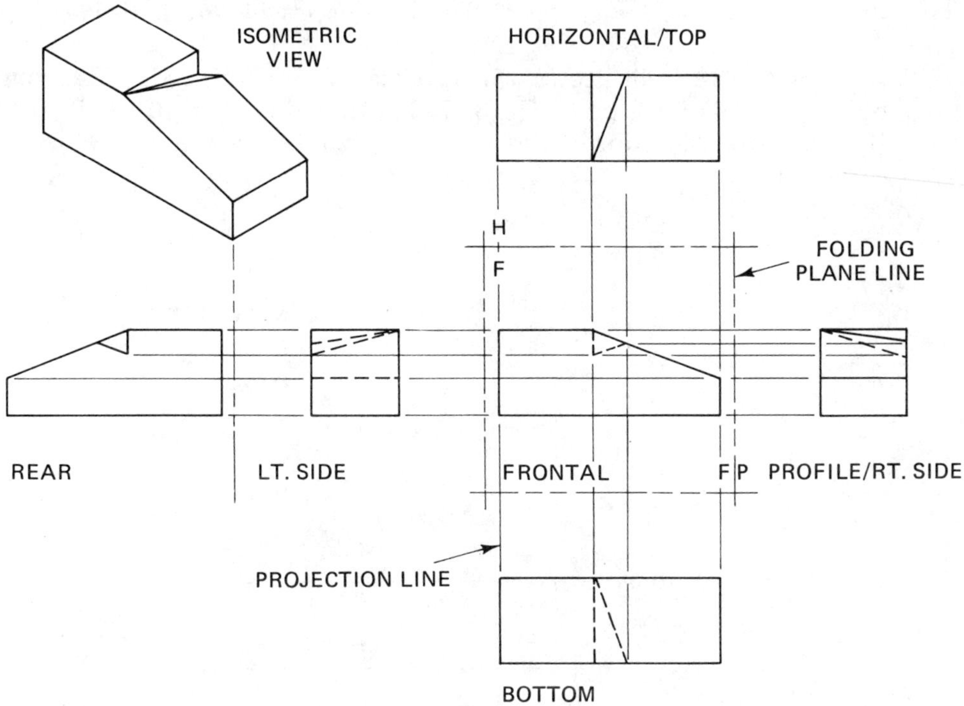

Figure 6-16 Third-angle orthographic projection.

6.6.2 Folding Lines

The view boundaries, called *folding lines*, are useful in descriptive geometry and normally remain throughout the development of the drawing. The line H/F in Fig. 6-16 denotes the folding line boundary between the horizontal and frontal views; F/P denotes the folding line boundary between the frontal and profile views. All folding lines represent the intersection of mutually perpendicular planes, which appear as adjacent views.

6.6.3 Projection of Points in Space

Descriptive geometry solves three-dimensional problems by successively solving two dimensional diagrams. Each view is a two-dimensional projection, with the folding plane lines acting as a rectangular coordinate system. The location of points in the horizontal or top view is achieved by locating the coordinates of the individual point just as we did in Chapter 4. To locate these same points in the profile on front view, simply project the points from the plan view to the profile view. The projection lines are perpendicular to the folding line. The point location in the front view requires (1) a reference elevation and (2) the elevations of all other points relative to that elevation.

Figure 6-17 illustrates the process of projecting the following data: point $H_1(2, 6)$ P(0), point $H_2(3, 7)$ P(-2), and point $H_3(6, 1)$ P (-6), where H represents the coordinates in the horizontal view and P represents the elevation.

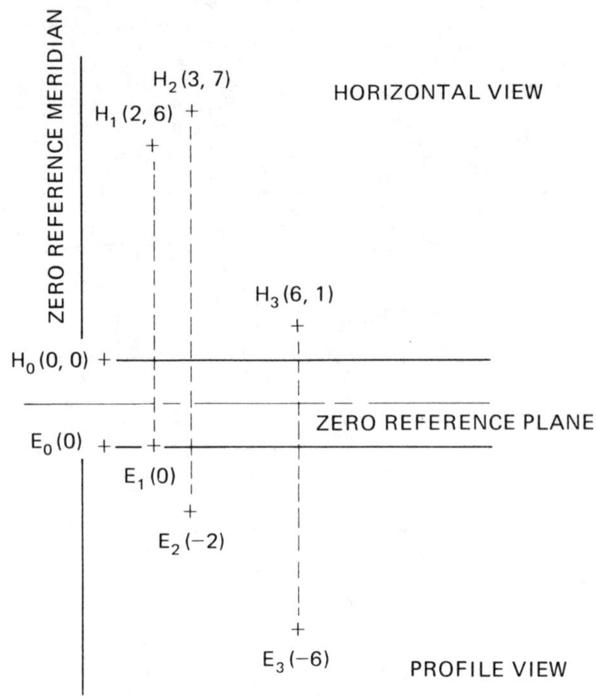

Figure 6-17 Projecting points in space.

6.6.4 Auxiliary View

An *auxiliary view* is a view of an object taken from any direction other than one of the principal views or planes. A primary auxiliary view is constructed perpendicular to one of the principal views and is separated from the principal view by a folding line. Auxiliary views are used to show a line, side, or area of an object as a "true view" or when it becomes necessary to complete the description of the object. A true view shows the actual, undistorted lengths, angles, and areas. Figure 6-18 shows an auxiliary view of a line never shown as a true view in any of the principal views.

6.6.5 True Length of Lines

The importance of determining the true length of a line cannot be overemphasized in descriptive geometry, for it is the basis of many subsequent

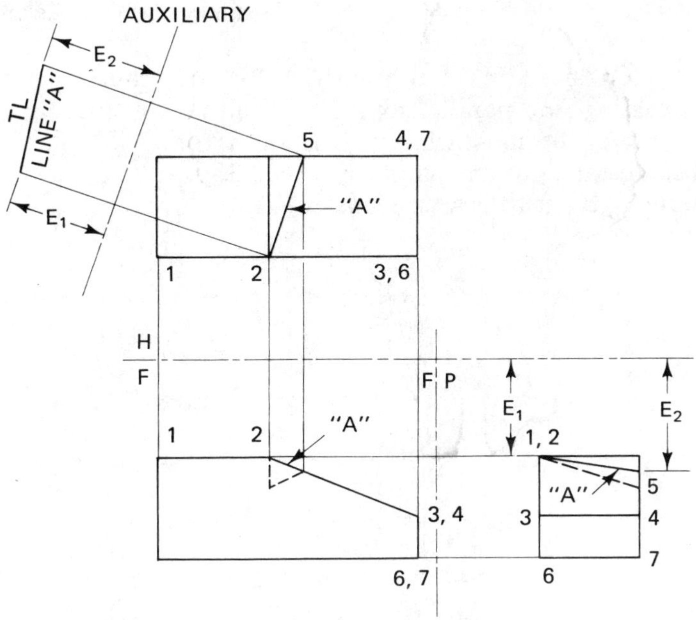

Figure 6-18 Determining the true length of a line using an auxiliary view projection.

techniques. For contrast you should recall that the true length of a line is analytically defined as

$$\text{TL} = \sqrt{(H)^2 + (L)^2 + (D)^2} \tag{6-1}$$

where TL = true length
H = height
L = length
D = depth

Several graphical methods can be used to determine the true length of a line. The first method uses the construction of a primary auxiliary view and the second requires the construction of a supplemental diagram.

The method of determining the true length that uses the construction of an auxiliary diagram is based on the fact that if a line is shown *parallel to a folding line in one view*, the line is shown as *true length in the adjacent view*. This is true whether or not primary auxiliary views are needed. For example, consider again the object and projected views shown in Fig. 6-18. Lines 1-2 and 2-3 in the horizontal view are shown parallel to folding line H/F; thus they should be and are shown as true length in the frontal projection. Similarly, line 3-6 in the frontal projection, shown parallel to the folding line F/P, is shown as a true length in the adjacent profile view. Note, that line 2-5 is shown as an oblique line to all principal folding planes and is never shown as a true length. An auxiliary view is used to show 2-5 in true view.

Consider line A-B Fig. 6-19. To construct an auxiliary view, first draw a folding line parallel to A-B in one of the principal views. Next, project point A_A by measuring the distance from H/F to A_F (E_1) and transferring this distance to the auxiliary view. Repeat to project B_A (distance E_2). The line A_A-B_A is thus a true length.

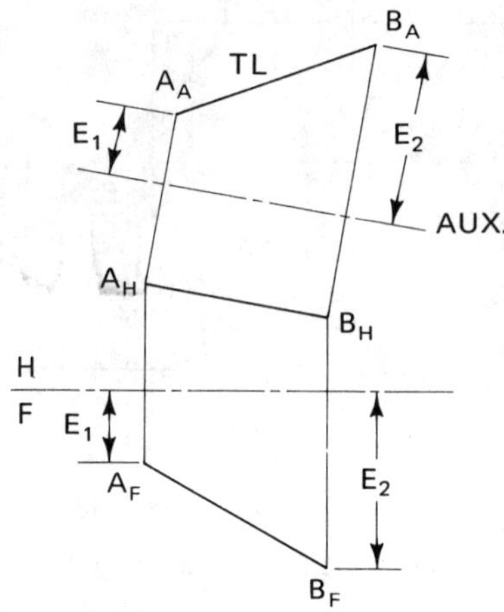

Figure 6-19 Procedure of constructing an auxiliary view projection.

The second method used to show the true length of a line is the construction of a supplemental true-length diagram. The construction uses the available data regarding the relative position of points in space to construct a diagram. For example, an imaginary surface containing the unknown length of 4–5, shown in Fig. 6-20, can be constructed by plotting the vertical and horizontal distances available from the frontal and horizontal views (see Section 6.6.3).

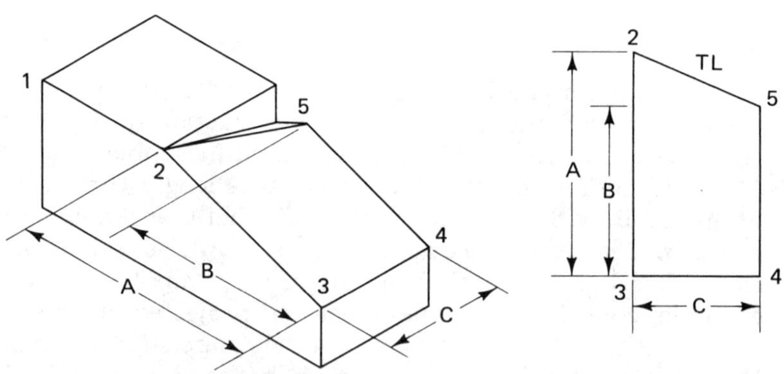

Figure 6-20 Determination of true length by construction of a supplemental diagram.

Either method may be used for determining the true length of a line; however, for most mining applications the use of auxiliary views provides additional information which is often pertinent to the solution.

6.6.6 Bearings and Grades

Entry projections, as noted in Chapter 4, are shown as simple lines on the mine map. The bearing of the projection is the direction of a vector relative to true meridian. Direction is defined by specifying its orientation and its sense. The *orientation* of a line is denoted by specifying its angular departure from a true north-south line (i.e., azimuth or bearing line), and the *sense* is noted by measurement from either the north or the south. Line *AB* in Fig. 6-21 has a 37°15′ angular departure measured from either the north moving eastward or from the south moving westward. The sense in this case is from point *A* moving toward point *B*. Thus the bearing (orientation and sense) of line *AB* can be specified as N37°15′E. Line *BA*, which would represent the same orientation but an opposite sense, would be specified as S37°15′W. The bearing of a line is shown only in the horizontal projection plane and is furthermore independent of the grade or slope of the line in space.

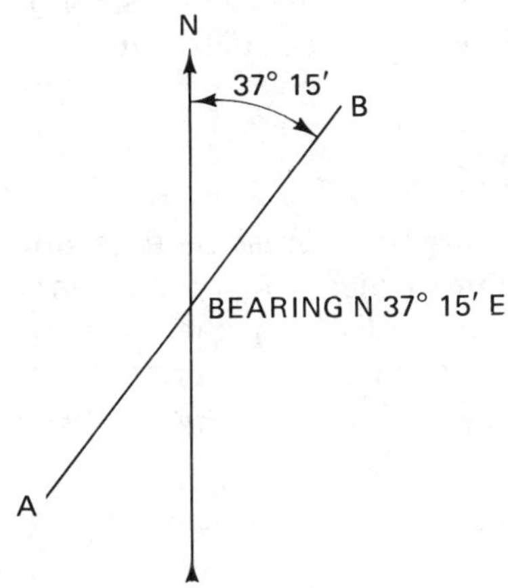

Figure 6–21 Bearing of a line.

The *grade* of a line defines the slope or inclination of a line in space. A horizontal grade has a zero slope; all other grades are measured according to the following equation:

$$G = \frac{\text{rise}}{\text{run}} \times 100 \qquad (6\text{-}2)$$

where G = percent grade, the vertical rise in feet per 100 ft of horizontal run
rise = vertical distance measured as the difference between any two points on the line
run = horizontal distance measured between the two same points on the line
100 = constant for converting decimal to percent

The *true grade* of a line is shown in either the frontal or profile view if and only if the line shown in the horizontal view appears parallel to the H/F folding plane line. The true grade of any line which appears in the horizontal

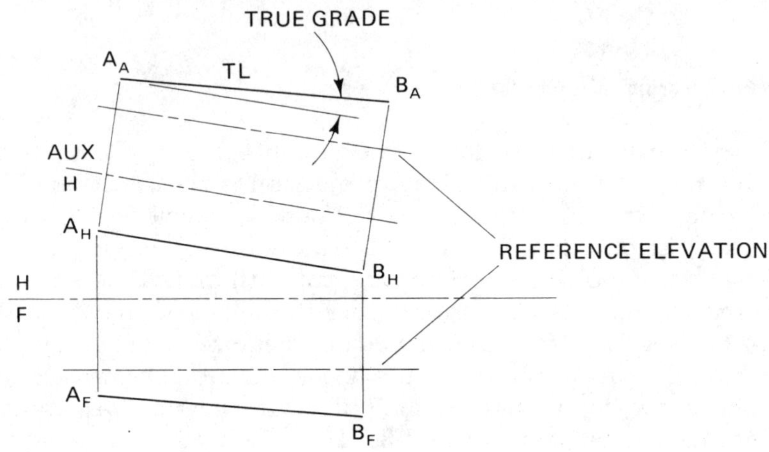

Figure 6-22 Determining true grade.

view as oblique can be shown in an auxiliary view by drawing a folding plane parallel to the line. The auxiliary view will also show the true length of the line. Figure 6-22 shows an example of such a line.

6.6.7 Projection of Points on a Line

Often points along a line segment must be determined in several views. By the nature of orthographic projection, points along any line as shown in adjacent views must coincide with the projector perpendicular to the folding line. Thus point C in Fig. 6-23 in the horizontal view must coincide with point C' in the frontal view as long as line a-a is perpendicular to the folding line H/F.

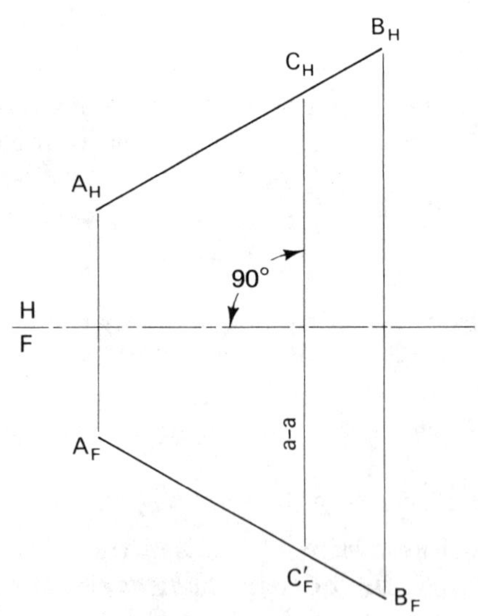

Figure 6-23 Projecting points on a line between views.

6.6.8 Strike and Dip

Most bedded deposits, coal seams being a typical example, present significantly less vertical relief and irregularity than does the surface topography that covers them. For exploratory purposes, the initial evaluations of the seam can be made treating the seam as a plane surface and using strike and dip designations, as opposed to a complete contour mapping of the coal seam.

The stratum is assumed to be a plane oriented in space. A plane is defined in one of several ways: (1) by locating the positions of three noncollinear points, each lying on the surface of the plane; (2) by defining a line and a point each existing on the surface of the plane; or (3) by defining two lines, which may or may not intersect. Since most exploratory data are point data (e.g., bore holes), the plane representing the seam is normally located by the first method—establishing three noncollinear points on the surface of the seam. Figure 6-24 shows the horizontal and frontal plotting of the raw data points *A*, *B*, and *C*.

The strike of the seam is the bearing of a line that is horizontal on the surface of the seam. The dip is the maximum angle the plane makes with a horizontal plane and is always perpendicular to the strike (when shown in the horizontal or plan view). Using these definitions, the strike and dip of plane *A*–*B*–*C* can be determined as follows:

1. (a) Determine the strike of the seam by constructing a horizontal line in the frontal view C_F-O. (Refer to Fig. 6-25.)

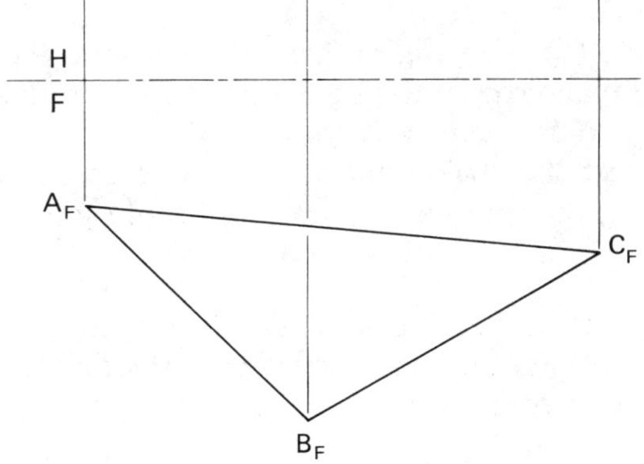

Figure 6-24 Frontal and horizontal views of plane *ABC*.

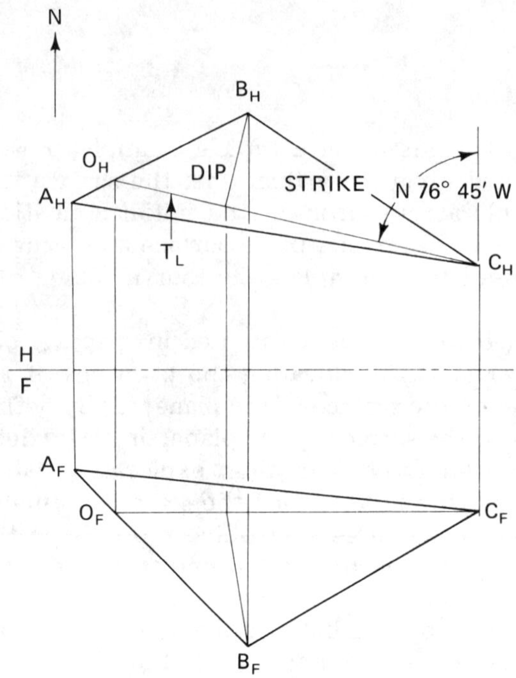

Figure 6-25 Determining the bearing of strike and dip of seam (plane) *ABC*.

(b) Project *CO* onto the horizontal view; since C_F-O_F is parallel to the folding line H/F, C_H-O_H is shown in true length and is the projection of the strike.

(c) Measure the angular departure of C_H-O_H from the meridian N76°45′W. (Normally, the strike bearing is measured from the N, due to the lack of true "sense" of the strike line.)

2. Determine the dip of the seam shown in Fig. 6-26 using the following procedure:
   (a) In the horizontal view, construct a perpendicular from any known point B_H to the strike line, B_H-P_H.
   (b) Draw a folding line parallel to B_H-P_H and project the points *A*, *B*, *C*, and *O* onto an auxiliary view. Note that all points along the strike line, namely O_H, P_H, and C_H, all appear as a single point in the auxiliary view.
   (c) Measure the angle (the dip angle) formed between the plane *A*-*B*-*C* and a horizontal plane that must be parallel to the folding line.

The information about the strike and dip can now be transferred onto the working maps by using the symbol $\underline{\quad | 54°30′}$ over N76°45′W, where the longer section of the tee represents the strike bearing and is oriented N76°45′W on the map, while the section perpendicular to it represents the downhill direction of the dip measured at a slope of 54°30′.

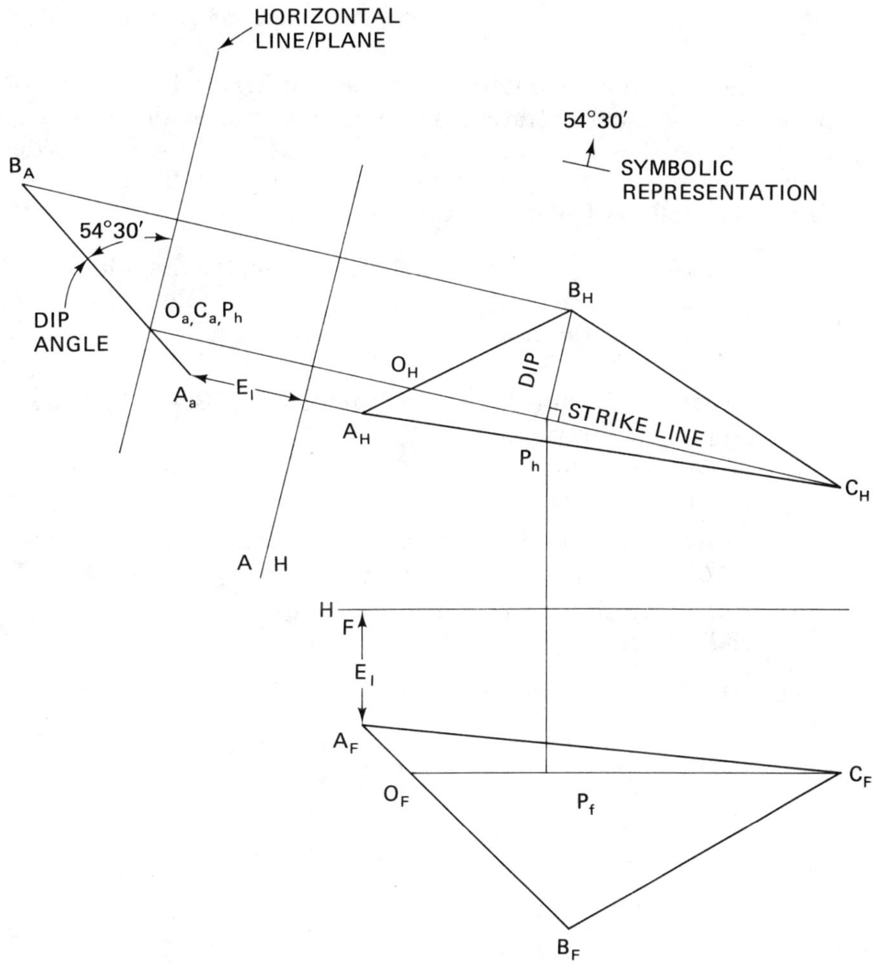

Figure 6-26 Determining the dip direction and angle of seam (plane) *ABC*.

6.6.9 Grade and Apparent Dip

In the example above, the greatest angle formed between the surface of the stratum and a horizontal plane corresponded with a line (shown on the horizontal or plane view) that was perpendicular to the strike. Any entry centerline corresponding with a line drawn between zero and 90° to the strike would obviously form a lesser dip angle with the horizontal plane. This lesser angle is called the *apparent dip* of an entry. In coal mining, the less the haulage grade, the more efficient haulage operations will be. Although a zero grade, that is, an entry centerline at the same heading as the strike, would be ideal for haulage, problems of water accumulation and drainage would occur in most coal mining operations. In general, the optimum range of grade for haulage that will maintain adequate drainage is between 0.75 and 2% grade.

Thus the problem of determining the bearing of haulage entries that will provide economical operation is one often solved by descriptive geometry.

Assume that points *C*, *D*, and *E* define a coal seam. To determine the bearing of the main entries which will provide a 2% grade from point *C*, proceed as follows (refer to Fig. 6-27):

1. Construct a line in the frontal view parallel to the folding line from point C_F to M_F. Since this line is parallel to the folding line H/F, projection of C_F-M_F in the horizontal view will appear in true length and be the strike line of *CED*.
2. Project C_F-M_F onto the horizontal view. Scale 100 ft along C_H-M_H and project this point onto C_F-M_F.
3. Perpendicular to C_F-M_F, measure 2 ft, scaled, below the point and mark the point *O*.
4. Construct a line from point C_F through point *O* intersecting line D_F-E_F. This line represents an entry from point *C* with a 2% grade.
5. Project this line on the horizontal view and measure the bearing of this heading—89°.

Figure 6-27 Determining the bearing of an entry at a 2% grade along seam *ABC*.

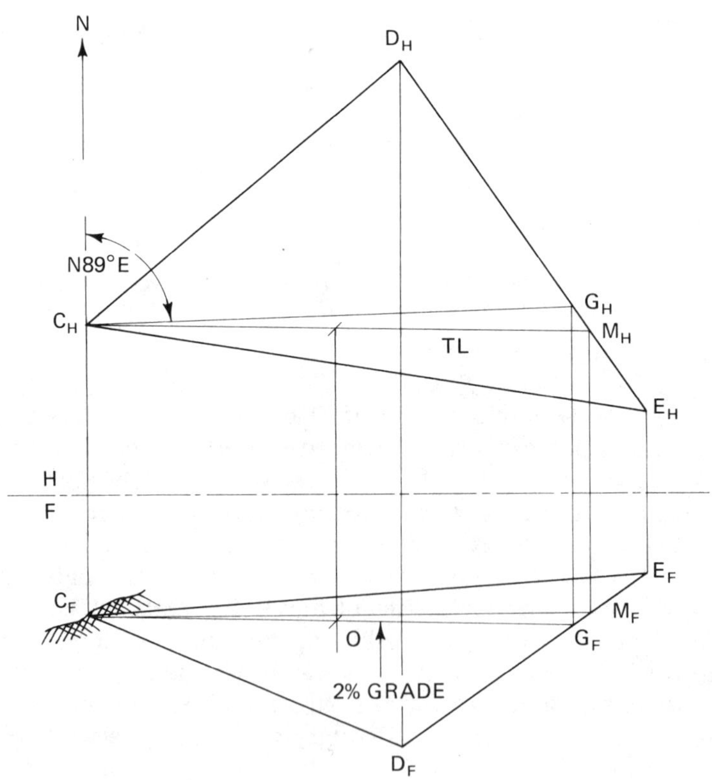

6.6.10 Determining Entry Bearing

Slope mines, which do not open on an outcrop, often establish the bearing of the mains based on the optimum grade and the location of the point of entry which is either most convenient for its operations or most stable in terms of rock mechanics. Therefore, the problem faced by the designer is to determine the shortest possible entry from the entry point (the portal or pit mouth) to the surface of the seam. This problem is solved as follows (refer to Fig. 6-28):

1. Construct an auxiliary view showing the plane *A–B–C* as a side view (thus a line as described in Section 6.6.7), showing *E*, the entry point, on a horizontal line (representative of the surface).
2. Construct an angle equivalent to the required grade from *E* to the plane.
3. Project the point of intersection onto the horizontal view (the point of intersection will be on a line from *E* perpendicular to the strike) and then onto the frontal view.
4. Measure the bearing angle.

Figure 6-28 Determining the bearing of the shortest entry from point *E* to seam *ABC* at a 12° slope.

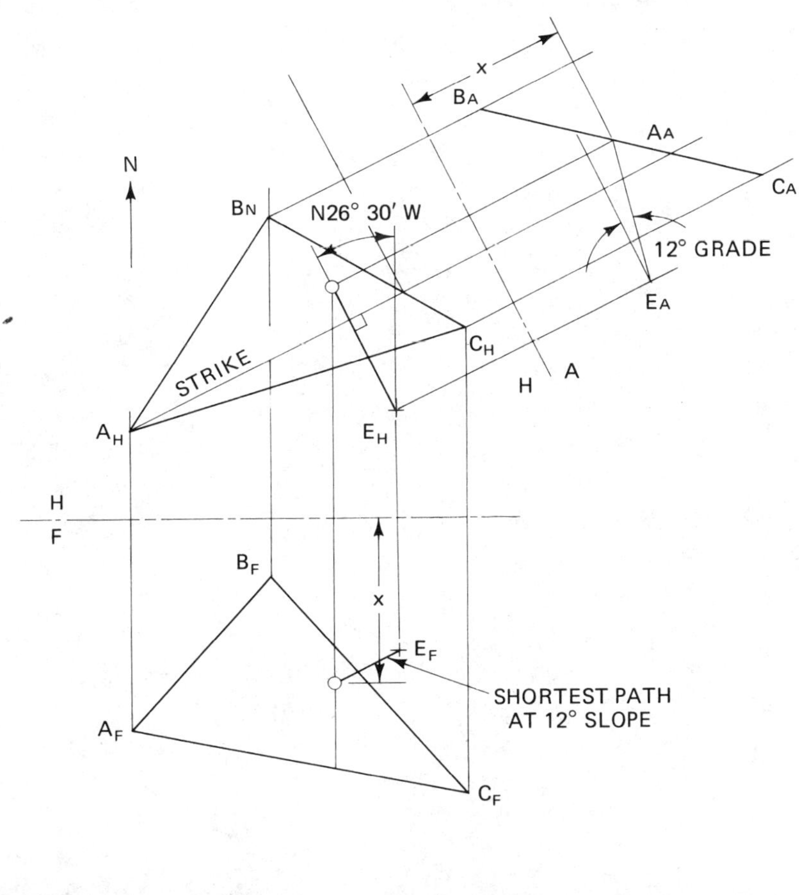

Note that any shortest grade from E to the plane will form a line perpendicular to the strike line. Thus the shortest entry from any specific portal location to the coal seam surface will have the same bearing regardless of the grade, the only difference being the point and angle of intersection.

6.6.11 Shortest Distance between Two Skewed Entries

The last basic technique to be presented which has application within the mining industry is the determination of the shortest path between two skewed entries. The shortest path between a point and a line is, of course, a perpendicular line from the point to the line. The shortest path between two skewed lines is a line that is mutually perpendicular to both lines. The graphic solution to this problem by the line method is shown in Fig. 6-29 and explained here.

1. Construct a folding line H/F parallel to AB and project AH and CD; AB will be in true length.
2. Construct a folding line $\frac{1}{2}$ perpendicular to AB shown in true length and project AB as a point. A line drawn from point AB perpendicular

Figure 6-29 Finding the bearing and grade of the shortest drift between two skewed entries.

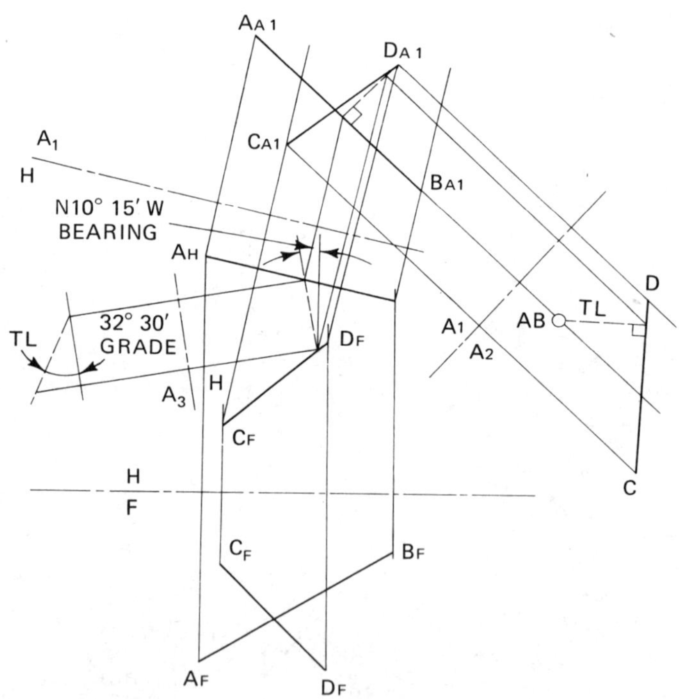

to line *CD* will show the shortest connection path *xy* from *AB* to *CD* in true length.

3. Project point *y* on *CD* back to view 1, and construct a perpendicular from point *y* to *AB*. This will locate point *x*.
4. Project *xy* back to the horizontal and frontal views. If the grade of *xy* is desired, you must show *xy* in true view in a third auxiliary view—3.

Several other methods can be used; however, for most purposes the line method should be adequate.

6.6.12 Summary of Descriptive Geometry

The techniques, descriptions, and applications discussed should supply you with the necessary information for solving most commonly encountered geometric problems. Chapter 7 discusses functional layouts that depend significantly on a mastery of descriptive geometry.

REFERENCES

1. John Hospers, *An Introduction to Philosophical Analysis*, 2nd ed., Prentice-Hall, Inc., Englewood Cliffs, N.J., 1967.
2. Kirk W. McCabe and John J. Jansky, "Satellite Imagery—A New Tool to Predict Hazardous Mine Roof," presented at 66th National Safety Congress and Symposium, October 2-5, 1978.
3. John J. Jansky and Kirk W. McCabe, "Evaluation of Remote Sensing Techniques for Planning Mining Operations," Pittsburgh Technical Support Center, MSHA, U.S. Dept. of Interior, 1978.
4. Paul R. Wolf, *Elements of Photogrammetry*, McGraw-Hill Book Company, New York, 1974.
5. Marland P. Billings, *Structural Geology*, 3rd ed., Prentice-Hall, Inc., Englewood Cliffs, N.J., 1972.
6. A.C. McLean and C. D. Gribble, *Geology for Civil Engineers*, George Allen & Unwin, London, 1979.
7. H. Dotson Cather, unpublished lecture notes, 1980.
8. A. Roberts, *Geotechnology*, Pergamon Press, Oxford, 1977.
9. Anthony L. Payne, "Exploration Drilling," in Arthur B. Cummins and Ivan A. Given, eds., *SME Mining Engineering Handbook*, Vol. 1, Society of Mining Engineers, AIME, New York, 1973.
10. Robert S. Lewis and George B. Clark, *Elements of Mining*, 3rd ed., John Wiley & Sons, Inc., New York, 1964.
11. Syd S. Peng, *Coal Mine Ground Control*, John Wiley & Sons, Inc., New York, 1978.
12. Khateeb M. Hussain, *Development of Information Systems for Education*, Prentice-Hall, Inc., Englewood Cliffs, N.J., 1973.

13. Arthur H. Robinson and Randall D. Sale, *Elements of Cartography*, 3rd ed., John Wiley & Sons, Inc., 1969.
14. G. C. Dickinson, *Statistical Mapping and the Presentation of Statistics*, 2nd ed., Edward Arnold, London, 1973.
15. W. K. Overbey, Jr., C. A. Komar, and J. Pasini III, *Predicting Probable Roof Face Areas in Advance of Mining by Geological Analysis*, U.S. Bureau of Mines, Health and Safety Research Program, TPR 70, May 1973.
16. S. P. Babu, J. A. Barlow, L. L. Craddock, R. V. Hidalgo, and E. Friel, *Suitability of West Virginia Coals to Coal-Conversion Processes*, West Virginia Geological and Economic Survey, Morgantown, W.Va., 1973.
17. John I. Platt and John Challinor, *Simple Geological Structures*, Thomas Murby and Co., London, 1974.
18. Clarence E. Douglass, Albert L. Hoag, and Daniel E. Alexander, *Descriptive Geometry*, 2nd ed., Rinehart Press, San Francisco, 1971.
19. E. G. Pare, R. O. Loving, and I. L. Hill, *Descriptive Geometry*, 4th ed., Macmillan Publishing Co., Inc., New York, 1971.

PROBLEMS

1. You are assigned the task of designing and scheduling the exploratory investigation of a nearby region. Prepare an overview of this project, prepare a listing of the specific places or people that should be consulted and prepare an overview of the data that must be obtained so that a mine can be planned.
2. Implement those portions of your plan that cost less than \$0.50/datum. What percent of your data base can be filled?
3. Evaluate your data (see Section 6.3). What percent of the data can be used without question?
4. Obtain population data about your county together with any scale planimetric map that contains the political boundaries of your region. Prepare a dot map depicting the population density.
5. A new mine in your county will require 650 workers. As a minimum they should have a 10th-grade education, be physically fit, and live within a 30-minute drive of the geometric center of the county. Can these requirements be met? (Prove your answer with data, not opinion.)
6. Obtain a $7\frac{1}{2}$-minute topographical map of your region. A coal seam located 550 ft below the highest elevation on the map is to be mined. Using an overlay, construct an isopleth (isopac) map of the overburden thickness. For thickness less than 50 ft, show the individual isopleths in red. How would the map change if the dip of the seam were 4% due south?
7. Again using the $7\frac{1}{2}$-minute topographical map as in Problem 6, a core drilling at the highest point on your map reveals a 40-ft thickness of sandstone 100 ft below the surface. Plot both the upper and lower surface outcrops within a 2-mi radius on an overlay sheet. Assume the bed to be perfectly horizontal.
8. Again using the $7\frac{1}{2}$-minute topographical map, draw a profile cross section between two core holes. The first, located at the highest elevation, has the following data:

sandstone, 25 ft thick; sandy-shale, 35 ft thick; shale, 25 ft thick; sandstone, 15 ft thick; and limestone, 40 ft thick. The second bore hole, located 1200 ft 515°30′W, has the following: sandstone, 15 ft thick; sandy-shale, 42 ft thick; shale, 35 ft thick; sandstone, 25 ft thick; and limestone, 55 ft thick.

9. In addition to the two bore holes in Problem 8, a third hole, located 1500 ft N38°14′N from the highest elevation, has the following: sandstone, 20 ft thick; sandy-shale, 25 ft thick; shale, 30 ft thick; sandstone, 20 ft thick; and limestone, 40 ft thick. Use descriptive geometry and determine the strike and dip of the limestone and the thickness of the limestone bed at each point.
10. Below the limestone is a seam of coal 48 in. thick. Find the bearing of an entry that has a 2% grade.
11. What is the bearing of the shortest sloped entry (10% grade) from the second bore hole to the first coal seam?

7

Ventilation System Layout

7.0 INTRODUCTION

Perhaps the most important and best known functional system needed for underground mining is the ventilation system. Ventilation is a term that encompasses the principles and practices governing the flow of air and other gases through underground entries. As underground operations have evolved from shallow pit and punch mines to the expansive multiseam mining of today, the need to *control* the *quantity* and *quality* of air has increased accordingly. Our ability to *control* the ventilation circuit depends on our knowledge of (1) the functions that need to be performed, (2) the principles of gas mechanics, and (3) the methods of laying out the ventilation devices to attain the desired effect.

The general requirements of the ventilation system underground differ significantly from those of surface installations. These differences can be summarized as follows:

1. The underground ventilation conduits are also the entries and work areas. Thus the flow path is large, irregular, and intermittently obstructed.

2. The ventilation circuit is continually advancing, retreating, or being modified together with mining activities.

3. The supply conduits are subject to an ever-present threat of ground movement, potentially causing a blockage of the airflow.

4. Local barometric changes cause gas to flow into and out of abandoned areas. This results in unexpected changes in the quality of air, thereby requiring corrective modification of the system.

5. The air quality changes continuously along its route. The coal seam liberates methane (CH_4) gas at various rates into the ventilation air circuit. The coal seam, ground control timbers, and haulage ties all add carbon dioxide (CO_2) gas into the ventilation air circuit. Finally, the surrounding coal and oxygen-deficient water absorb the oxygen (O_2) from the air stream.

6. The ventilation circuit can adversely and directly affect the stability of the ground control system.[1] The amount of water vapor the air will hold (humidity) is proportional to the temperature of the air. In the winter the temperature is cold and correspondingly, the air is dry. Once underground the constant 60° F temperature eventually heats the air, increasing its ability to hold water vapor. Water from the top, ribs, and bottom is thus absorbed. The opposite occurs in the summer. The hotter, more humid air cools as it flows through the mine, causing water vapor to condense on the entry surfaces. The periodic drying out (fall and winter) and wetting (spring and summer) of these surfaces causes certain materials, such as shale, to expand with moisture absorption and contract with moisture escape. The cyclic nature of the expansion and contraction can cause accumulative microscopic failures in the material, which can produce larger cracks, leading to roof falls.

7.1 FUNCTIONAL ANALYSIS

To control the ventilation network you must first understand the functions that must be performed. Regardless of whether or not a ventilation system is above or below ground, the functions performed are:

1. Provide ventilation when and where needed
2. Ensure safety (first of personnel and then of the system and surrounding property)
3. Maximize reliability
4. Improve maintainability
5. Optimize value

7.1.1 Provide Ventilation

The primary function of the ventilation system is to supply air when and where needed. This function can be subdivided into component purposes which also fall into a relative importance ranking. These subfunctions include:

1. Supply air to workers in the quantity and quality needed.
2. Supply the quantity of air needed to mix, dilute, and make harmless any explosive or noxious gases.

3. Supply air with sufficient velocity to carry away dusts and gases.
4. Control the humidity and temperature of the air.

Air for Workers. The most important aspect of the ventilation system is to supply to workers the quality and quantity of air needed to sustain life. These two characteristics, *quality* and *quantity*, are the bases of mine ventilation.

The quality of air refers to the type and percentage of the gaseous compounds that make up air. Dry, uncontaminated air is a mixture of various gases, including:

Nitrogen (N_2)	78.09% (by volume)
Oxygen (O_2)	20.95% (by volume)
Carbon dioxide (CO_2)	0.03% (by volume)
Argon and other rare gases	0.93% (by volume)

The two active components in air affecting workers are oxygen and carbon dioxide. Oxygen is essential to ensure the conversion of energy and the maintenance of tissue performance (life), while the carbon dioxide percentage in the air is linked to the physiological triggering of respiration. The quality and properties of air and mine gases are discussed in greater detail in Section 7.2.

The quantity of air refers to the volumetric amounts needed by human beings to sustain normal bodily functioning. Hartman[2] calculates that 20 cubic feet per minute (ft^3/min) is needed for active men when the oxygen content is 19.5%. However, the 1971 mining law in the state of West Virginia[3] mandates that 100 ft^3/min per worker be supplied in a mine classified as nongaseous, and 150 ft^3/min for each person working in a gassy mine (CH_4 0.25% at the face).

Mix, Dilute, and Make Harmless. The second most important function of providing ventilation is to supply the air in such a way that harmful concentrations of gas will be diluted and thus rendered harmless. Methane gas suspended in the air is highly explosive in concentrations ranging from 5 to 15%. To reduce the percentage to a safe level, we must either increase the quantity of air or we must mix the local regions of high methane concentration with those of lower concentrations (i.e., distribute the methane more evenly within an area). To achieve the first solution, *dilution*, we simply increase the amount of airflow within an entry. However, to achieve *mixing*, we must cause the flow to "go turbulent" and we must be able to direct the air along routes that it would normally not take, for instance, into corners. These solutions for making gas concentrations harmless fall within the domain of fluid mechanics and are discussed in Section 7.3.

Carry Away Contaminants. After the dangerous concentrations of gas or dust have been lowered, the ventilation system must supply the velocity to carry out the harmful agents. If the velocity is too low, some of the heavier, more harmful gases may stratify in layers along the bottom and present a danger. If not carried away, the coal dust would linger in the entry, presenting a visual and respiratory danger for the worker and posing a threat of an explosive ignition.

Control Humidity and Temperature. As noted earlier, the varying air humidity adversely affects the ground control system. Additionally, high humidity increases worker effort and general discomfort. In most mines the temperature underground is constant (approximately 60° F). However, the outside air temperature can present extreme operating problems in the winter, such as the freezing of conveyor belt heads and the icing of escapeways. The practice of humidity and temperature conditioning of the air is commonly ignored in coal mine ventilation.

State and federal laws mandating functional control. The laws that govern the specification of air quality and quantity are explicitly stated in Sections 75.300 through 75.330-1 of the federal specifications[4] and by most state control boards. These specifications are summarized in Table 7-1.

TABLE 7-1 SUMMARY OF MINE LAWS

Quality controls			
Component	Location in mine	Specification	Reference
Oxygen (O_2)	Working sections	19.5%	(75.301) (22-2-4a)
Carbon dioxide (CO_2)	Working sections	0.5%	(75.301) (22-2-4a) (75.301-2)
Noxious poisonous gases	Working sections	TLV TLV	(25.301) (22-2-4a)
Explosive gases	Working sections	CO, 2.5% H_2, 0.8% H_2S, 0.8% C_2H_2, 0.4% C_3H_8, 0.4% MaPP, 0.3%	(75.301-5) (75.301-5) (75.301-5) (75.301.5) (75.301-5) (75.301-5)
Methane (CH_4)	Working sections	1%	(75.307)
	Most returns	1%	(75.309, 75.310)
	Beltways/ trolleyway	1%	(75.327-1)
	Air to working sections from pillar section	0.25%	(22-2-5(b)) (75-312) (75-311)

TABLE 7-1 *(continued)*

Quantity controls			
Nature of flow	Location in mine	Specification	Reference
Volumetric flow rate	Each face	3,000 ft^3/min	(75.301) (75.301-1)
	Last open crosscut	9,000 ft^3/min	(22-2-4) (75.301)
	Intake of pillar line	9,000 ft^3/min	(22-2-4) (75.301)
	80 ℄ Crosscut 200	12,000 ft^3/min	(22-2-4(f))
	Intake of pillar line When bleeder closed	20,000 ft^3/min	(22-2-4)(i))
Velocity	Working sections	60 ft/min	(75.301-4)
	Trolley haulage entry	250 ft/min	(75.327-1)

Ventilation overview. The subfunctions described above are achieved by using the principles of fluid mechanics (Sections 7.2 and 7.3) and by implementing those principles underground (Section 7.4). As shown in Fig. 7-1, the energy needed to cause the airflow throughout the underground workings is, by law, supplied by a large mechanical fan. Intakes deliver fresh air to the working sections, the network of main, submain, and other entries. Auxiliary fans and check curtains are used to divert air to the face, pillar areas, and other working regions. From these utilization areas, the air exits the mine along a second network of entries designated as *returns*. The intake and return air courses are separated from one another by coal pillars and block structures called *permanent stoppings*. (Refer to the Appendix for a detailed presentation of the ventilation devices, features, and symbolizations.)

7.1.2 Ensure Safety

The primary function of the ventilation system is to ensure the supply of air to the workers; therefore, the safety function and the ventilation function share a common purpose. The insurance of a safe ventilation system is complicated because (1) we cannot physically observe good versus bad air; (2) we cannot easily observe the streamlines of the airflow, thus we cannot always be sure that air is flowing the way it should; and (3) the quality and quantity of the air changes continuously along its circuit. As a result, the

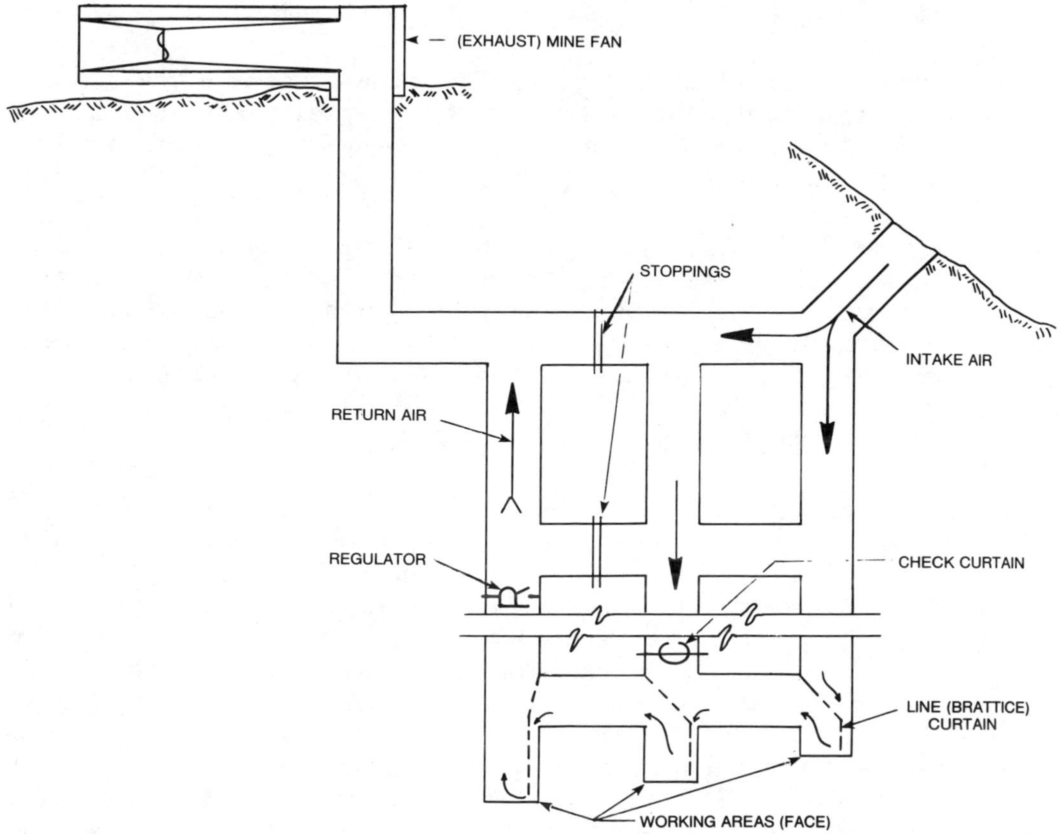

Figure 7-1 Overview of mine ventilation.

insurance of safety is based on identifying the state of the system and preventing workers from contacting dangerous parts of the ventilation network. This is done using the principles of safety outlined in Chapter 5 and using the following safety schemes.

Isolation schemes. The intake air enters the mine and is circulated to the workers. Return air is circulated away from the working sections and eventually exhausted out of the mine. Intake air is considered fresh and uncontaminated, whereas return air is considered to be gaseous, dusty, and otherwise contaminated.

Isolation attempts to locate all working places on the intake air. Further, isolation schemes physically separate safe intake air courses from the return air courses or abandoned sections by using pillars and concrete structures (stoppings). Isolation is also used to separate gaseous sections from nongaseous sections by using barrier pillars and by splitting the air circuit.

Protected agent schemes. As a product of combustion, methane and coal dust explosions create carbon monoxide gas, which is highly toxic to human beings. Although the risk of methane explosion is low, it is always present. As part of the protected agent scheme, each coal miner is required to carry a self-rescue breather, which is designed to chemically convert the toxic carbon monoxide into relatively harmless carbon dioxide.

Protected equipment schemes. One protected equipment scheme involves enclosing electrical equipment in such a way that if methane is ignited, the explosion will be contained within the equipment enclosure and will not propagate to the ventilation circuit. This scheme is known as *permissibility*. Another protected equipment scheme is the use of tamperproof locks on ventilation regulators to prevent unqualified personnel from altering the circuit parameters.

Deenergized contact. Perhaps the greatest advancement in ventilation safety in the past twenty years has been the development of the small gas chromatograph, which monitors the component percentages of different gases in the airstream. In the past and still today, the flame safety lamp has been the principal method by which oxygen and methane amounts have been "observed." In the hands of an experienced professional, very accurate oxygen approximations are possible; however, the flame safety lamp does not lend itself to recording instrumentation or automatic interlocking with equipment. Electronic methane monitors and gas chromatographs do permit equipment interlocking, so that electrical power can be deenergized in the event of high methane percentages, and also continuous recording of methane and other gases, as may be desired within sealed portions of the mine.

Alerted contact schemes. Alerted contact schemes notify a person of potential danger but do not deenergize the system. Again, methane monitors or gas chromatographs can be linked to buzzers, lights, or other alarm systems.

Environmental exclusion schemes. Several methods are employed to reduce the hazards of methane explosion. One of these is degasification or predrainage of the methane gas in advance of mining. Using degasification, high, unexpected concentrations of methane gas liberated at the face can be avoided, thus making the face safer. Another method of environmental exclusion of methane is the use of protection pillars to surround gas wells or, in the case of wells located in the path of a longwall panel, the plugging (sealing) of the well far below the seam level.

These safety schemes, together with other activities for ensuring safety, are described in more detail in Section 7.4.

7.1.3 Maximize Reliability

Ventilation reliability is extremely critical to the well-being of miners and to the profitability of the company. Various types of failure modes threaten the ventilation system. Following the classification scheme introduced in Chapter 5, these failures include:

Defect failures. Defect failures occur as a result of improperly installed ventilation control devices, stopping, overcasts, and so forth. These failures result in misdirected airflows or in "short circuiting" of the air from the intake air circuit to the return air circuit. More often than not, such "failures" do not interrupt the service, but they are extremely costly and should be corrected.

Degradation failures. Degradation or a general wearing out of the components is present only in devices frequently moved or used, such as face equipment.

Improper use, abuse, or maintenance. Failures resulting from deliberate or unintentional abuse of ventilation equipment, such as cutting of line brattice, removal of stopping blocks, and pinning back of check curtains.

Macroscopic failure. The major macroscopic failures within the ventilation system are mine fires and methane explosions. Both of these related occurrences cause major downtime effects to the ventilation system.

Spillover failure. Many major failures in the ventilation system are those caused by failures of neighboring systems. These include air circuit blockage resulting from roof failures and main fan failures resulting from electrical circuit failures.

These and other failures all contribute to lowering the reliability of the ventilation system. Layout tactics for improving reliability which have significant influence on the ventilation system are:

Use of redundant circuits. A minimum of two entries on intake and a return are needed for establishing the ventilation system. Since at least three entries are needed for a continuous miner section and five for a conventional miner section, redundant air circuits are a common practice.

Overdesign. The greatest operational cost of the ventilation system results from leakage of air from the intake to the return across the permanent stoppings. Leakage, which can be as much as 80% in some large mines,[2] also requires larger fans, greater fan speeds, and higher intake velocities, all of which adversely influence the reliability of the system. To avoid these problems, some mines have used double-wall stopping construction, particularly in mains in which the pressure difference between the intake and return is greatest.

Control of the environment. The greatest environmental threat is the presence of explosive methane gas at the face. Predrainage of methane gas by drilling bore holes in advance of mining has proved successful in reducing the amount of methane as much as 75% at the face.[5] Another environmental threat comes from ground movement, most notably roof falls. To reduce the probability of roof falls, great care is taken to position entries in locations of maximum stability. Also, intersections between entries and crosscuts (usually the greatest unsupported spans) are provided additional support.

Proper installation. Perhaps the most important activity which can increase the continuity of the ventilation system is proper installation of control components: the fan, regulators, stoppings, and overcasts.

These factors are the most important for extending the operation life of the ventilation system. The other reliability factors, noted in Chapter 5, also influence the continuity of the system, but, not in the same proportion.

7.1.4 Assure Maintainability

Maintainability of the ventilation system involves:

1. Time estimates and the probability of restoring service following failure.
2. The ability to advance retreat or expand the system.
3. The ability and requirements for inspecting and maintaining the system.

Restoration. Three major failures that can cause interruption of the ventilation service are (1) fan stoppage, (2) explosion and fire, and (3) major roof fall blocking an entry. Unless the service is restored within 15 minutes, the mine must be evacuated. A discussion of time and probability estimates associated with restoration of service following each failure is beyond the scope of this text.

Modifications. The major form of modification of the ventilation system is the advancement of the system. This extension is principally dependent on the number of entries, the degree to which the present entries are utilized, and the size of the mine fan. As the length of the intakes increases, the flow through the intake entries approaches a maximum, the leakage increases, the frictional losses increase, and the speed and pressure capabilities of the mine fan approach their limits. Eventually, these conditions necessitate installation of an additional mine fan.

Inspection and maintenance. Most inspection and subsequent maintenance activities applied to the ventilation system center about "surveys" of the airflow parameters, quality, quantity, and pressure. These surveys are used to indicate leakage, measure air particle contamination, monitor explosive gas percentages, and indicate inappropriate pressure drops along the ventilation circuits. The survey data are compared to acceptable norms and used to guide further inspection and maintenance to improve the variances. For example, leakage greater than 30% of the total is usually indicative of cracked or otherwise defective stoppings. A closer inspection of airflow across stoppings within the area surveyed would then reveal which specific structures require repair. Failure-cued maintenance would then be applied to correct the defect.

7.1.5 Optimize Value

The value of the ventilation system can be measured as the ratio of quantity of air delivered to the utilization regions (e.g., face, gob area, belt) per dollar expended. The major items contributing to dollar expenditures include capital costs (principally fan installation, which is a fixed cost), device construction costs (see the Appendix), electrical power costs, and maintenance costs. The higher the ratio, the more optimum the value. The value ratio is subject to wide variation, depending on (1) leakage, (2) quality of device construction, (3) amount of gas liberated by the seam, (4) type of mining, and (5) size of entries. For a more detailed analysis of the economics of the ventilation system, see Hartman.[2]

7.2 AIR AND OTHER GASES

Most matter can be broadly classified as being solid, liquid, or gas. The classification depends somewhat on conditions of time, temperature, and loading, but in general, solids, liquids, and gases can be differentiated as follows: A *solid* has the properties of mass, volume, and definite shape; a *liquid* has the properties of mass and volume but no definite shape; a *gas* has the property of mass but neither definite volume nor shape.

The importance of differentiating solids, liquids, and gases is that each state is governed by different laws of physics. Liquids and gases are called fluids and their actions are defined by the principles of fluid mechanics, whereas the actions of solids are defined by the theory of solid mechanics. This section presents properties of gases and the following section presents the basics of fluid mechanics.

7.2.1 Mine Gases

As noted earlier, pure, dry air is a mixture of various compounds, including oxygen (20.95%), nitrogen (78.09%), carbon dioxide (0.03%), and argon (0.93). As air is coursed underground, deletions (e.g., the absorption of oxygen) and additions to the air change the quality. All these changes are of an adverse nature to the workers underground. Table 7-2 is a tabulation of the source, effect, and properties of major mine gases that can be introduced to the air.

TABLE 7-2 PROPERTIES OF TYPICAL MINE CASES

Gas	Specific* gravity	Density† (lb/ft^3)	TLV‡ (ppm)	Hazard
Air	1.0	0.075	—	(At elevated pressure)
Nitrogen (N_2)	0.967	0.073	—	Simple asphyxiant
Oxygen (O_2)	1.105	0.083	—	(At elevated pressure)
Carbon dioxide (CO_2)	1.529	0.115	5000	Affects respiration
Methane (CH_4)	0.554	0.042	—	Flammable and a simple asphyxiant
Carbon monoxide (CO)	0.967	0.073	50	Very toxic
Nitric oxide (NO)	1.036	0.078	25	Very toxic
Nitrogen dioxide (NO_2)	1.589	0.119	5§	Very toxic
Sulfide dioxide (SO_2)	2.264	0.170	5	Very toxic
Hydrogen sulfide (H_2S)	1.191	0.089	10	Very toxic
Hydrogen (H_2)	0.0695	0.0052	—	Flammable and a simple asphyxiant

*Ratio of density of gas to that of air at sea level and 70°F.

†Determined at sea level and 70°F.

‡Threshold limit value: average (time-weighted) concentration for a daily 7- or 8-hour workday that will not produce an adverse effect in most workers.

§Ceiling value (maximum allowable concentration).

Source: Data courtesy of Mesa. M. G. Zabetakis, Safety Manual No. 2 Mine Gases, United States Department of the Interior, Mining Enforcement and Safety Administration, 1976. pp. 40–41.

7.2.2 Properties

All gases share certain common physical properties. A basic property shared by all materials, including gases, is mass. A mass when acted upon by gravity produces a force or weight. The weight of a unit volume of gas is measured by specific weight or density (lb/ft^3).

The relative densities of equal volumes of gases are as important as the absolute values (densities) since lighter gases tend to rise and heavier gases tend to fall. The specific gravity is the ratio of the weight of a unit volume of one substance compared to the weight of an equal volume of another substance assumed to be a standard. For liquids and solids the assumed standard is *air*. The specific gravity (sp. gr.) can be symbolically stated by the formula

$$\text{sp. gr.} = \frac{W_a}{W_s}$$

where W_a = weight of a unit volume of fluid
W_s = weight of an equal volume of standard fluid (either air or water)

The specific gravities of some of the more significant mine gases are compiled in Table 7-2.

Gas is composed of independent molecules that tend to move or redistribute when a force is exerted on them. If laterally contained, the molecules move closer together proportional to the force and tend to equally distribute the force on the containing surfaces. The force distributed upon a given unit area is called pressure (e.g., $lb/in.^2$). The weight of a given volume of gas depends on the density of the gas. If the given volume is spread over a wide area, the pressure is small. However, if the same volume were stood on end so that only a small area at the bottom of the column was covered, the pressure would be much larger. If a unit area is considered, the pressure exerted by a column of gas is thus dependent upon density and the height of the column according to Eq. 7-1.

$$P = \gamma h \tag{7-1}$$

where P = pressure (lb/ft^2)
γ = specific weight (lb/ft^3)
h = height of column of gas (ft)

Pressure and the volume occupied by a gas are directly proportional to the temperature according to Eq. 7-2a.

$$PV = f(T) \tag{7-2a}$$

where P = pressure
V = volume
$f(T)$ = functional statement of temperature

or, stated as the perfect gas law,

$$pv = RT \tag{7-2b}$$

where p = pressure (lb/ft^2)
v = specific volume (ft^3/lb)
R = constant of proportionality for the gas
T = temperature (°R)

7.2.3 Applications

The specific gravity, Table 7-2, shows the relative weights of equal volumes of gases compared to air. The application of this information aids ventilation surveying since lighter, generally more volatile gases, tend to migrate toward the entry top, while the heavier toxic gases tend to migrate toward the floor.

One application of the perfect gas law aids in calculating the effects of changing barometric pressure on the mine loads. A major source of added mine gas is from inactive or abandoned workings which are considered "sealed" in that they are not regularly ventilated. The introduction of these gases into the mine ventilation system is governed by the perfect gas law and more specifically Boyle's modification of that law. If a gas can be held at a constant temperature, then pressure times volume is a constant. An increase in the pressure will produce a decrease in the volume. Similarly, a decrease in the pressure produces a corresponding proportional increase in the volume. This condition is often stated as Boyle's law:

> The volume occupied by a given body of gas varies inversely with the pressure.

Symbolically,

$$PV = K \tag{7-3}$$

where P = pressure
V = volume
K = constant

Within a mine the temperature varies slightly if at all; however, the pressure, which corresponds to the atmospheric or barometric pressure, varies measurably. If a portion of a mine has been abandoned and is sealed (not ventilated), the changes in atmospheric pressure will affect the volume of air contained within that abandoned section. From Eq. 7-3 any decrease in the barometric pressure will produce a corresponding expansion in the volume of air contained within the sealed sections. This expansion must overflow into operating ventilation circuits, producing an additional load on those circuits. It should be noted that coal has an affinity for absorbing oxygen. Thus after a period of time the gas within the sealed area will be mostly nitrogen and carbon dioxide.

7.3 FLUID MECHANICS

7.3.1 Conservation of Mass

Perhaps the simplest and most critical fluid principle in mine ventilation theory is the conservation of mass or continuity. The continuity principle states that the amount of fluid (air) input into a mine (or entry) minus the amount of fluid that leaves must be equal to the amount of fluid that has accumulated (or has been lost) inside the mine. Simply, then: "in - out = accumulated" for any given system. Mathematically, this is written

$$M_{\text{in}} - M_{\text{out}} = M_{\text{acc}} \tag{7-4}$$

where M_{in} = mass input over a given time, $(\Delta M/\Delta t)$ (e.g., lb/min)
M_{out} = mass that leaves over that same period of time (lb/min)
M_{acc} = mass that has increased or decreased inside the system within that same period of time (lb/min)

For a gas flowing through an entry, the mass that passes any given point (per unit time) is dependent on the area of the opening, the velocity of the moving gas, and the density of the gas (at the temperature, pressure, and velocity of movement) such that

$$M = \rho VA \tag{7-5}$$

where M = mass of gas per unit time (lb/min)
ρ = density of gas (temperature, pressure, velocity) (lb/ft^3)
V = velocity of gas (ft/min)
A = area of opening (ft^2)

Equation 7-5 defines the *mass rate of flow* within a conduit. This equation can be substituted into Eq. 7-4 to yield

$$(\rho VA)_{\text{in}} - (\rho VA)_{\text{out}} = M_{\text{acc}} \tag{7-6}$$

If for any given inlet and outlet in a mine, the temperature, pressure, and velocity are, for all practical purposes, constant; then the density, ρ, can also be considered constant. Equation 7-6 may be simplified by dividing all factors by the density to yield

$$(VA)_{\text{in}} - (VA)_{\text{out}} = (\text{volume})_{\text{acc}} \tag{7-7}$$

The term VA is called the volumetric flow rate, Q. For the English system of units (where V = ft/min and A = ft^2), Q is qual to cubic feet per minute. Note that in Table 7-1 all quantity controls regulated by federal and state specifications are volumetric flow rates currently stated in English units.

For any given opening, the volumetric flow rate Q is defined as the product of velocity V and area A, or

$$Q = VA \tag{7-8}$$

where V = air velocity (ft/min)
A = stope area or flow area (ft^2)

7.3.2 Steady Flow/Steady State

A special, but common, condition for the flow within a given mine or entry is called steady flow/steady state. Steady flow/steady state assumes that all volumetric input and output flows are constant over time and that no significant changes are occurring within the system. Specifically, the volumetric changes inside the system are zero. Thus Eq. 7-7 can be written as

$$(VA)_{\text{in}} = (VA)_{\text{out}}$$

Equation 7-7 written for the condition shown in Fig. 7-1 would appear as

$$V_1A_1 = V_2A_2 = V_3A_3 = V_4A_4 = Q \tag{7-9}$$

Thus for any given volumetric flow rate, the velocity increases for smaller areas and decreases for larger areas. Figure 7-2 shows a small-section view of an air course. The volumetric flow rate through areas 1 to 4 is a constant 9000 ft^3/min; however, the velocity at each corresponding point will vary depending on (and only on) the area at each point. Using Eq. 7-8, the velocities at points 1, 2, 3, and 4 are 90, 60, 100, and 91.82 ft/min, respectively.

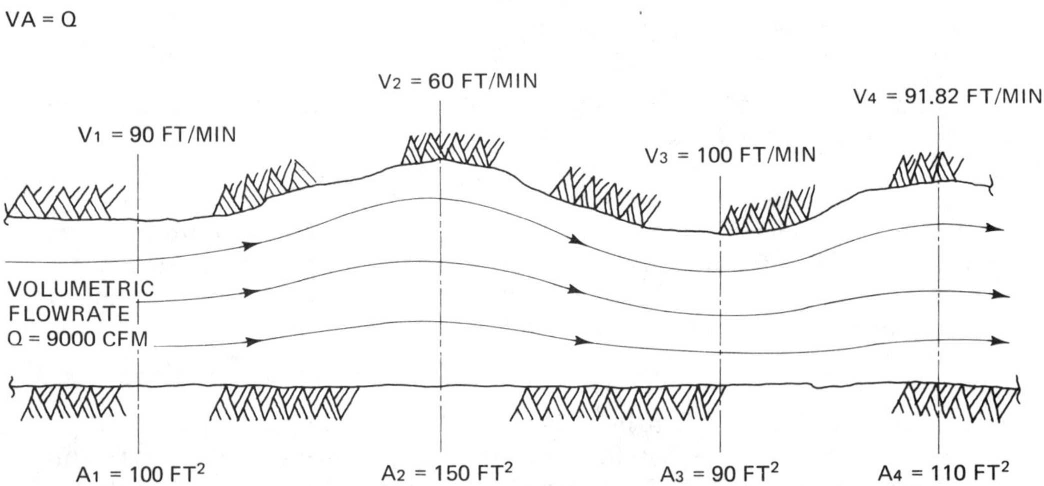

Figure 7-2 Profile of mine entry showing the airstream lines and velocity changes.

7.3.3 Pressure

Pressure, a fluid (thermodynamic) property, is a quantified measure of the stored or potential energy of a fluid. As a rough analogy, fluid pressure can be likened to voltage in an electrical system. Several different types of pressure concepts occur in fluid mechanics, including atmospheric pressure, absolute pressure, gage pressure, static pressure, and total pressure.

Atmospheric pressure, or *barometric pressure*, is the pressure of air that results principally from a column of air (a given volume) being acted upon by gravity. Obviously, the higher the column of air, the greater will be the weight of air pushing down and the higher will be the atmospheric pressure (see Eq. 7-1). The normal or standard atmospheric pressure measured in English units is 14.7 psi (lb/in.^{2}). Atmospheric pressure is measured relative to absolute pressure. *Absolute pressure* is the pressure that would exist in a perfect vacuum and is assumed to be 0 psi.

Gage pressure is the pressure a fluid exerts on its surroundings measured relative to (above or below) the surrounding atmospheric pressure. In the English system of units, the values of gage pressure are recorded as psig (pounds per square inch-gage). The absolute equivalent to psig can be determined by adding the atmospheric pressure to the gage pressure or

$$\text{psia} = \text{psi of atmosphere} + \text{psig} \tag{7-10}$$

Thus 20 psig would be equivalent to 34.7 psia if the atmospheric pressure were 14.7 psi.

Static pressure is the gage pressure of a fluid measured perpendicular to the direction of flow or measured when no flow exists. Figure 7-3 (top) shows the arrangement of an instrument measuring static pressure. *Total head*, is the gage pressure of a fluid measured parallel and against the direction of flow. Figure 7-3 (bottom) shows the arrangement of an instrument measuring dynamic pressure. As discussed in the next section, both total and static pressures are effected by velocity; however, total pressure head has a velocity component that directly influences its measurement.

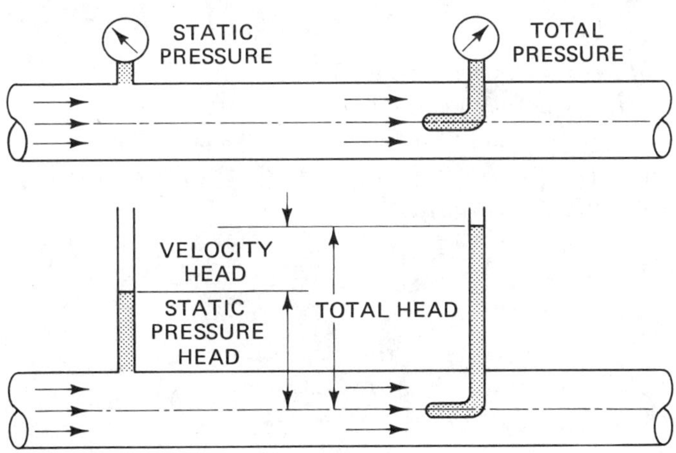

Figure 7-3 Schematic showing static and total pressure measurement.

7.3.4 Conservation of Energy: The Ideal Fluid

The energy changes that occur throughout the ventilation system from the inlet to the outlet are governed by the first law of thermodynamics, or the conservation of energy. For any system the first law can be stated:

$$(\text{energy})_{\text{in}} - (\text{energy})_{\text{out}} = (\Delta\,\text{energy})_{\text{system}} \tag{7-11}$$

For an ideal fluid at steady flow/steady state with no heat or work exchanges across the system boundary, the first law can be reduced to what is commonly called *Bernoulli's equation:*

$$\text{(static pressure energy)} + \text{(kinetic energy)} + \text{(potential energy)} = \text{constant} \tag{7-12a}$$

or

$$\frac{P}{\gamma} + \frac{V^2}{2g_c} + H = K \tag{7-12b}$$

where P = static pressure (lb/ft^2)
γ = specific weight of fluid (lb/ft^3. 0.075 lb/ft^3 for air)
V = velocity of fluid (feet per second)
g_c = 32.17 ft/sec^2 (a constant)
H = height above a reference plane (ft)
K = constant value of all energies at any given point in the same flow circuit (ft)

Note that the English unit for each energy term is lb-ft/lb or simply ft. Historically, fluid energy has been measured as the height in feet above a datum plane and termed *head.* Thus each term often includes the word "head": for example, pressure head, velocity head, and vertical head. Thus Eq. 7-12b is often written

$$H_s + H_v + H_z = K \tag{7-12c}$$

where H_s = static pressure head (ft)
H_v = velocity head (ft)
H_z = head above a reference elevation (ft)

The total energy at points 1, 2, 3, and 4 in Fig. 7-2 can be determined from Eq. 7-12b as follows:

$$\frac{P_1}{\gamma_1} + \frac{V_1^2}{2g_c} + H_1 = \frac{P_2}{\gamma_2} + \frac{V_2^2}{2g_c} + H_2 = \frac{P_3}{\gamma_3} + \frac{V_3^2}{2g_c} + H_3$$

$$= \frac{P_4}{\gamma_4} + \frac{V_4^2}{2g_c} + H_4 = K \tag{7-13a}$$

For most cases the values of the specific weight, δ, will be constant, and are thus written without the corresponding subscripts. For Fig. 7-2 each point lies more or less on the same horizontal reference plane; thus the potential energies, H_z, of each will be the same. With these considerations in mind. Eq. 7-13a can be simplified to

$$\frac{P_1}{\gamma} + \frac{V_1^2}{2g_c} = \frac{P_2}{\gamma} + \frac{V_2^2}{2g_c} = \frac{P_3}{\gamma} + \frac{V_3^2}{2g_c} = \frac{P_4}{\gamma} + \frac{V_4^2}{2g_c} = K' \tag{7-13b}$$

If the volumetric flow rate and areas are known, each of the velocities can be calculated from Eq. 7-9. Thus, by knowing the pressure at any one point, all pressures can be simply calculated.

The values of $(P/\delta + V^2/2g_c)$ are the mathematical representation of the *total dynamic head.* Note that the total head is constant when height remains unchanged. If the velocity decreases as a result of the area of flow increasing, the velocity head also decreases.

The result is an increase in the static head or, more specifically, in the static pressure head. Thus kinetic energy is *converted* to stored pressure energy. Energy is not lost, just changed in form.

7.3.5 Real Flow: Head Loss

In ideal flow, energy is converted from velocity to pressure and back, depending on changes in the area of flow without any change in the total energy. In real flow, such conservation is not the case. Flow involves the movement of particles of air through a passage. This movement results in a rubbing friction between the air particles and between the air and the walls of the entry. This friction, in turn, produces the conversion of some of the energy into heat energy. However, heat energy is not recoverable; it is "lost." This loss of energy is illustrated in Fig. 7-4. The amount of heat produced is proportional to the square of the velocity and is measured as the difference between the total energy at point 1 and that at point 2 such that

$$\frac{P_1}{\gamma} + \frac{V_1^2}{2g_c} + H_1 - \frac{P_2}{\gamma} + \frac{V_2^2}{2g_c} + H_2 = \text{energy (heat) lost} = \text{head loss} \quad (7\text{-}14)$$

Head loss, that energy which is not available for work, can be divided into two categories, *frictional losses*, due to air rubbing along the length of

Figure 7-4 Schematic showing the relative effects of frictional flow.

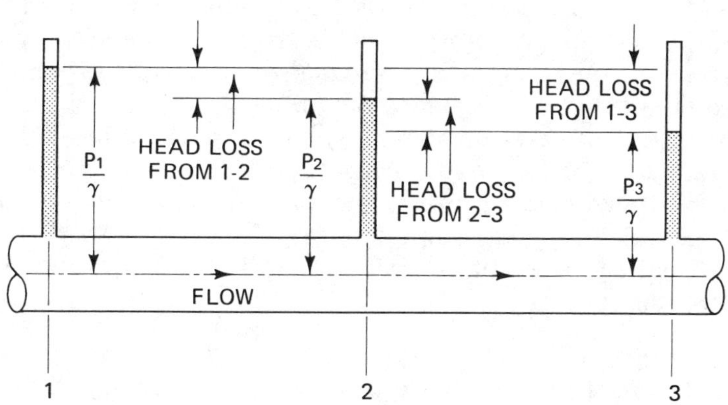

the flow passage, and *shock* or *minor losses*, due to friction produced from changes in the direction of the air stream. These two losses can be stated as

$$H_L = H_f + H_x \tag{7-15}$$

where H_L = total head loss (ft)
H_f = frictional head losses (ft)
H_x = shock head losses (ft)

Frictional head. The frictional head losses can be calculated as a function of the velocity squared, the length of the passage, the perimeter of the passage, the area of opening, and the relative roughness of the mine walls. Symbolically, these relationships can be stated as follows:

$$H_f = \frac{KLOV^2}{5.2A} \tag{7-16a}$$

where H_f = frictional head loss (inches of water)
K = constant (includes relative roughness, type of mine, type of roof support, type of rock, straightness of entry)
L = length (ft)
O = perimeter of the opening (ft)
V = velocity (ft/min)
A = area (ft^2)
5.2 = conversion constant ($lb/ft^2/in.$ of H_2O)

Substituting Q/A for V (from Eq. 7-8), Eq. 7-16a can be rewritten using the volumetric flow rate and area as follows:

$$H_f = \frac{KLOQ^2}{5.2A^3} \tag{7-16b}$$

The geometric shape of the entry that provides the smallest ratio of perimeter length to area (O/A) is a circular opening. Any other shape (e.g., rectangle, square) contributes to greater rubbing surface and thus greater head losses. As the length of flow increases, the head loss proportionally increases, since the product LO is the rubbing surface. However, as the area increases, the head loss decreases by the *cube* of this area $(A)^3$. As the requirements for volumetric flow rate, Q, increase, the head loss increases by the *square* of these increases $(Q)^2$. Finally, the value of K is an empirical factor that is related to the irregularities of the rib surface, type of roof support, straightness of entry, and type of strata. The values of K shown in Table 7-3 have been found to vary from 30×10^{-10} for smooth, straight, roof bolted, clean air courses to 145×10^{-10} for timbered, highly irregular, rough air courses.[6]

Shock (minor) losses. Shock losses, H_x, are those internal or fluid frictional losses that result from the air stream changing direction. Any time a

TABLE 7-3 SCHEDULE OF FRICTIONAL FACTORS (K) FOR DIFFERING MINE ENTRY CONDITIONS

Type of Airway	Irregularities of surfaces, areas, and alignment	Values of $K \times 10^{-10}$*											
		Straight			Sinuous or curved								
					Slightly			Moderately			High degree		
		Clean (basic values)	Slightly obstructed	Moderately obstructed	Clean	Slightly obstructed	Moderately obstructed	Clean	Slightly obstructed	Moderately obstructed	Clean	Slightly obstructed	Moderately obstructed
Smooth lined	Minimum	10	15	25	20	25	35	25	30	40	35	40	50
	Average	15	*20*	30	25	30	40	30	35	45	40	45	55
	Maximum	20	25	35	30	35	45	35	40	50	45	50	60
Sedimentary rock (or coal)	Minimum	30	35	45	40	45	55	45	50	60	55	60	70
	Average	55	*60*	70	65	70	80	70	75	85	80	85	95
	Maximum	70	75	85	80	85	95	85	95	100	95	100	110
Timbered (5-ft centers)	Minimum	80	85	95	90	95	105	95	100	110	105	110	120
	Average	95	*100*	110	105	110	120	110	115	125	120	125	135
	Maximum	105	110	120	115	120	130	120	125	135	130	135	145
Igneous rock	Minimum	90	95	105	100	105	115	105	110	120	115	120	130
	Average	145	*150*	160	155	160	165	160	165	175	170	175	195
	Maximum	195	200	210	205	210	220	210	215	225	220	225	235

*All values are for air weighing 0.075 lb/ft^3.

Source: G. E. McElroy, "Engineering Factors in the Ventilation of Metal Mines," *U.S. Bur. Mines Bull. 385* (1935), p. 43.

fluid particle is forced to alter a straight-line path, it interferes with adjacent particles, causing a loss of energy. These losses are termed "minor losses" only to differentiate them from the frictional losses that occur at the wall-fluid interface. In fact, these minor losses can be a major source of "lost energy" (up to 30% for certain mines).

Shock losses are present at every turn, overcast, regulator, line brattice, and face and also result from the presence of workers and equipment operating in the flow path. The amount of lost energy depends on the severity of flow direction change and the square of the velocity of the fluid. This can be symbolically represented as

$$H_x = \frac{KV^2}{2g_c} = K'V^2 \tag{7-17}$$

where H_x = shock loss (in.)
K' = constant that depends on the geometry of flow path and conversion of units (in.-sec^2/ft^2)
V^2 = velocity of flow (ft/sec)

The total head loss represented by Eq. 7-15 can now be rewritten

$$H_L = \frac{K_1 LOV^2}{5.2A} + K_2 V^2 = \left(\frac{K_1 LO}{5.2A} + K_2\right) V^2 \tag{7-18}$$

The effect of minor head loss in a fluid is shown graphically in Fig. 7-5. (See Hartman[2] and McElroy[6] for a more detailed analysis.)

For any given mine, the values of K, O, A, and L are constant. Minor losses can also be considered constant. Thus for a given mine the head loss can be written

$$H_f = RQ^2 \tag{7-19}$$

where H = pressure loss (inches of water)
R = total resistance factor of the mine [H_2O/100,000 ft^3/min^2]
Q = volumetric flow rate of the mine fan (100,000 ft^3/min)

7.3.6 Energy Input: Mine Fans

All coal mines, by law, use mechanical methods (fans) to cause a pressure differential between the inlet and outlet. To obtain the needed quantities of air, Q, the total pressure or head developed by the mine fan, H_{fan}, must as a minimum equal the total head loss throughout the mine plus the velocity head energy, $V^2/2g_c$, since this is also "lost" energy. Thus the amount of airflow forced through any mine or mine segment depends on the characteristics of the mine and those of the fan. For any two points within the flow stream and on either side of the mine fan, Bernoulli's equation can be written:

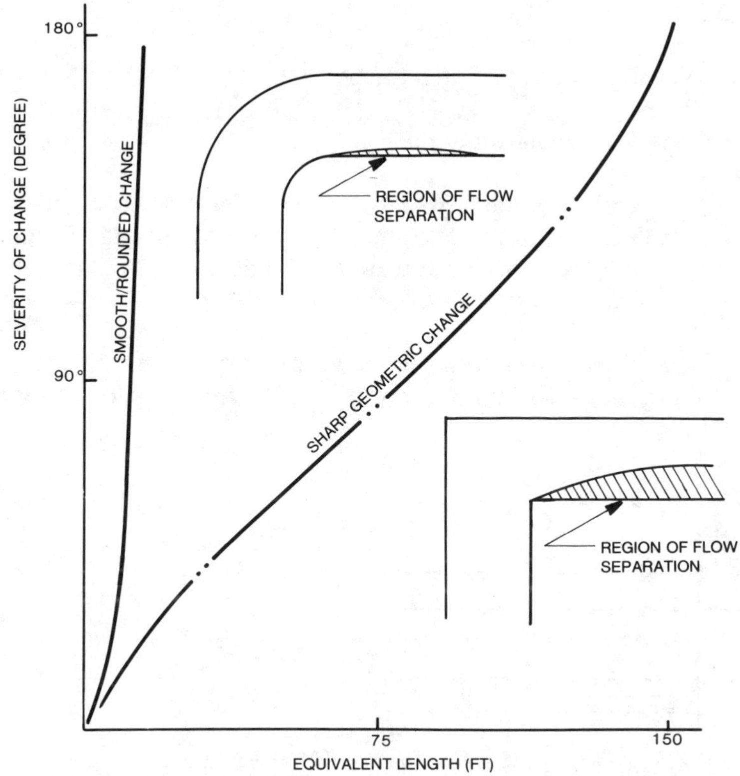

Figure 7-5 Effect of flow geometry on minor head loss.

$$\left(\frac{P_1}{\gamma} + \frac{V_1^2}{2g_c}\right) + H_1 + H_{\text{fan}} = \left(\frac{P_2}{\gamma} + \frac{V_2^2}{2g_c} + H_2\right) + H_{L\,1\text{-}2} \qquad (7\text{-}20)$$

In general, certain characteristics are shared by all types of fans. These can be summarized,[2,7,8] as follows:

1. The *volumetric flow rate Q varies directly* as the *speed S* (rpm) of the fan:

$$Q = f(S) \qquad (7\text{-}21)$$

2. The *power input* hp, (horse power) to the fan *varies directly* as the *cube* of the *speed S* and the *density* γ of the air:

$$\text{hp} = f(S^3, \gamma_{\text{air}}) \qquad (7\text{-}22)$$

3. The *pressure P* increase induced by the fan varies directly as the *square* of the fan *speed S* and the *density* γ of the air:

$$\Delta P = f(S^2, \gamma_{\text{air}}) \qquad (7\text{-}23)$$

For a more detailed analysis of fan characteristics and fan types, see Osborne[8] or contact specific mine fan manufacturers.

7.4 VENTILATION PRACTICES AND PROBLEMS

7.4.1 Arrangement of the Fan

Several arrangements of the mine fan affect the nature of ventilation. These include exhaust systems, blower systems, and combinations of exhaust/blower systems operating in direct series or parallel installations. The most common system is the exhaust system.

Exhaust system. In the exhaust system shown in Fig. 7-6(a), the mine fan pulls the stale air out of the mine by creating a lower pressure at the

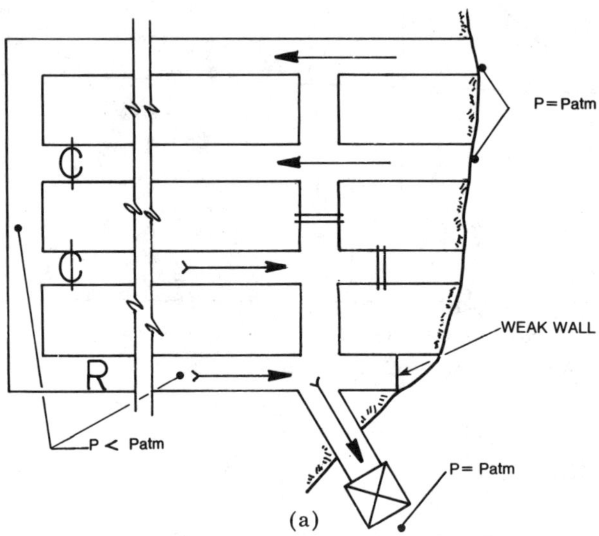

Figure 7-6 (a) Exhausting type of main fan. (b) Blowing type of main fan.

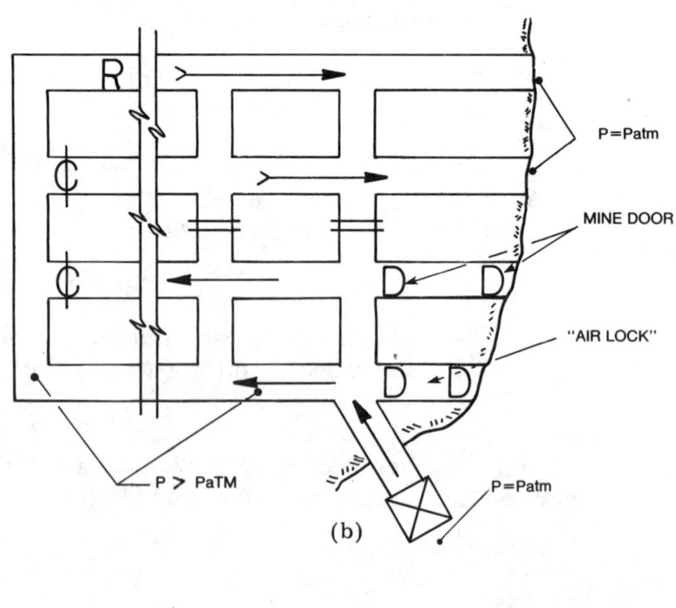

intake end of the fan. The pressure at the intake to the mine is, of course, atmospheric pressure. Thus the pressures inside the mine vary below atmospheric.

The exhaust system has several advantages over other arrangements:

1. Main haulage can be established simply on the intake air courses, thus making track haulage safer.
2. If the exhaust fans fail, the increasing pressure inside the mine will tend to retard methane seepage from the virgin coal. This makes evacuation safer.
3. Power savings are usually greater with the exhaust system.
4. In the event of fire, hazardous gases move out the return airways.

The advantages of the exhausting methods of ventilation are countered by the following limitations:

1. Exhaust system fans handle contaminated (dust-laden) air.
2. Exhaust systems are *pressure limited;* that is, they may not be as easily adjusted for increased mine resistance as an equivalent blower system might.
3. The lower pressures within the mine cause greater methane liberation than with the higher-pressure blower system.
4. During winter months, icing of entries becomes a problem.
5. The effects of dry air in winter and humid air in summer will have the greatest effect on the mine top in the critical mains.

Blower system. With the blower system shown in Fig. 7-6(b), the fan discharges air into the mine at greater than atmospheric pressure. The static pressure at the outlet is, of course, atmospheric; thus the interior pressure of the mine is always greater than atmospheric pressure.

Traditionally, the blower system would course air from the fan to the face and return the "stale" air through the mains. Haulage would then be established on the return air. This offered the advantages that (1) in winter, icing would be prevented, and in summer, the haulage routes would be cooler; and (2) there was less wall sweating and drying in the mains and therefore a more stable top. Such practices were acceptable as long as the mine was not a gassy mine or as long as the operations did not create too much dust.

Presently, the establishment of haulage on intake is required because all coal mines are now treated as being gassy and increased mechanization has increased particulate content. To provide for haulage on intake using a blower system, a series of mine doors are used to establish an airlock, as illustrated in Fig. 7-6(b).

Like the exhaust system, the blower system has certain advantages and limitations. The major advantages are that the higher static pressure of the air course causes less methane liberation, and bore holes drilled from the surface into gob or pillared areas can be used as direct returns.

Some of the disadvantages are similar to those for the exhausting system; however, the major limitation is that mine doors create a costly and inconvenient burden on the mine operation.

7.4.2 Leakage

A common problem of all types of mine layouts is leakage. Leakage refers to the "short circuiting" of air from the intake into the return (or the outside in the case of blower systems) before the air can be utilized by the working sections. Leakage, in effect, is an unregulated natural division in the air intake. The amount of leakage at any one point depends on the pressure drop across a ventilation device between the intake and return. Thus greater leakage would be expected between the intake and return courses near the main fan (where P is greatest), and less leakage would be expected across devices closer to the working sections.

Leakage does have one valuable quality—it establishes a small rate of flow in crosscuts which would otherwise be a "dead" airspace. Thus a small leakage aids in preventing the accumulation of explosive gases in these areas.

Air leakage is costly to the general operation of the mine since the fan is required to operate at higher speeds [recall that $Q \approx f(S)$], supplying more airflow than would normally be needed if leakage were zero. The effect on power is therefore cubed [since hp = $f(S^3)$, hp = $f(Q)^3$]. Air leakage for typical coal mines can be as high as 70 to 80%. At only 50% leakage the power increase would be 3.375 times that at no leakage. The total amount of leakage depends on the size of the mine, as leakage is directly related to the total number of stoppings. The quality of stopping construction is an additional factor. Poor quality means greater leakage and more costly operation.

7.4.3 Methods of Layout

Although numerous arrangements of air course circuits are possible, only three basic layouts are used: (1) the simple series [Fig. 7-7(a)], (2) radial splitting [Fig. 7-7(b)], and (3) redundant splitting [Fig. 7-7(c)].

Simple series. The *simple series* layout is characterized by a single fan that moves air successively from one working section to another. Thus the

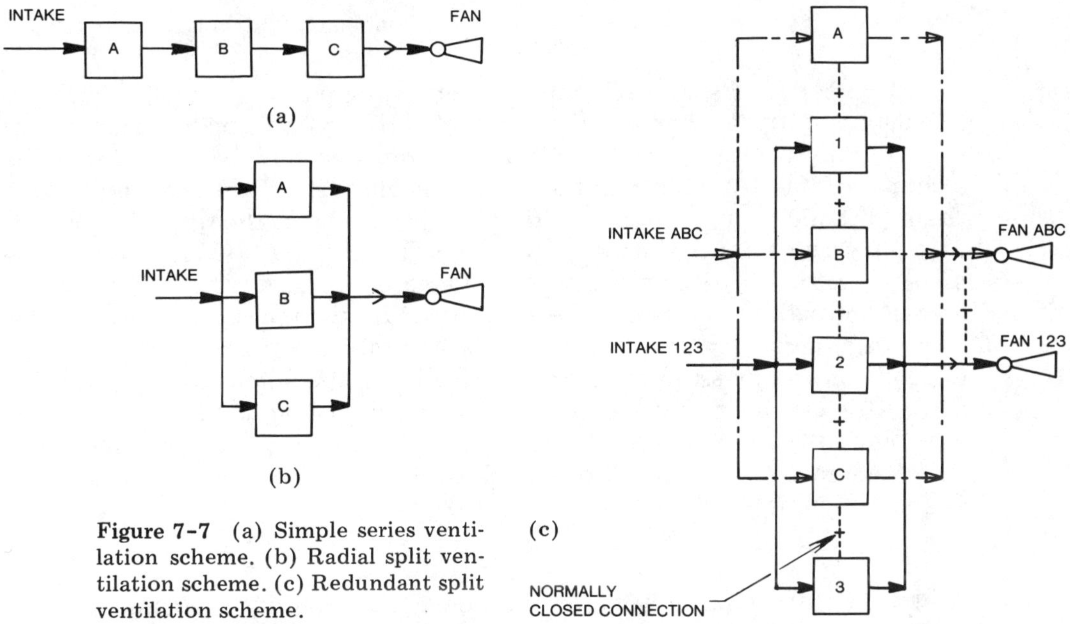

Figure 7-7 (a) Simple series ventilation scheme. (b) Radial split ventilation scheme. (c) Redundant split ventilation scheme.

intake air for section 2 is the return air from section 1. The fluid mechanic characteristics of flow and head loss can be summarized as

$$Q_1 = Q_2 = Q_3 \tag{7-24}$$

and

$$H_T = H_1 + H_2 + H_3 \tag{7-25a}$$

$$R_T = R_1 + R_2 + R_3 \tag{7-25b}$$

For small mines the simple series network *can* provide the primary function of supplying air when and where needed. However, as the mine increases in size, the quality of the air cannot be maintained. Even for small mines the series layout suffers in relation to the other systems when secondary functions of safety, reliability, maintainability, and cost are considered. The health and safety of all sections downstream from the first section are intolerable. The reliability is poor since a roof fall anywhere along the flow path would effectively stop the ventilation to all sections. Although the simplicity of the system would facilitate inspection and maintenance activities, it inhibits advancement of the system. Finally the cost of power to ventilate a series network is reflected by the additive effect of the individual head losses, Eq. 7-25a.

Radial splitting. A *radial splitting* layout is characterized by a single fan supplying power to an entry, from which several branches or splits are taken to the various working faces. A *split*, then, is a division or offtake from an air course to form a parallel circuit.

Two types of splits are commonly encountered: natural splits and regulated splits. *Natural splits* are those in which the airflow divides as a result of the naturally occurring resistances of each possible flow path. Leakage is an example of an unwanted natural split. A small portion of the air splits from the intake and flows around or through the stopping to the return air course. *Regulated splits* are desired splits which occur as a result of the construction of a regulator. By decreasing the flow area of the regulator, the resistance across that path increases and the quantity of flow decreases proportionately.

Regardless of whether the split is natural or regulated, the flow rate depends on the pressure drop or resulting head loss across each path of the split. This can be grasped more easily using the example illustrated in Fig. 7-8. Since the fluid energies at points *a* and *b* are equal for all three splits, the pressure drop resulting from the head losses must be equal for splits 1, 2, and 3. Symbolically,

$$H_{L1} = H_{L2} = H_{L3} \tag{7-26a}$$

Now substituting Eq. 7-16b for the above and simplifying (assuming that each air course has equal perimeter and areas), we obtain

$$L_1 Q_1^2 = L_2 Q_2^2 = L_3 Q_3^2 \tag{7-26b}$$

or for paths 1 and 2,

$$\frac{Q_1}{Q_2} = \sqrt{\frac{L_2}{L_1}} \tag{7-26c}$$

Thus the relative flow rate through each air course is equal to the square root of their respective lengths. Note that this is only an approximation since shock losses are not considered.

Since the flow rate depends on the H_L from point *A* to point *B*, another method often used to approximate the effect of splitting is to simplify Eq. 7-16b as follows:

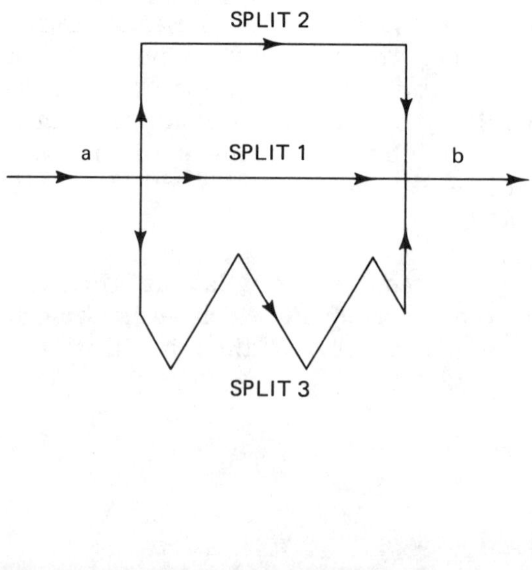

Figure 7-8 Split effects.

$$H_L = \frac{KLOQ^2}{5.2A^3}$$

$$Q^2 = \frac{5.2A^3H_L}{KLO} = \frac{A^3}{LO}\left(\frac{5.2H_L}{K}\right)$$

Since 5.2, K, and H_L are the same for each split, then for any split the *relative* flow rate can be defined as

$$Q_1 = A_1\sqrt{\frac{A_1}{L_1O_1}} \tag{7-27}$$

where $Q_1 \simeq$ relative flow rate (cfm)
$A \simeq$ area (ft^2)
L = length (per 1000 ft)
O = perimeter (ft)

Additionally, for parallel flow:

$$Q_{total} = Q_1 + Q_2 + Q_3 \tag{7-28}$$

An example of the use of Eqs. 7-25 and 7-26 follows.

Example

Find the flow rate in splits 1, 2, and 3 if the height of all air courses is 5 ft; the width is 15 ft; the length of the splits are 3500, 4200, and 2200, respectively; and Q total = 120,000 cfm.

$$Q_1 = A_1\sqrt{\frac{A_1}{L_1O_1}} = 75\sqrt{\frac{75}{3.5(40)}} = 54.89$$

$$Q_2 = A_2\sqrt{\frac{A_2}{L_2O_2}} = 50.11$$

$$Q_3 = 69.24$$

$$Q_1 + Q_2 + Q_3 = 174.24$$

Then

$$Q_1 = \frac{54.89}{17.24}(120{,}000) = 37{,}800 \text{ cfm}$$

$$Q_2 = 34{,}510 \text{ cfm}$$

$$Q_3 = 47{,}685 \text{ cfm}$$

The fluid mechanic characteristics are summarized as

$$Q_T = Q_1 + Q_2 + Q_3 \tag{7-28}$$

$$H_L = H_1 = H_2 = H_3 \tag{7-28a}$$

and

$$\frac{1}{\sqrt{R_T}} = \frac{1}{\sqrt{R_1}} + \frac{1}{\sqrt{R_2}} + \frac{1}{\sqrt{R_3}} \tag{7-29}$$

The radial splitting layout is the most common. This layout accomplishes the primary function for all types and sizes of mines. The safety of a radial splitting system is greatly enhanced over that of the simple series, since each working section can:

1. Better control of the air flow within each section is achieved.
2. Simplify the distribution of air in each working section.
3. Lessen the probability of propagating the effects of an explosion toward other working sections.
4. Cause the dust or gas produced in one section to enter a return directly.

The reliability of the system is improved because roof falls cannot cause wide-ranging failure of the system. The maintainability is slightly more complex in inspection and maintenance; however, the ability to advance, retreat, or modify the system is greatly improved. Finally, because the head loss is less, the cost of operating the radial split system is significantly less than for an equally complex series system.

Redundant splitting. The *redundant splitting* arrangement is characterized by two or more radial splitting networks which can be linked together in the event of a main fan failure of either system. Under normal operations each system functions as a radial splitting system, independent of one another, separated by stoppings and mine doors. If a failure occurs in one system, the entries linking the two are opened or closed as needed to create a single large radial splitting system.

The advantage of such a system is that it increases the reliability of the mining operation. If each mine fan or system had a reliability index of 95%, then using Eq. 5-2, the improved reliability of the redundant layout would be calculated as

$$\begin{aligned} R &= 0.95 + 0.95 - (0.95)^2 \\ &= 0.9975 \end{aligned}$$

A truly redundant splitting system requires that each mine fan be capable of assuming the total load. In general practice, this necessitates that each fan be sized 80 to 100% larger than would be specified normally. The value of such a layout ultimately depends on (1) the actual reliabilities of the fan, (2) the lifetime cost of fan downtime, and (3) the cost of supplying the redundant capacity.

7.5 SPECIAL SITUATIONS

7.5.1 Face Ventilation

The major functions of the ventilation system are to prevent the accumulation of dangerous quantities of methane gas and to deliver needed oxygen to workers. Since methane is most prevalent at the face and since the greatest numbers of workers are concentrated at or near the face, ventilation of the face region becomes a critical problem.

As with any field flow system, an inlet and outlet must be defined. To provide this delineation two methods are commonly employed: line brattice or ventilation fan and tubing. Each method operates to separate the inlet and outlet physically from the face and working room. The line brattice has the advantage of being economical, but the ventilation tubing provides greater maneuverability of workers and machinery and is easier to advance as the mining advances.

Regardless of which method is used, the equipment can be positioned so as to create two possible systems for delivering air to the face: (1) the blower system and (2) the exhaust system. (*Note:* There exists no correlation between the face distribution system and the type of main fan system. Thus a mine may use a blower distribution system at the face while the main fan is positioned to provide an exhausting flow.) Each system of face ventilation has distinct advantages and limitations. Figure 7-9 shows a typical exhaust system and the air streamlines that result. From this diagram the following advantages and disadvantages of the exhaust system are illustrated:

1. The dust, methane, and stale air flow away from the workers and equipment while fresh air is circulated to them.
2. Air flow tends to recirculate in corners *A* and *B* (particularly *A*), creating a potential for explosive accumulations of methane and dust.
3. Rigid support of the brattice cloth is required along its entire length; otherwise, the curtain would collapse toward the rib.
4. The distance from the end of the curtain or tube to the face is more critical with the exhaust system than with the blowing system.

Figure 7-10 shows a typical blower system and the resulting air streamlines. The following characteristics are illustrated:

1. The face, including corners *A* and *B*, is swept to a greater extent than in the exhaust system, providing better gas and dust control.
2. The dust, methane, and stale air flow over the workers and equipment, contributing to health and safety problems (not to mention poorer morale).

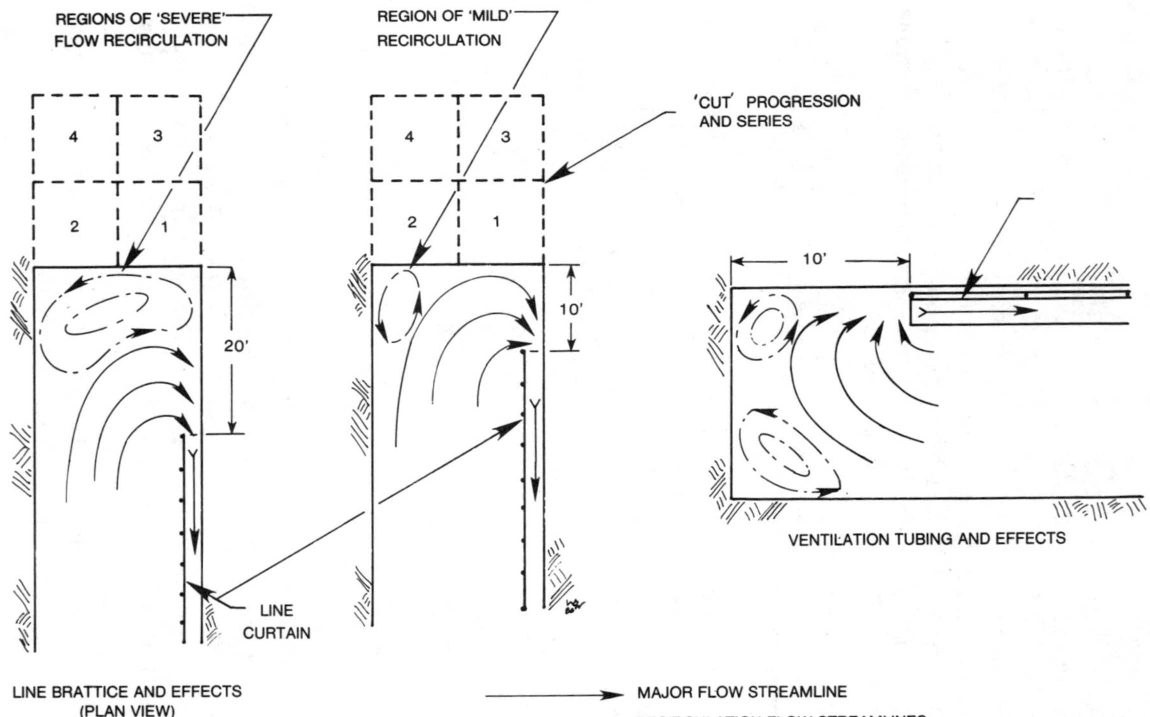

Figure 7-9 Exhaust system of face ventilation and effects.

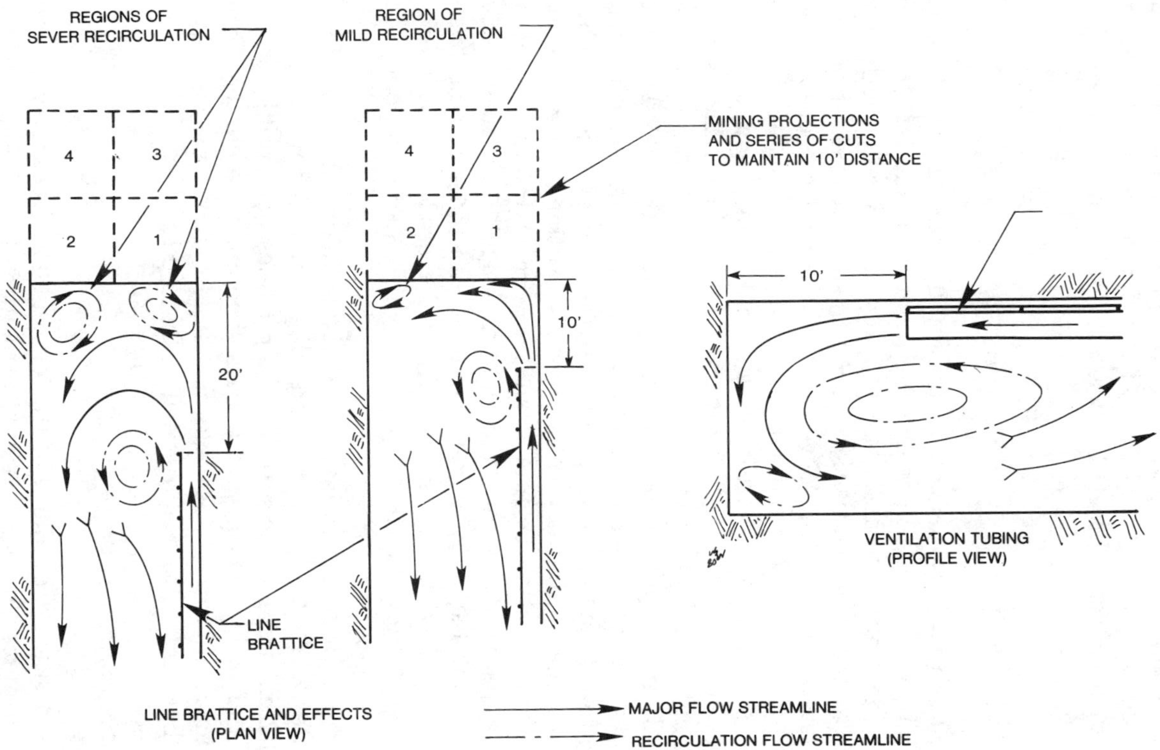

Figure 7-10 Blower system of face ventilation and effects.

In general, use of the blower system for face ventilation is rare, the exceptions being those "hot" mines or faces where the methane percentages are continuously greater than 0.75%. For the typical gassy mine (methane consistently greater than 0.25%) the exhaust system is used in conjunction with diffuser fans directed toward those corners where air tends to recirculate.

7.5.2 Pillar Section

Retreat mining (also called pillaring, pillar recovery, and second mining) is that phase of mining in which the pillars, initially left to support the roof during development, are systematically mined and the immediate top allowed to fall. Two ventilation conditions are of upmost concern during the pillaring operation: (1) delivering the needed fresh air to the working sections, (2) diluting and removing stale air in and around the abandoned and mined out sections. (Note that concern for methane immediately adjacent to the face is not usually a primary concern since pillar extraction does not occur in "virgin coal.")

All active pillaring areas use a system of "bleeder" entries. Bleeder entries are used to remove air–methane mixtures from the working sections along the pillar line across the GOB, and to eventually deliver such mixtures to the appropriate return circuit. Normally, the bleeder entries surround the panel to be mined and are driven in advance of other mining, thereby permitting some "methane drainage" prior to subsequent development of the panel. After development and pillar mining has depleted the panel, the bleeder entries remain and are used to ventilate the GOB continuously to prevent the buildup of high methane concentrations in the old workings.

Fresh air from the intake is delivered along the pillar line and allowed to enter the GOB. The pressure differential from the pillar line to the bleeder entries must be sufficient to provide "perceptible airflow" across the GOB and within the bleeder entries. The methane content from the bleeder entering a return split should not exceed 2.0% methane. An example of pillar line ventilation is illustrated in Fig. 7-11.

7.5.3 Longwall/Shortwall

The ventilation procedures of the longwall and the shortwall are simpler than those required for conventional or continuous-mining face ventilation. The development of retreating longwall panels requires at least three parallel headings, located along either side of the panel and bleeder entries driven across the top, parallel to the intended face. (Greater concern for more intake escapeways is creating a "need" for four entry developments.) Figure 7-12 shows a plan view of a typical shortwall panel. In general, entries 1, 2,

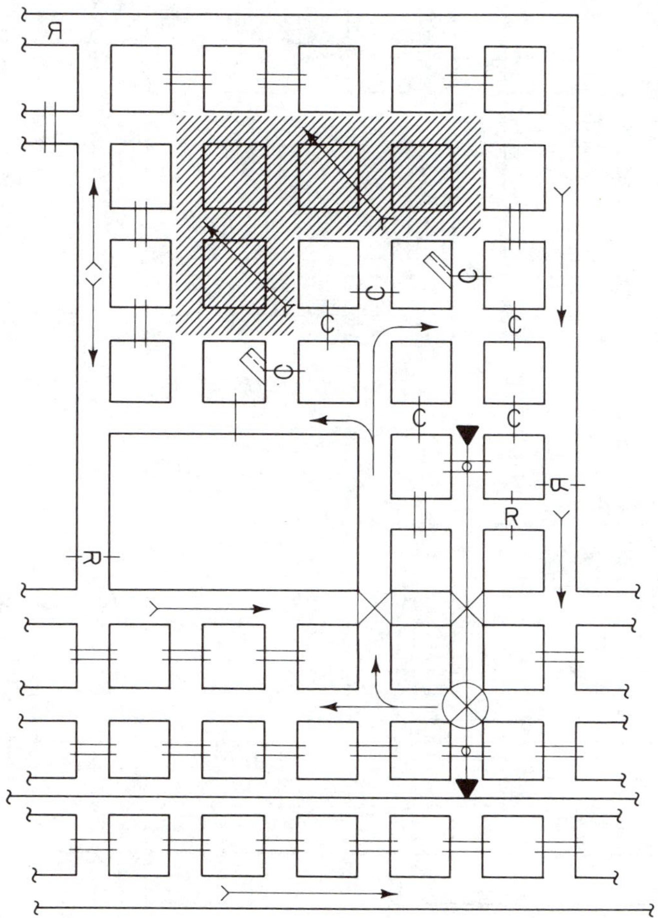

Figure 7-11 Pillar line ventilation.

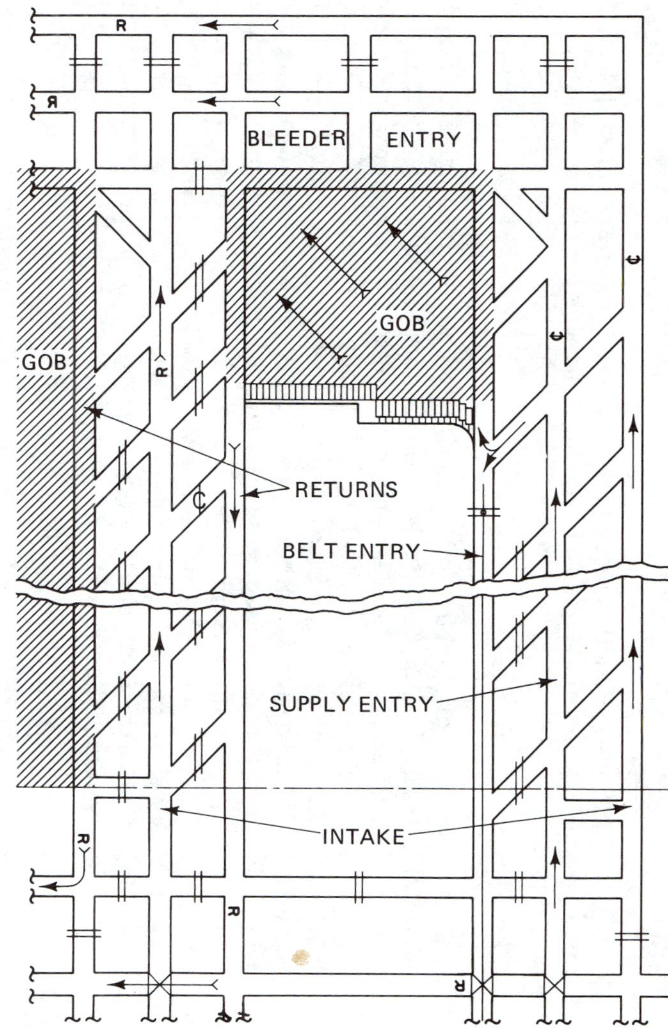

Figure 7-12 Possible shortwall ventilation layout.

and 3 on the headgate side are employed as intake, supply entry, and belt entry, respectively, while 4 and 5 on the opposite side are used as returns.

7.5.4 Belt Ventilation

Beltlines are used principally for intermediate and mainline haulage. As of March 30, 1970 (75.326 Fed. Spec.), entries used as intake airways must be physically separated from entries using belt haulage. This requires that each belt entry be on a separate split of air. (Normally, the air used to ventilate a beltline should immediately enter a return. However, if the mine design so requires and the design is approved by the federal representative, the air from a belt entry may be used to ventilate an active working place. Such a practice is not generally recommended.)

To place each belt entry on a separate split, a ventilation box check is used at either end (see the Appendix). A regulator is then placed as close to the tailpiece or drive as possible to regulate airflow in the belt entry. Such a ventilation system is shown in Fig. 7-13.

Figure 7-13 Belt ventilation techniques.

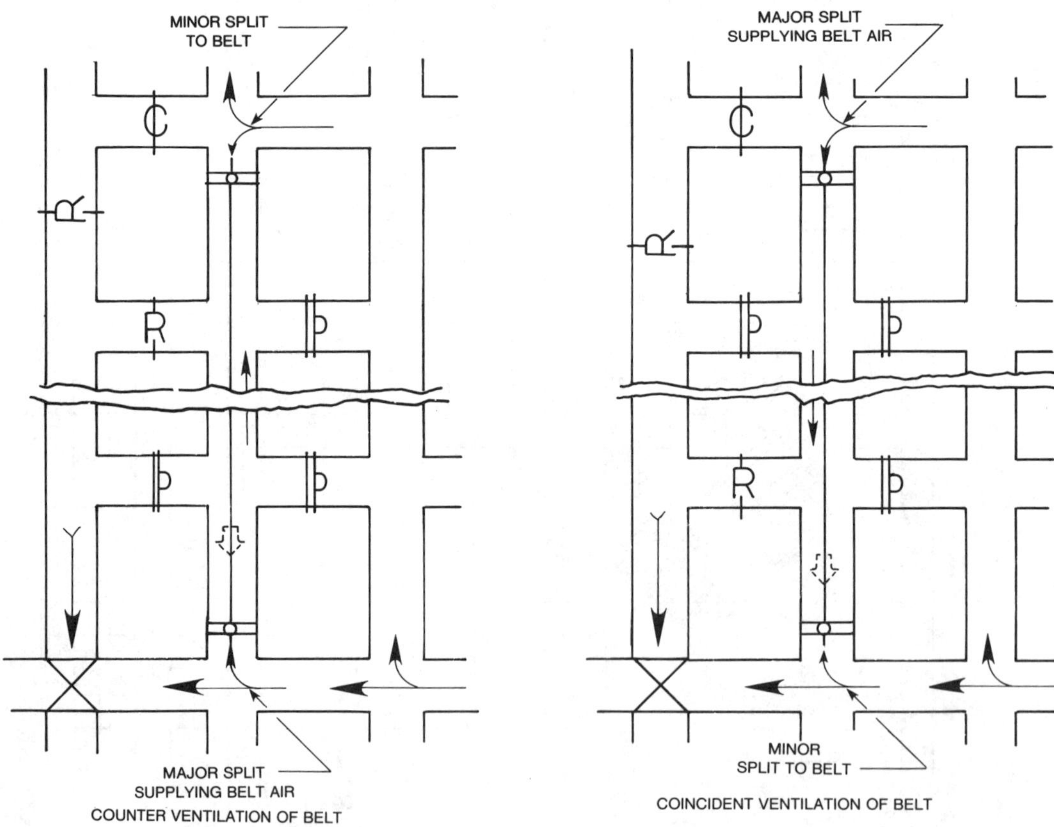

7.5.5 Electrical Equipment

Certain permanently mounted transformers, rectifiers, battery-charging stations, and belt power centers must be ventilated by a separate split of air. To facilitate this requirement, the electrical equipment requiring ventilation should be installed within a crosscut linking an intake and return. By removing several blocks of the permanent stopping, the split requirements can be achieved with a minimum of trouble and expense.

7.5.6 Layout of Main Entries

The best layout scheme depends on the needs of the system and the costs associated with supplying those needs. Although the actual cost associated with a ventilation system depends on many factors, the relative cost difference between two alternatives depends primarily on the number and type of ventilation structures needed to accomplish the ventilation tasks. For example, if a series of six main entries were used for supplying the intake and return needs of a mine, what system of layout would provide the lowest cost? Obviously, the least number of stoppings would be a single row equally dividing four intakes and four returns as shown in Fig. 7-14(a).

The simplicity of the example above is complicated when additional constraints are placed on the system. For instance, let us assume that the seam is gassy and that development occurs equally to either side of the main entries centerline. The single row of stoppings is still the least costly [Fig. 7-14(b)]; however, the gas requirements necessitate the use of "outboard" returns to protect the intake entries from methane contamination, as shown in Fig. 7-14(c). Outboard returns also provide greater flexibility in isolating

Figure 7-14 Various layout strategies designed to minimize cost and maximize function: (a) least-cost ventilation layout; (b) addition of submain service; (c) spreading returns.

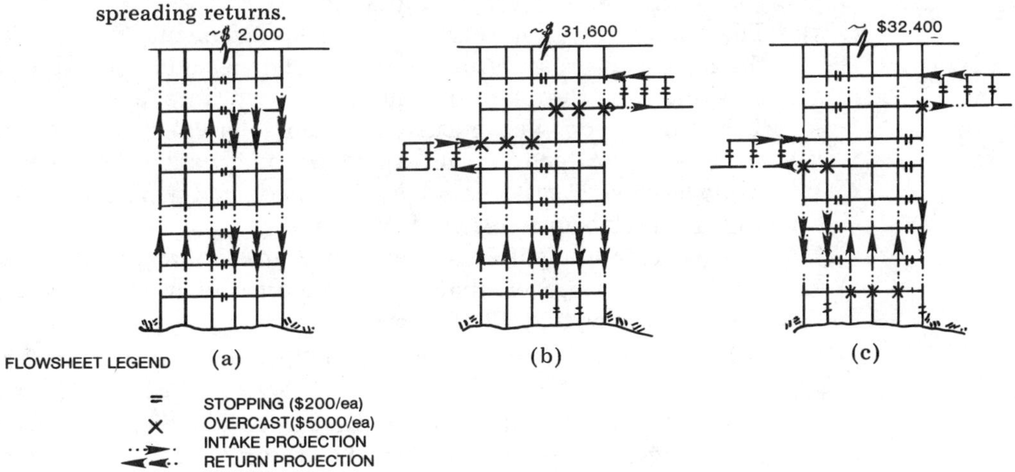

or bisecting each side. Because of these added benefits, the outboard return layout is more valuable and therefore the most common intake/return layout used underground.

7.6 VENTILATION LAYOUT REQUIREMENTS

7.6.1 Federal Requirements

The 1969 Health and Safety Act[3] specifically mandates that "a ventilation system and methane and dust control plan" be submitted to government agencies which meets or contains the following criteria:

1. An accurate, up-to-date map of the coal mine at a scale of not more than 500 ft to the inch and supporting data that shall include:
   (a) The limits of the mine property, including all known underground workings bordering the mine above, below, and on adjacent properties.
   (b) The location of all oil and gas wells.
   (c) The location of all surface-installed fans, type of fan, manufacturer's name, size of fan, and complete current operating specifications.
   (d) Location of all surface mine openings.
   (e) Any abnormal conditions or reservations, such as faults, which may affect mine ventilation system design.
   (f) Projections of anticipated mine development for at least 1 year.
   (g) Direction and volume of air at each surface mine opening.
   (h) All underground workings, with the active working sections delineated.
   (i) Location of all stoppings, overcasts, undercasts, regulators, seals, and ventilating and access doors.
   (j) The volume of air entering and leaving each split, passing through the last open crosscut in each set of entries and rooms, at the intake end of each pillar line, and at each working face.
   (k) The velocity of the air current when such velocity is required at each working face, in all conveyor belt haulage entries, where trolley haulage systems are maintained, and where trolley wires and trolley feeder wires are installed.
   (l) Average entry height in conveyor belt and trolley haulage systems.
   (m) Areas that have been abandoned and areas from which pillars have been wholly or partially removed.
2. A ventilation system and methane and dust control plan which shall show in detail:
   (a) The methane and dust control practices along all haulageways and travelways, at all transfer points, at underground crushers and

dumps, in all active working places, and in such other areas as may be required by the district manager.

(b) All face ventilation systems used and drawings illustrating system use, anticipated air quantities and velocities in the work place, and the use and application of the system under all anticipated mining conditions.

(c) When auxiliary face ventilation systems are used, a detailed plan of the system, including equipment specifications, fan capacity, method of application, and methods to be used for maintaining continuous airflow to the working face in the event of auxiliary equipment failure.

(d) The bleeder entry system, when such a system is used, including:

(1) Methods for maintaining the bleeder entries free of obstructions such as roof falls and standing water.

(2) Ventilating devices, such as regulators, stoppings, and bleeder connectors, used to control air movement through the GOB bleeder entries.

7.6.2 Additional Guidelines in the Preparation of Layouts

In addition to the government regulations, layouts should correspond to the following guidelines:

1. The ventilation system and the mining methods should be designed simultaneously, as opposed to the layout of entry projections and the subsequent overlay of the ventilation system needed to supply the required air quality and quantity.
2. Approximately every five crosscuts, stoppings should be equipped with access doors to facilitate movement of personnel between the intakes and returns.
3. Because of their expense ($4800+ in 1978), overcasts should be carefully planned to prevent the need to move them shortly after their installation.
4. When possible, use intake for haulageways.
5. Exhausting mine fans should use all available surface openings as intake entries, thus limiting the total mine resistance and subsequent energy losses.
6. Split the airflow whenever needed to lower losses.
7. Unite splits for discharge as near the surface as possible (exhausting system) to avoid higher energy losses.
8. Minimize the total distance traveled to avoid high losses.
9. Balance the resistance offered by the intake and return air courses (usually, the number of intakes equals the number of returns).

10. Minimize obstructions to lower head losses.
11. As much as possible, avoid leakage by ensuring high-quality construction of ventilation devices. Make periodic ventilation surveys for these purposes.
12. Avoid recirculation of air and dead spots.
13. Minimize the total number of ventilation devices.
14. Each split must be regulated; avoid placing regulators in the intake or close to the face, as haulage would be adversely affected.
15. All active workings should be on separate splits of air.
16. All pillar sections should be on separate splits of air from development sections.
17. Inactive workings should be ventilated by return from active sections when possible.
18. Ventilation development (e.g., the mining of bleeder entries for the purpose of methane drainage) should be completed prior to pillar recovery.
19. Use the largest entry cross section as possible to achieve the greatest reduction of total head loss.
20. Position slope or drift entries at least 50 ft apart and shaft entries 300 ft apart (greater concentration of flow) to avoid recirculation of return and intake air.

7.6.3 Sequence of Layouts

Two types of ventilation layouts are used in mining operations: (1) the projection or planning layout, nd (2) the existing excavation layout. The *projection layout* is designed in conjunction with the design and layout of projected entries. The *existing excavation layout* is the scaled representation of mine entries and the locations and descriptions of the ventilation courses, devices, and structures relative to those entries. Ideally, the two will coincide.

For purposes of instruction, this text presents the necessary sequences for completing a projected ventilation layout based on the assumption that the initial excavation projections have been completed. Figure 7-15 represents a completed ventilation layout (*Note:* This assumption is imposed only as a learning strategem. Theoretically, the design of ventilation layout, as well as all other operation designs, is undertaken in conjunction with the general mine design and projection.)

In general, the ventilation sequence proceeds as follows:

1. Ventilate the mains by placing the main fan and showing the directions and symbols for intake and return air courses. (Do not place the symbols for ventilation devices or structures until all air courses have been delineated.)
2. Ventilate the submains, beginning with the submain farthest away and

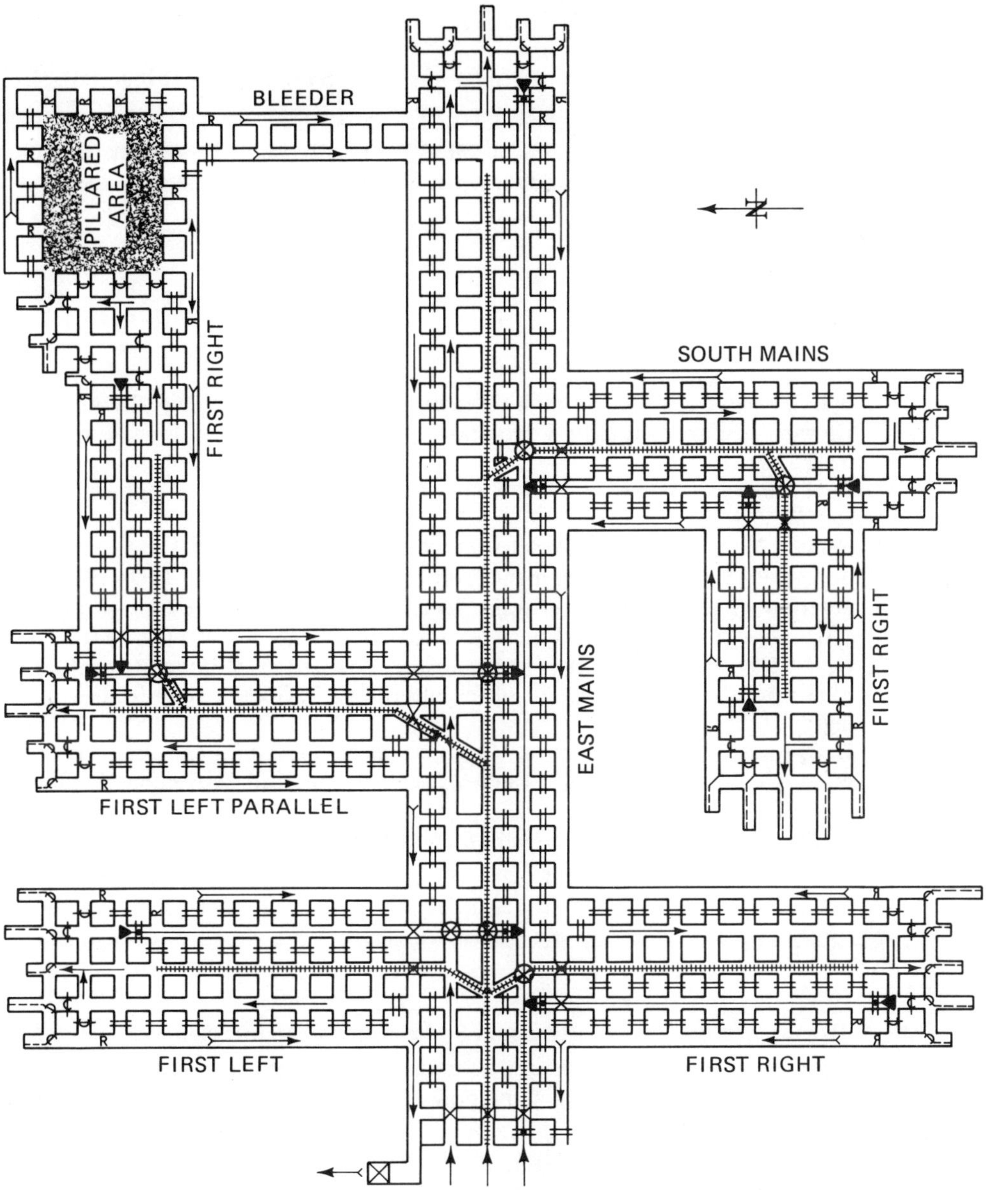

Figure 7-15 Completed mine ventilation map (WVU Mine Extension Service base map).

progressively work back to the outcrop, shaft, or main portal, Again, just delineate the return and intake air courses.

3. Ventilate the panel entries and development sections, showing the needed intake and return airways.
4. Ventilate the last crosscut and face for each section. Place the face structures in either exhausting or blowing configurations.
5. Working from the face toward the panel entries, butt entries, and so

forth, show the approximate symbols to maintain the ventilation directions and segregation shown.

6. Calculate needed flow rates, air velocities, resistances, and so forth, placing those specifications on the layout as needed. Side-note the remaining calculations.

REFERENCES

1. Syd S. Peng, *Coal Mine Ground Control*, John Wiley & Sons, Inc., New York, 1978.
2. Howard L. Hartman, *Mine Ventilation and Air Conditioning*, John Wiley & Sons, New York, 1961.
3. Coal Mine Safety Law of 1971, State of West Virginia, W. Va. Dept. of Mines, Charleston, W.Va., 1971.
4. Federal Mine Safety and Health Act of 1977, Public Law 91-173 as amended by Public Law 95-164, U.S. Government Printing Office, Washington, D.C., 1977.
5. Maurice Deul and Ann G. Kim, "Methane Drainage—An Update," presentation at 1978 Coal Convention, American Mining Congress, St. Louis, Mo., 1978.
6. G. E. McElroy, *Engineering Factors in the Ventilation of Metal Mines*, U.S. Bureau of Mines Bulletin 385, U.S. Dept. of the Interior, U.S. Government Printing Office, Washington, D.C., 1935.
7. D. S. Kingery, *Introduction to Mine Ventilating Principles and Practices*, U.S. Bureau of Mines Bulletin 589, U.S. Dept. of the Interior, U.S. Government Printing Office, Washington, D.C., 1960.
8. William C. Osborne, *Fans*, Pergamon Press, Oxford, 1966.

PROBLEMS

1. What is the methane percentage if 60 ft^3 of methane is mixed with 650 ft^3 of air?
2. How much air would be needed to reduce the percentage in Problem 1 to 1%?
3. The flow rate in a return is 60,000, with a methane content of 1.7%. How much methane gas is present?
4. The carbon dioxide content at the face is 0.2%, with a flow rate of 10,720 ft^3/min. What must be done, and in what amount, to correct this?
5. A working section liberates 250 ft^3/min of methane. Calculate the air flow rate needed to provide 0.25%.
6. The volume of an abandoned portion of a mine is 300,000 ft^3. The pressure on day 1 is 14.79 psia (30.11 in. Hg). On day 2 the pressure is 14.2 psia (28.91 in. Hg). What will be the total effect to the ventilation system as a result of this change? What assumption is needed to calculate this effect?
7. From Problem 6, what is the effect if several days later the pressure increases to 31.2 in. Hg?

8. Assume in Problem 6 that the pressure changed in a linear relationship from day 1 to day 2. Write the continuity equation for this.
9. Assume for Fig. 7-1 that the flow rate Q_1 was equal to 12,200 cfm and that A_1 = 121 ft^2, A_2 = 115 ft^2, A_3 = 112 ft^2, and A_4= 124 ft^2. Find the velocities at each point.
10. What is the total dynamic head of flow if the static pressure is 14.9 psia, the flow rate is 9200 ft^3/min, and the area is 105 ft^2? (Assume 0.075 lb/ft^2.)
11. If the point in Problem 10 were point 1 and another point in series with it had the characteristics P_2 = 14.7 psia and A = 110 ft^2, determine the velocity at 2 and the head loss from 1 to 2.
12. Calculate the difference in head loss between one entry (A = 100 ft^2, 2000 ft long, supplying 15,000 ft^3/min) and two identical entries supplying the same quantity. What happens if four entries are used?
13. Plot or sketch the effects to head loss due to independently changing Q, L, and K.
14. Research the literature and determine the range through which Q, O, K, and A vary between mines. (For instance, Q should provide a velocity range of 60 to 500 ft/sec.) What is the corresponding range of head loss effects?
15. Assume the 600 hp is required to supply a mine with no leakage. What is the required hp if leakage were 30%? 60%?
16. Determine the total mine resistance if 250,000 ft^3/min produced a 1.72-in. loss.
17. Determine the length of entries 2 and 3 if L_1 = 200 ft, Q_1 = 9100 ft^3/min, Q_2 = 9600 ft^3/min and Q_3 = 12,000 ft^3/min.
18. Find the flow rate in splits 1, 2, 3, and 4 if the entry height is 4 ft and the entry width is 16 ft; if L_1 = 2200 ft, L_2 = 3200 ft, L_3 = 1900 ft, and L_4 = 3000 ft, Q_T = 190,000 ft^3/min.
19. Determine the head loss for six entries (three intake and three return) 2100 ft in length, supplying 19,000 ft^3/min when the entries are straight, moderately obstructed, of height 3.5 ft, and of width 16 ft. What is the effect if a 30% leakage occurs halfway? What is the effect if that same 30% were to occur uniformly at each stopping? (The stoppings are on 100-ft centers.)

8

Haulage System Layout

8.0 INTRODUCTION

Great changes have occurred in the methods used for mine haulage. Originally, haulage depended upon human beings carrying, pulling, or pushing containers filled with the mined ore. Later, animals were used to supply haulage power. Various animals have been employed, including dogs, ponies, mules, oxen, and horses. As the technology of power conversion advanced, steam engines and electrical motors replaced the less efficient animal systems. Today various devices are being used to transport workers, supplies, and materials, including belt conveyors, shuttle cars, mine locomotives and mine cars, and fluid entrainment pipelines. The most appropriate selection and layout of these methods forms an important aspect of design.

Underground haulage system layout and operation requirements differ significantly from those found on the surface, for a variety of reasons:

1. The route layout of the haulage system is less flexible underground, being dependent on nonhaulage factors such as optimum ground control, property bisection philosophy, existing surface facilities, hydrology, property ownership, and seam extent.
2. The haulage system must conform to the extraction operation; thus it continually advances and retreats, requiring both permanent and temporary forms of haulage.
3. The ever-present danger of ground movement (e.g., roof falls) threatens the continuity of the haulage service.

4. The ground stability directly affects the width of an entry. Poor stability decreases entry size, thereby limiting the track gage and ultimately the capacity or layout of the system.
5. The bottom is usually shale or clay, both becoming soft (exhibit plastic behavior) and gummy when wet and exposed to high bearing loads.
6. The mine environment includes highly corrosive water, erosive particles, and explosive dusts and gases, all adversely affecting system reliability.
7. The length of the system and duty cycle (frequency of trips) depend on location within the mine.
8. The underground haulage system must be capable of transporting workers, materials, supplies, and equipment upon the same system.

These differences, together with haulage principles and traditional practices, influence the eventual methods used to achieve haulage. The purpose of Section 8.1 is to introduce the common functions of the haulage system. Section 8.2 reviews the various mine layout characteristics as they influence the different methods of haulage. Section 8.3 presents some of the principles and graphical techniques used in track haulage design. Section 8.4 discusses conveyor belt haulage. Section 8.5 reviews the fluid entrainment systems of haulage.

8.1 FUNCTIONAL ANALYSIS

The first task needed to produce any layout is to understand the functions performed by the system. The haulage system underground must be designed to achieve the following functions:

1. Provide transportation
2. Ensure safety
3. Maximize reliability
4. Improve maintainability
5. Optimize value

8.1.1 Provide Transportation

The primary function of the haulage system is to provide necessary transportation of the following:

1. Mined materials from the face to the surface.
2. Personnel from the surface to locations within the mine and back to the surface.
3. Supplies from the surface to locations within the mine.

4. Equipment from the surface to locations within the mine and occasionally back to the surface.
5. Waste material (gob) from the face to locations within the mine or on the surface.

As used here, the haulage system is comprised of any equipment used to transport or move any of the items listed above over any distance. The major methods and types of transport equipment presently used to achieve these purposes are shown in Fig. 8-1.

For the purposes of overview we can divide the haulage system into four categories (Fig. 8-2): (1) face or primary haulage, (2) intermediate or

Figure 8-1 Major methods of haulage and transportation used in mining.

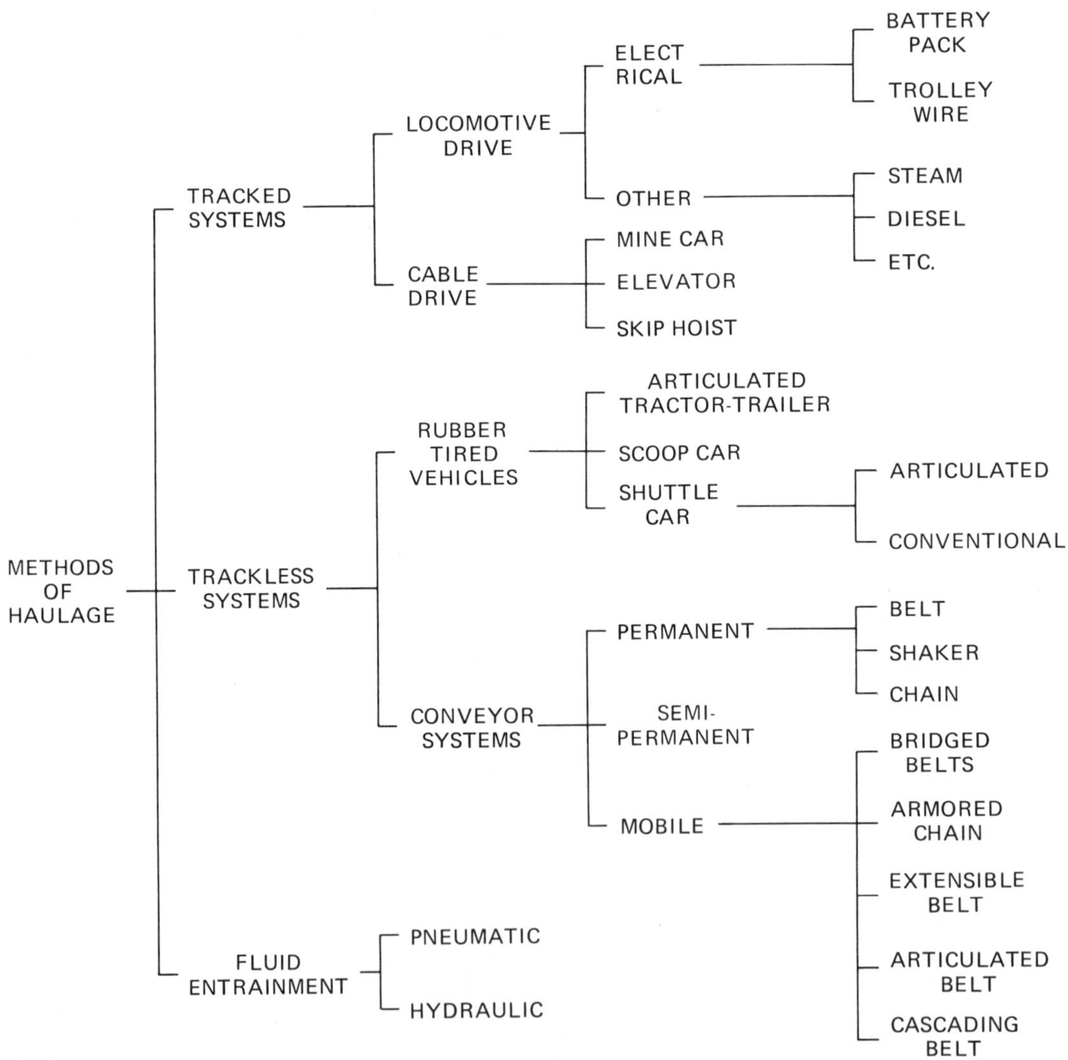

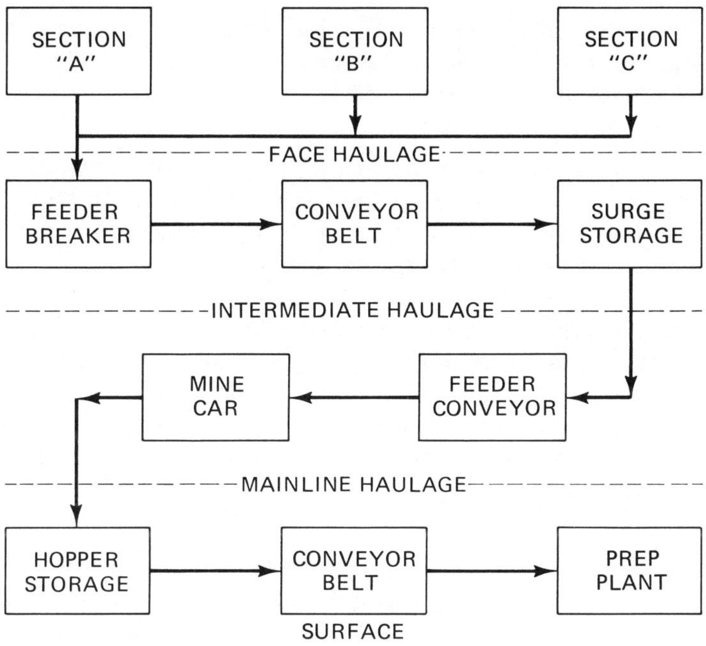

(a)

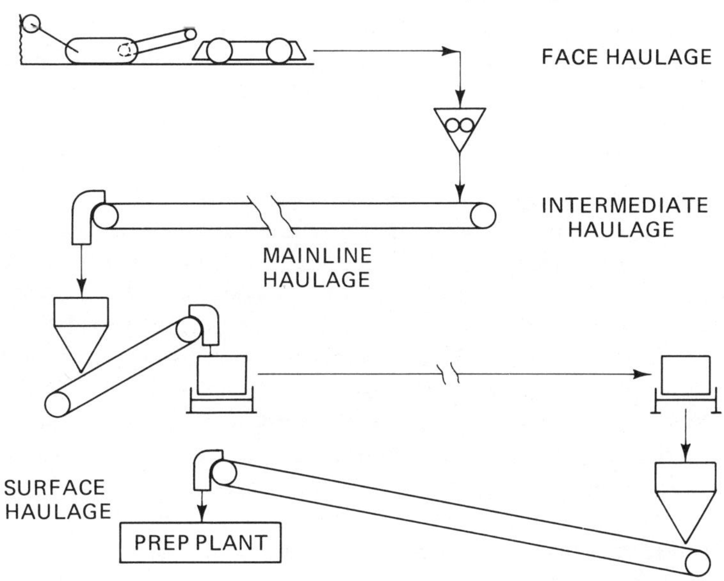

(b)

Figure 8-2 (a) Block flowchart showing the coal haulage pattern. (b) Symbolic flow chart showing the coal haulage flow.

secondary haulage, (3) mainline haulage, and (4) hoisting. (For further simplification, the following discussion centers on the transport of coal, leaving the other items for Section 8.2)

Face haulage. *Face haulage* is that phase of the total system which transports the mineral and waste from the face to either the secondary or mainline phase. The methods used for face haulage are usually characterized by frequent but short trips from the face to an unloading station. Because of the layout of the face, most equipment requires a high degree of maneuverability. Also, since the face is usually the most environmentally unstable area with the greatest bottom variability, highest dust and gas content, and least stable ground, the equipment used must have the greatest speed and power control and have the most elaborate safety provisions. These requirements are most often met by rubber-tired vehicles and by some of the mobile face conveyor systems.

Secondary haulage. *Secondary haulage* is that intermediate stage of the haulage system which is used to transport the output from several adjacent working sections to mainline haulage points. Secondary haulage phases are included when the travel time from the face to the mainline locations lowers the productivity of the sections. Secondary haulage is characterized by infrequent trips, usually less than 1 mile in length. The secondary haulage equipment is more permanent, requires less speed and power control, and does not require the stringent safety provisions needed for face equipment. Typically, either conveyor belts or locomotive and mine cars are used as secondary haulage.

Mainline haulage. *Mainline haulage* is usually the final stage in any haulage system, terminating on the surface at some centralized storage point or at the coal preparation facility near the mine property. Mainline haulage is characterized by scheduled trips of length 8 miles or more. Mainlines methods are permanent methods which must be highly reliable and adaptable to supplies, materials, and personnel transport. Conveyor belts and mine locomotive systems have both been used, with the tracked system being the traditional favorite. The longer the trip, the more favorable will be the use of tracked haulage. The favorability of track over conveyor belt usually has an economic breakpoint between 4 and 5 miles.[1]

Mainline haulage systems that provide only material transport are fluid entrainment systems using air, water, or oil as the transport fluid. Because of the "containment of the system" within a pipe, many of the disadvantages of underground haulage systems mentioned in Section 8.0 disappear. Another advantage of the fluid entrainment system is that it does not require transfer to a hoisting system when used in a shaft mine.

Hoisting. Shaft mines and steeply pitching slope mines require an additional system for connecting the surface transportation system with the subsurface mainline system. Track systems are usually limited to long grades of less than $4\frac{1}{2}\%$ (although some special rack and pinion designs have been tested for grades of up to 25%[2]). Conveyor belts are limited to less than 30% grade, with the limit being the slippage of the material on the belt. Hoisting is accomplished by either directly lifting the mine car to the surface in cage elevators, or, more commonly, by using a large-capacity skip (bucketlike container) filled from mine cars at a central underground dumping station and lifted to the surface, where it is automatically dumped.

8.1.2 Ensure Safety

Haulage system safety is first concerned with the safety of the personnel and second with property safety. Each of the methods of haulage has its own set of safety-related problems. Only three hazards are commonly shared between the methods: (1) overspeed, (2) confined spaces, and (3) probability of explosion. Like other functional systems, the safety of the haulage system is controlled using the principles and schemes presented in Chapter 5. Some of the major safety schemes include the isolation scheme, fail-safe scheme, alerted contact scheme, and environmental exclusion.

Isolation scheme. Isolation schemes attempt to maintain a physical space between the system and personnel or other functional systems. In the haulage system this is achieved in a number of ways. For instance, track haulage requires a minimum of 24 in. of clearance on the traffic side of the haulage road and a minimum of 12 in. on the opposite or "tight" side. Additionally, "shelter holes" at least 5 ft deep and 4 ft wide must be provided every 105 ft. For belt conveyors transporting personnel, a minimum vertical clearance of 18 in. is required, with 36 in. of clearance provided on both sides of the belt.

Fail-safe scheme. With the exception of belt conveyors and fluid entrainment methods, human operators control both the speed and direction of the haulage vehicles. If the operator were incapacitated for one reason or another, the system is designed to fail-safe: that is, to stop. For belt and fluid systems operation, parameters are monitored and the system automatically stops when valves exceed certain predetermined limits. Further, fail-safe schemes are exemplified by hoisting equipment that must incorporate automatic brakes and catches to stop the cage in the event of cable failure.

Alerted contact scheme. The most organized safety scheme used with the haulage system is that of alerted contact. Alerted contact includes audible warnings given by mobile equipment operators when passing through check curtains, safety lights installed on the track to indicate "in-use" status, and a central dispatcher to control the scheduling of track haulage trips.

Environmental exclusion scheme. All haulage equipment that is used in the last open crosscut must be rated "permissible" by the U.S. Bureau of Mines. Permissibility is described in more detail under electrical safety in Section 10.1.2.

These schemes combine with safety audits, adequate safety management, automation, and system layout to ensure the safety of personnel, equipment, and property. However, since most haulage methods are controlled directly by an operator, the most significant factor in haulage system safety is control over the operator's behavior.

8.1.3 Maximize Reliability

Reliability of the haulage system depends on either eliminating failures or providing a layout that can be adjusted to minimize the effect of failures. Downtime of the haulage system directly influences the profitability of the mine. Like most underground functional systems, the severity of the environment, the high load demands, and the complexity of the system all contribute to an estimated reliability index ranging from 1 to 10% of an equivalent surface haulage system.

Various failure modes were presented in Chapter 5. All of these adversely limit the performance of the haulage system. As a review, the various failures modes are:

Defect failure. Defect failures occur as a result of poor design, improper material selection, or poor quality control. These failures can be lessened by adequate inspection of equipment before use, selection of a reliable manufacturer, and use of standard proven designs.

Aging failure. Aging failures are linked to environmental effects over an extended period of time. The most severe aging failures are the rotting of crossties, corrosion of metal components, and ozone embrittlement of rubber tires and conveyor belt materials.

Degradation failure. The most frequent failure mode of the haulage system is degradation, particularly mechanical wear. High loads, steep grades, infrequent lubrication, and improper adjustment all contribute to higher-than-normal wear. Erosion and abrasion resulting from an excessively dusty environment are also degradation problems experienced by haulage systems. Frequent lubrication, derating speed, and selection of particle exclusion designs usually control erosion and abrasion.

Improper use, abuse, or maintenance. Many components of the haulage system that fail and are sent to rebuilding shops require bearing replacement. In 85% of these cases the bearing failure resulted from a lack of routine lubrication. Additional "abuse" failures include cut power cables, improperly aligned conveyor belts, derailments due to excessive speed, and deliberate vandalism of equipment.

Accumulative macroscopic failures. Mechanically overloading the system causes macroscopic failures of the haulage system. Such failures are usually predictable and avoidable.

Spillover failure. Spillover failures are caused by the dependence of the haulage system on another system or as a result of the proximal relationship of the haulage system to another system that experiences a failure. An example of the first type of spillover failure is a failure of the haulage system resulting from a failure within the electrical system. The electrical system failures thus "spill over" to affect the haulage system. An example of the latter is a roof fall that blocks track haulage. The neighboring relationship of the ground control system to the haulage system causes the effects of the roof failure to spill over, thereby interrupting the haulage system.

These specific failures and many others contribute to lower reliability values than may be desirable. To optimize the reliability of any haulage system, the following layout principles should be observed:

1. Simplify the design.
2. Select reliable components.
3. Use redundant designs.
4. Employ overdesign and derating tactics.
5. Control the operating environment.
6. Use human factors engineering.
7. Install equipment properly.
8. Provide proper training.
9. Inspect and maintain equipment.

Each of these reliability improvement measures depends on the individual method being analyzed. Thus the expansion of each measure is reserved for later sections.

8.1.4 Maintainability

System maintainability is measured by:

1. The time and probability of restoring service to a failed system
2. The ability to advance, retreat, or modify the haulage system
3. The ability and requirements for inspecting and maintaining the haulage system

Service restoration. Major failures resulting in haulage system downtime include roof falls, motor failures, mine car derailment, and electrical power outages. The probability of restoring haulage service in a given period of time depends on the severity of the failure, the accessibility of the failure, and the nature of the repairs. With many haulage system components, such as failed motors, direct replacement has proved to be more cost effective than trying to repair the equipment under the normally adverse mine conditions. Other factors affecting service restoration that were discussed in Chapter 5 apply to the haulage system, including simplicity of the system, control of environmental conditions, and the selection of reliable equipment.

Modifications. The ease of advancement, retreat, or change to the haulage system depends on the methods of haulage being used, the conditions of the mine, and the region being modified. The two major methods most frequently contrasted are track and belt systems. Belt systems are limited in application by their requirement of straight alignment, whereas track systems are limited in bottom competency and grades less than $4\frac{1}{2}\%$. Given straight and level entries with regular and competent bottom, track haulage is probably the *simplest* to advance, although it may not necessarily be the most *cost effective.* The requirements and problems of advancing each are discussed in later sections.

Inspection and maintenance. Adequate inspection and maintenance are perhaps the most overlooked activities of the haulage system. Most designers of haulage equipment recognize this and overdesign the components accordingly. The costs of this neglect are difficult to estimate because they depend on the size and condition of the mine as well as the type of haulage system. Specific inspection and maintenance guidelines are developed in later sections.

Value. The most valuable or cost-effective haulage system is the one that will provide the aforementioned functions at the least cost. Each system has operating advantages and disadvantages which determine their suitability for a particular application and, of course, their corresponding cost. These traits will be discussed in detail for each specific class of transport. Assuming, however, that all factors are roughly equal, the value of any method of haulage will be proportional to the tons of product transported divided by the number of miles. Obviously, to increase the value you either increase the quantity of product transported or minimize the number of miles it must be transported.

The number of tons that can be transported is linked to mining conditions, mining methods, and economic conditions, as well as haulage factors such as the size of mine car and locomotive or the width and speed of the belt. Note that these latter haulage factors increase the initial capital outlay needed and further complicate the value analysis.

Portal location effects. The number of haulage miles is a simple calculation that depends only on the location of the faces relative to the existing portal. Assuming that the coal property is uniform relative to seam thickness, strike and dip, roof conditions, and proximate analysis, the minimum number of total transport miles is achieved by bisecting the coal property (equal-area balancing) and locating the existing portal in the geometric center of the property.

Several mathematical and graphical methods exist for locating the geometric center. Graphical methods are simplest and quickest for common geometric shapes, in which the coal property can be approximated as a combination of circles, ellipses, rectangles, squares, or triangles. The geometric center of squares and rectangles is the intersection of the principal diagonals. The center of an ellipse is the intersection of the principal axis. For a circle the intersection of any two diameters delineates the geometric center. The geometric center of the circle or the ellipse can also be found by first enscribing the property outline with a square or rectangle, then drawing the principal diagonals. The geometric center of a triangle is the intersection of the lines that bisect each angle. These lines also bisect each opposite side. Thus the geometric center is found by bisecting each side with your divider and connecting these points with the vertex of the opposite angle. These techniques are shown in Fig. 8-3. Note that moderately irregular shapes can

Figure 8-3 Graphical methods of determining the geometric center of common shapes.

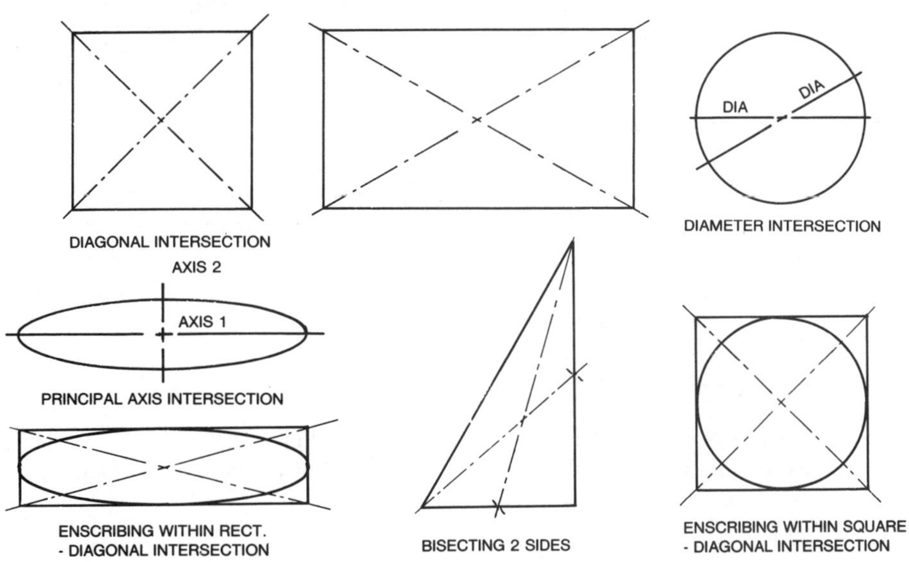

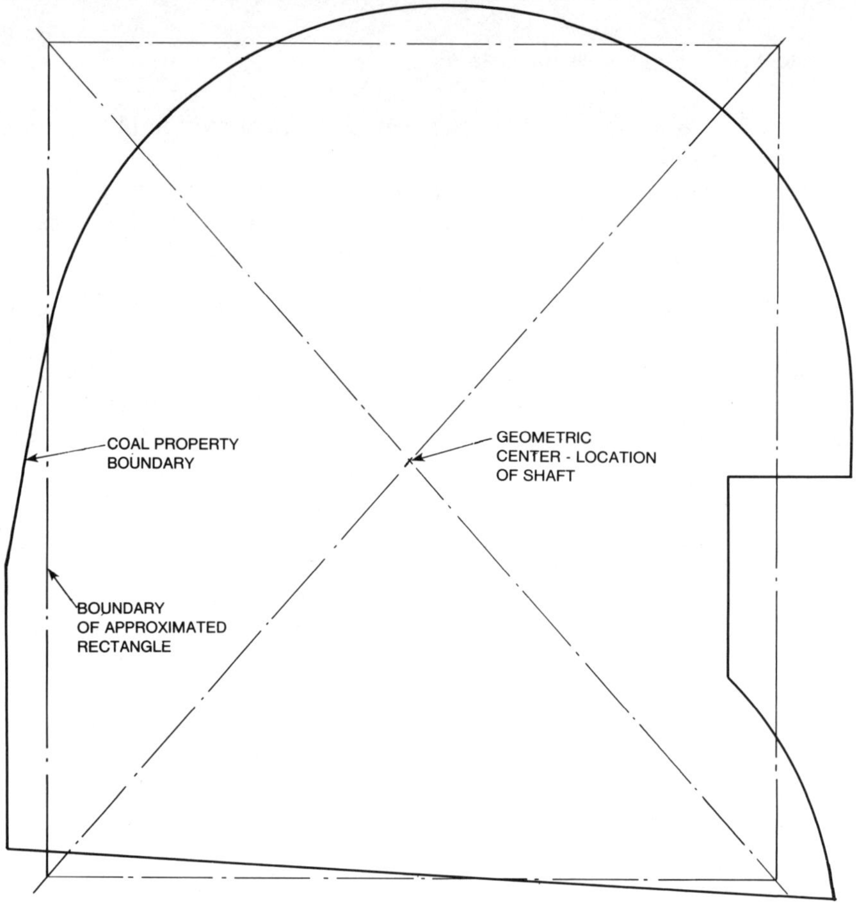

Figure 8-4 Using an approximated rectangular boundary to find the geometric center.

be approximated by these standard shapes and the geometric center easily found as long as the amount of property "added" by the approximation is offset by property excluded from the shape, as shown in Fig. 8-4.

8.2 MINE LAYOUT CHARACTERISTICS

Various mine characteristics influence the layout of the haulage system and the eventual selection of the best transportation mode (see Fig. 8-1). Some of these factors include (1) the mining method, (2) the seam and environmental factors, (3) the location or transport duty cycle, and (4) other less categorized factors.

8.2.1 Type of Mining System

The types of mining systems include (1) conventional, (2) continuous, (3) longwall, and (4) shortwall systems. The choice of the optimum transportation system to match the type of mining system is predicated upon the *principle of balance*, which can be stated as follows:

> No method or intermediate phase of the transportation system used should result in an accumulation of material at any point throughout the total mining operation which causes the retardation or stoppage of any other phase.

Theoretically, the capacity per unit time of the face transportation phase should be equal to the rate of the extraction process. Similarly, the capacity of any secondary transportation system should equal the combined rate of the contributing face systems.

The capacity of the conventional mining method can be adequately met by a face haulage system using shuttle cars. Both the mining method and the haulage method operate as "batch processes" and are thus well matched. As the capacity of extraction increases and as the method of extraction becomes more continuous, say for longwall mining, the method of haulage needs to change to the use of conveyors. Basically, the principle of balance holds that the total output of a mine can be only as efficient as the capacity of the poorest phase of the transportation system.

8.2.2 Seam and Environmental Characteristics

Major seam and environmental characteristics affecting the selection of the best transportation system are:

1. Height of entry
2. Orientation of strike and magnitude of dip
3. Regularity of the seam bottom
4. Density of oil and gas wells
5. Nature of the bottom
6. Nature of the top

(Since belt and track haulage are the two most common methods of haulage away from the immediate face area, the following discussion is limited to these two alternatives.)

Height of entry. Height of entry is usually governed by the thickness of the seam. The total height of a tracked system includes the track ballast (usually a minimum of 6 in. for competent bottoms), tie extensions, height of rail (depends on weight of rail), the total height of the mine car or locomotive, and the head clearance needed for safe trolley wire tracking. Conveyor belts, on the other hand, require only the roller clearances and clearance for the coal. Although "low-coal" mine cars are produced and used in many mines, the recent tendency is to use the more compact, permanent belt systems for low coal. Future trends would dictate that as the higher (more accessible) coal seams are mined out, the height of mine headings will continue to decrease, thereby necessitating the use of beltline haulage.

Strike and dip. The heading of the strike and the orientation and magnitude of the corresponding dip play a direct role in the grade of the mainline and secondary haulage entries. The optimum grade required to maintain adequate drainage while minimizing haulage operation and maintenance cost is between 0.75 and 2.0% grade. This ideal grade may not be possible due to (1) the necessity of locating mainline portals in the most stable rock, (2) the density of oil and gas wells, or (3) the method of bisecting the property. Thus the actual haulage grade may vary from a minimum of 0%, where haulage is along the strike, to a maximum grade equal to the dip angle. For tracked haulage the maximum grade that does not adversely affect efficiency is $4\frac{1}{2}$%. Beltline haulage has a much wider range of operational grade, with 23% an effective limit.[3]

Grade also affects the best layout position for the portal. Ideally, loaded haulage trips would be downhill, and the return of empties would be uphill. The value, or ton/mile cost, must be considered. If all haulage were downhill, the portal would be located at the lowest seam elevation and the mining distance would continually increase, thus decreasing the value. The optimum portal location is no longer the geometric center of the property but is rather some location between the center and the lowest point. The exact point requires a weighted balancing of the cost of uphill transport over short distances with the downhill transport over longer distances.

Bottom irregularity. To maintain smooth, efficient track haulage, the bottom must be smooth and regular. This requirement necessitates cutting the high spots (brushing) and filling in the depressions. Brushing is an expensive, time-consuming operation; thus if much brushing is required, the best haulage alternative might be the use of beltline haulage.

Density of oil and gas wells. The location and density of oil and gas wells will dictate the location and directions of headings. Mine cars, because of their articulated nature, are able to turn through small radii, whereas most belt systems must maintain perfectly straight alignments. Thus, as the density of wells increases, the use of tracked haulage must also increase.

Nature of the bottom. The nature of the bottom includes the reaction of the bottom to tectonic stresses (heaving or swelling), reaction to haulage loads (local bearing stress), amount and sources of water (drainage problems), and so forth. As these problems increase in severity, the advantages of tracked haulage decrease and those of rope suspension beltline haulage increase.

Nature of the top. The stability of the immediate roof to the minor and major tectonic loads influences the total width of the entry. As the width decreases to account for an unstable top, the size of mine car, the field side clearances between the mine cars and the rib, the possible turning radii, and the rail gage are all adversely affected. As the entry width decreases, the production capacity of track systems decreases. However, the productivity of the belt system is only partially affected since the capacity depends on the belt width, and speed is independent of entry width. Thus if entry width must decrease due to an adverse top, the acceptability of beltline methods increase.

8.2.3 Transport Duty Cycle

The transport duty cycle, which is defined as the length and frequency of travel, is dictated by location within the mine. This factor was developed in Section 8.1.1 by dividing the mine into production regions, including (1) face or primary haulage, (2) intermediate or secondary haulage, (3) mainline haulage, and (4) hoisting. To review briefly, face haulage was characterized as frequent but short trips from face loading to a surge or unloading station. Secondary haulage was characterized by less frequent trips, less than 1 mile in length. Mainline haulage, the final stage in any haulage system terminating either on the surface or at hoisting tipples, was characterized by scheduled trips up to 8 miles or more.

8.2.4 Other Factors

Certain other, less well categorized factors influence the optimum selection of the most appropriate haulage system. These include personnel requirements, ventilation ease, safety, maintainability, initial cost, generation or propagation of dust, ability to automate, and system reliability. These factors are discussed in later sections.

The categories listed above are summarized in the selection matrix shown in Fig. 8-5. The actual procedures for the selection of the optimum system of transportation are governed by principles of operations research and modeling which are outside the scope of this presentation.

8.3 TRACK HAULAGE LAYOUT

Thus far, certain general guidelines for track use and layout have been presented. These guidelines have included discussion of grade, radius of curvatures, regularity of bottom, and necessary clearances. This section presents the information needed to produce engineering layouts of track positioning for rail curves and turnouts.

METHOD OF HAULAGE	DUTY CYCLE AND CAPACITY	TRAVEL DISTANCE	ENTRY (SEAM) HEIGHT	LOCATION IN MINE	ENVIRONMENTAL FACTORS	COMMENTS
SHUTTLE CAR	FREQUENT: 2.0–1.2 TONS/MIN	200′–600′ FT	HIGH OR LOW COAL	IMMEDIATE FACE HAULAGE	WATER EFFECTS TRACTION GRADE LIMITED	MOST FLEXIBLE FACE SYSTEM
TRACTOR TRAILER (a) 2 UNITS (b) 4 UNITS	(a) 2.0–1.35 TONS/MIN (b) 2.0–1.5 TONS/MIN	(a) 500′–1500′ (b) 2250′–2800′	HIGH OR LOW COAL	SECTION HAULAGE	WATER EFFECTS TRACTION-GRADE LIMITED	—
MINE CAR	INTERMITTENT: >350 TONS/SHIFT	>5 MI (GENERALLY)	PRINCIPALLY HIGH COAL, LOW COAL LIMITS CAPACITY	INTERMEDIATE AND MAINLINE HAULAGE	COMPETENT BOTTOM, GRADE $<3.5\%$,	MAY REQUIRE EXTENSIVE BRUSHING, MOST COMMON
CONVEYOR BELT	CONTINUOUS: 1068–14400 CU FT/HR (100 F.P.M.)	<4 MI (GENERALLY)	HIGH OR LOW "PREFERRED" FOR LOW COAL	FACE INTERMEDIATE AND MAINLINE HAULAGE	STRAIGHT ENTRIES, 18° SLOPE, NO BOTTOM LIMITATIONS	INSTALLATION CRITICAL, LABOR INTENSIVE
FLUID TRANSPORT	CONTINUOUS DEPENDENT ON: PIPE SIZE PUMP CAPACITY INJECTION DENSITY	2 MI (GENERALLY)	HIGH OR LOW, REQUIRES LITTLE HEAD ROOM	INTERMEDIATE AND MAINLINE HAULAGE	NO OBVIOUS LIMITS	"NEW TECHNOLOGY" LABOR INTENSIVE

Figure 8-5 Some selection factors affecting the method of haulage.

The procedures used to position track underground follow those used to locate railroads above ground[4]

1. Reconnaissance
2. Preliminary survey
3. Location survey

Above ground, reconnaissance entails an examination of the topography of an area to determine the optimum location for a railroad line. Underground, reconnaissance is based on the treatment and analysis of information gathered from three sources: (1) *literature searches and historical data* available about the seam to be mined, (2) *surface exploration data*, and (3) *subsurface data*. The purpose of reconnaissance is to outline the most attractive routes for the haulage system, including possible portal locations and entry orientations.

The preliminary survey above ground is an extension of the data-gathering phase, intended to provide specific information about the possible alternatives to allow mapping and selection. The preliminary survey of underground haulage has the same goals, differing only in the indirect geophysical methods needed to gather the data.

The leeway available to the mine designer for selecting the "best" line is not as great as it is for the surface railroad designer, since the former must consider the stability of permanent portals, methods of property bisection, and so forth.

Certain guidelines can be outlined for the preliminary layout and positioning of mainline track:

1. Longer haulage distances, with grades greater than $1\frac{1}{2}\%$, should be avoided whenever possible.
2. High density of faults, low overburden, and other known roof anomalies should be avoided.
3. "Valley lines" and low spots in the topography of the coal stratum should be avoided because of the difficulty of drainage.
4. Regular and competent bottom should be sought.

Once the position of the mains is established, the mining engineer can make preliminary calculations to estimate the width of mainline entry and the stability of the bottom. These calculations in part establish the size of track, size of car, size of wheels, gage of track and tie, and construction schedules. These specifications in turn establish the minimum radius of curvature (depends on the length of the car wheel base, the diameter of the wheels, and the flange clearances), the curve grade, and the elevation of the outer rail (an easement needed to account for centrifugal forces).

8.3.1 Track Components

Various components combine to produce an effective track layout. Figure 8-6 shows a profile view of a track layout. The major components include ballast, ties, track, and trolley wire. The track ballast is the stone bed that elevates and secures the individual ties. Ideally, ballast is composed of limestone or other competent stone capable of withstanding the high loads and poor environmental conditions experienced underground. In reality, however, ballast is any easily attainable rock, such as gob, high in shale content, which rapidly degrades and must be replaced frequently. When possible, stone high in shale, sandstone, or coal content should be avoided because of the maintenance costs and safety hazards associated with their degradation.

The purposes of ballast are (1) to elevate the tie-track assembly to avoid water and subsequent corrosion, (2) to secure the ties both laterally and transversely, (3) to distribute and dump the bearing loads from the ties across the mine floor, (4) to smooth out the minor irregularities of the floor (major high spots should be brushed and dips filled to establish an initial uniformity), and (5) to provide the height for superelevation of the outer rail when curves and turnouts are needed.

Ties are used to position and secure the two track rails relative to one another and to provide structural support of the rail. Ties are constructed of wood or steel. Wooden ties are more traditional, providing lower initial cost and greater availability. Steel ties are rolled sections offering the maximum strength-to-weight ratio. Contrasted with wooden ties, steel ties provide minimum head room; do not rot (although when exposed to acidic mine water, they corrode rapidly); are uniform in quality; can be more rapidly ad-

Figure 8-6 Track profile.

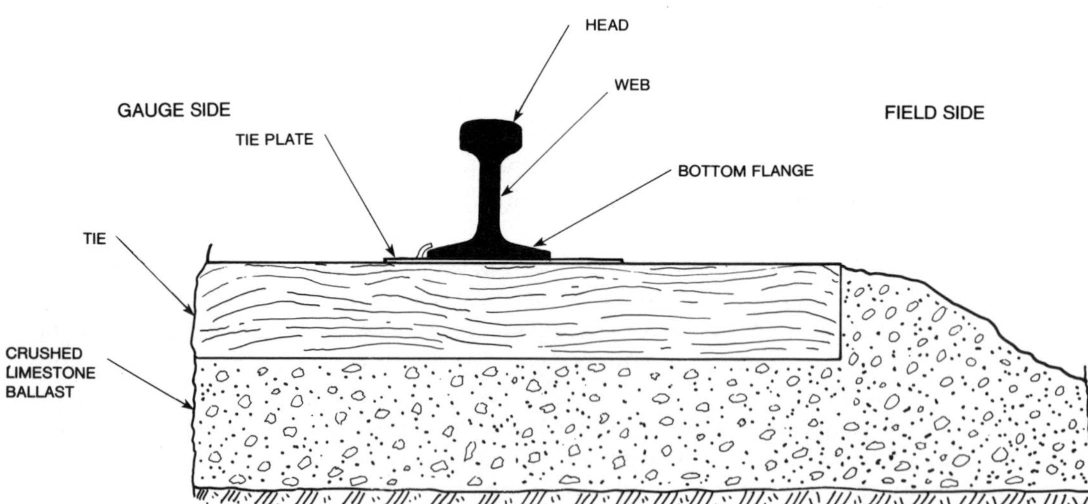

vanced, retreated, or permit rail replacement; are fireproof; do not require separate tie plates; are reusable; provide the strongest support; and cost the most.

The most variable component in track haulage is the rail. Rails vary in configuration and weight. The standard rail dimensions specified by the American Society of Civil Engineers are the most common rails used underground (Table 8-1). Since roughly the same profile characteristics are used for "light" as for "heavy" rail, the load-carrying structural characteristics depend on the unit weight of the rail. The resistance of the rail to bending (stiffness) is proportional to the square of the unit weight. The weight of the rail needed depends on (1) location, (2) tie spacing, and (3) load (weight per wheel). Recommended rail weights for coal mine haulage vary from 10 to 15 times the ton per wheel weight.[5,6] Because of this wide variability in size and cost, you should consult specific manufacturers who are familiar with your regional mining conditions and problems before selecting a specific rail weight.

The *gauge* of the track is defined as the perpendicular distance between a line $\frac{5}{8}$ in. below the top of the center of one rail head to a similar line on the opposite rail head. Standard railroad gauge is 4 ft $8\frac{1}{2}$ in. Mainline gage underground will vary from the AMC standard 42 to 30 in., with a major intermediate size of 36 in. The widest possible gage should be used since stability, capacity, and availability all increase as the gage size approaches the standard (42 in.) size. Gage on curves must increase to allow for wheel flange clearance and proper tracking. The increase is approximately $\frac{1}{8}$ in. per 2° for curves greater than 8°. (But the increase should not exceed 1 in.)[7] From these requirements the selection of the most appropriate gage depends on (1) the entry cross section and the necessary production capacity (both define the width of the mine car), and (2) the shortest radius of curvature. These requirements have been reduced to equation form as follows:[5,7]

$$G \mathrel{\dot{\geqslant}} \frac{1}{2} W \tag{8-1a}$$

$$G \mathrel{\dot{<}} \frac{R}{4} \tag{8-1b}$$

where G = track gage (in.)
W = width of mine car (in.)
R = shortest radius within mine (in.)

The trolley line is a hard copper extrusion suspended from the top by insulated tie anchors. The function of the trolley wire is to supply the necessary electrical power to the dc drive motors of the locomotive. The trolley line is hung at a 6-in. field-side offset from the return rail. Because of the high dc current density, a *feeder line* is commonly installed parallel to the trolley line to reduce electrical resistance and improve reliability.

TABLE 8-1 MAJOR RAIL STANDARDS FOR TRACK HAULAGE

Nominal Weight Per Yard	Type of Rail	DIMENSIONS IN INCHES								SECTION DESIGNATION		
		HT	BW	HW	W	HD	FD	BD	E			
12 lb.	ASCE	2	2	1	3/16	9/16	1 3/32	11/32	57/64	—	—	—
16 lb.	ASCE	2 3/8	2 3/8	1 11/64	7/32	41/64	1 23/64	3/8	1 1/16	—	—	—
20 lb.	ASCE	2 5/8	2 5/8	1 11/32	1/4	23/32	1 15/32	7/16	1 11/64	—	—	—
25 lb.	ASCE	2 3/4	2 3/4	1 1/2	19/64	25/32	1 31/64	31/64	1 15/64	—	—	—
30 lb.	ASCE	3 1/8	3 1/8	1 11/16	21/64	7/8	1 23/32	17/32	1 25/64	—	—	—
35 lb.	ASCE	3 5/16	3 5/16	1 3/4	23/64	61/64	1 25/32	37/64	1 15/32	—	—	—
40 lb.	ASCE	3 1/2	3 1/2	1 7/8	25/64	1 1/64	1 55/64	5/8	1 9/16	—	—	—
45 lb.	ASCE	3 11/16	3 11/16	2	27/64	1 1/16	1 31/32	21/32	1 41/64	—	—	—
50 lb.	ASCE	3 7/8	3 7/8	2 1/8	7/16	1 1/8	2 1/16	11/16	1 23/32	—	—	—
55 lb.	ASCE	4 1/16	4 1/16	2 1/4	15/32	1 11/64	2 11/64	23/32	1 103/128	—	—	—
60 lb.	ASCE	4 1/4	4 1/4	2 3/8	31/64	1 7/32	2 17/64	49/64	1 115/128	6040	60 AS	603
	MISC.	4 1/4	4 1/16	2 5/16	1/2	1 7/16	2 1/8	11/16	1 3/4	6051	—	—
65 lb.	ASCE	4 7/16	4 7/16	2 13/32	1/2	1 9/32	2 3/8	25/32	1 31/32	6540	65 AS	653
70 lb.	ASCE	4 5/8	4 5/8	2 7/16	33/64	1 11/32	2 15/32	13/16	2 3/64	7040	70 AS	701
75 lb.	ASCE	4 13/16	4 13/16	2 15/32	17/32	1 27/64	2 35/64	27/32	2 15/128	7540	75 AS	753
	MO. PAC.	4 3/4	4 3/4	2 9/16	9/16	1 7/16	2 15/32	27/32	2 5/64	7550	75 MP	—
	S. PAC.	4 15/16	4 7/16	2 7/16	33/64	1 3/8	2 5/8	15/16	2 1/4	7524	75 SP	757
80 lb.	ASCE	5	5	2 1/2	35/64	1 1/2	2 5/8	7/8	2 3/16	8040	80 AS	800
	DUDLEY	5 1/8	5	2 21/32	17/32	1 1/2	2 3/4	7/8	2 1/4	8022	80 DY	—
85 lb.	ASCE	5 3/16	5 3/16	2 9/16	9/16	1 35/64	2 3/4	57/64	2 17/64	8540	85 AS	851
	CAN. PAC.	5 1/8	5	2 1/2	9/16	1 7/16	2 11/16	1	2 11/32	8524	85 CP	—
	CB&Q	5 3/16	5 3/16	2 21/32	9/16	1 35/64	2 3/4	57/64	2 17/64	8543	85 CB	852
	MO. PAC.	5 7/32	5 1/4	2 15/32	75/128	1 3/4	2 39/64	55/64	2 21/128	8550	—	—
	PS	5 1/8	4 5/8	2 1/2	17/32	1 21/32	2 15/32	1	2 15/64	8531	85 PS	—
	PRR	5	5	2 9/16	17/32	1 3/4	2 3/8	7/8	2 1/16	8533	85 PR	—
	SOO LINE	5 3/8	4 7/8	2 1/2	9/16	1 15/32	2 29/32	1	2 29/64	8520	—	—
90 lb.	ASCE	5 3/8	5 3/8	2 5/8	9/16	1 19/32	2 55/64	59/64	2 45/128	9040	90 AS	—
	ARA-A	5 5/8	5 1/8	2 9/16	9/16	1 15/32	3 5/32	1	2 37/64	9020	90 RA	902
	ARA-B	5 17/64	4 49/64	2 9/16	9/16	1 39/64	2 5/8	1 1/32	2 11/32	9030	90 RB	905
	AT&SF	5 5/8	5 3/16	2 9/16	9/16	1 15/32	3 5/32	1	2 37/64	9021	90 SF	903
	C&NW	5 17/32	5 3/32	2 1/2	1/2	1 17/32	2 31/32	1 1/32	2 23/64	9035	90 OM	—
	D&RG	5 1/2	5 1/8	2 9/16	9/16	1 5/8	2 7/8	1	2 5/8	—	—	906
	GRT. NO.	5 3/8	5	2 5/8	9/16	1 15/32	2 7/8	1 1/32	2 15/32	9024	—	—
	INTRBGH	5	5	2 7/8	11/16	1 25/32	2 11/32	7/8	2 3/64	9050	90 RT	—
	U. PAC.	5 3/4	5 3/8	2 3/4	17/32	1 1/2	3 3/8	7/8	2 9/16	9023	—	901
	DUDLEY	5 1/2	5	2 21/32	9/16	1 1/2	3 1/32	31/32	2 3/8	—	90 DY	—

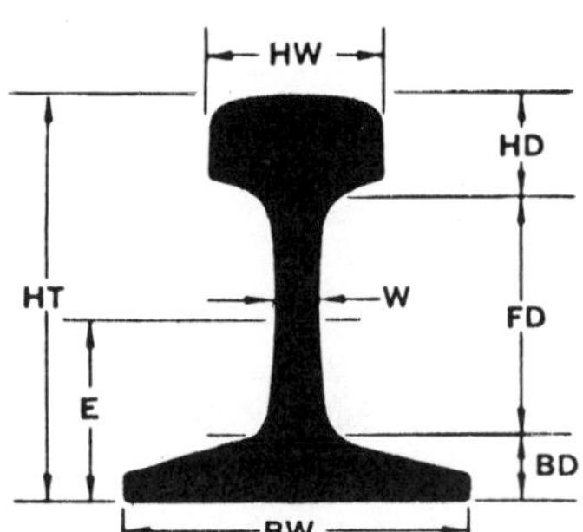

Tee Rail Data—Diagram

(Dimensions for Tee Rail Data Table)

HT—Height

BW—Width of Base

HW—Width of Head

W—Web (at center point)

HD—Depth of Head

FD—Fishing

BD—Depth of Base

E—Bolt Hole Elevation

Source: *Trackwork: Frogs, Switches, Turnouts, Light Trackwork and Fastrak*, L. B. Foster Co., Catalog, Pittsburgh, Pa., 1979.

8.3.2 Curvature and Heading Layout

Of extreme importance to the layout of the track is the layout of curves. Mine track curves usually have a much shorter curvature radius than does surface track. Mine track curves are usually specified in terms of radius, whereas railroad track curves are circular arcs with standard cords of 100 ft which are designated by a specific center angle. Underground curves with a radius greater than 200 ft are often specified similar to railroad curves. Compound curves that are provided above ground for speed easement are not needed below ground. Thus this section discusses only the layout of "simple" curves.

Curve terminology.[4,7] A *simple curve* can be defined as a circular arc extending from one tangent to the next. Underground the two tangents are any two skewed entries that could intersect. Figure 8-7 shows a simple curve. The point where the curve begins, that is, the point where the curve leaves the first tangent, is called the *tangent to the curve*, T.C. (or *point of curvature*, P.C.). The point where the curve intersects and joins the second tangent is called the *curve to tangent*, C.T. (or *point of tangency*, P.T.). The point where both entries would intersect if they were extended is called the *point of intersection*, P.I. (or *vertex*, *V*). The distance between P.I. and either T.C. or C.T. is the *tangent distance*, *T*. The distance from P.I. to the curve is called the *external distance*, *E*. The line from T.C. to C.T. is called the

Figure 8-7 Descriptive nomenclature of a simple curve.

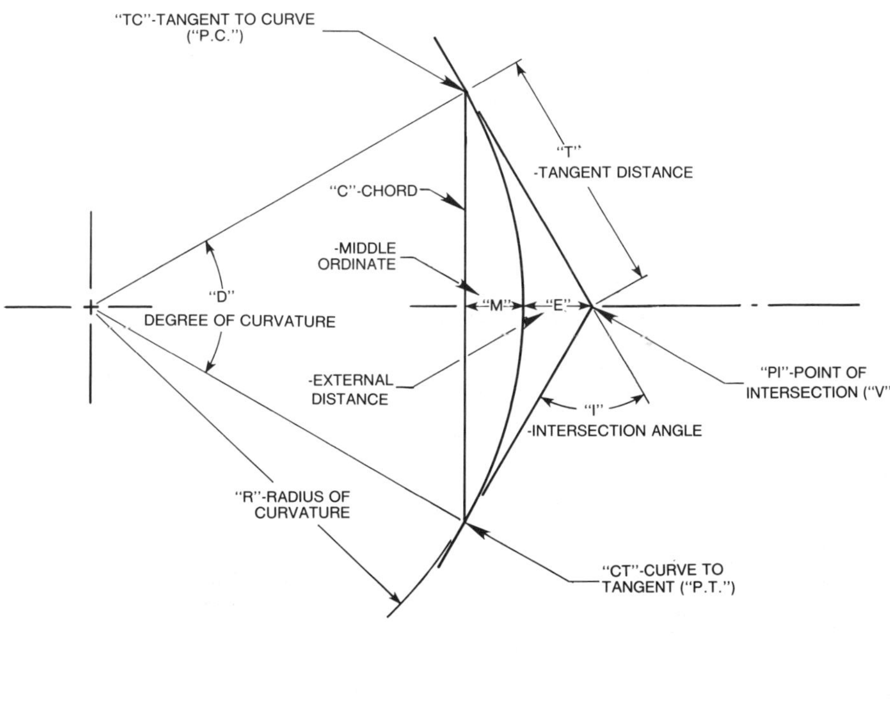

chord, C. The line drawn from the middle of the chord to the curve is called the *middle ordinate*, M. The angle of deflection between the two tangents is called the *intersection angle*, I. The radius of the curve is, of course, called the *radius*, R. The angle at the center, subtended by a chord of 100 ft (20 m in metric measure) is called the *degree of curvature*, D. (A chord less than 100 ft is termed a *subchord*, c, and its central angle is called a *subangle*, d.) The *length of curve*, L, is the sum of chords beginning at T.C. moving toward C.T.

In the United States, railroad curves are laid off using the designation of degree of curve, D. Underground most radii shorter than 200 ft in radius are designated by specifying the radius, R. The mathematical relationship between D and R can be stated as

$$\sin\frac{D}{2} = \frac{50}{R} \tag{8-2}$$

where D = degree of curve
R = radius

Example 1

Find the degree of curvature for a 65-ft radius.

$$\sin\frac{D}{2} = \frac{50}{R} = \frac{50}{65}$$

$$D = 100.6^\circ$$

(Check this result graphically by plotting 100.6° and a 100-ft chord.)

The radius selected is not an arbitrary choice of the designer but depends on the track gage (the wheelbase of the cars), the diameter of the wheel, and the flange-rail clearance. Figure 8-8 shows the minimum safe

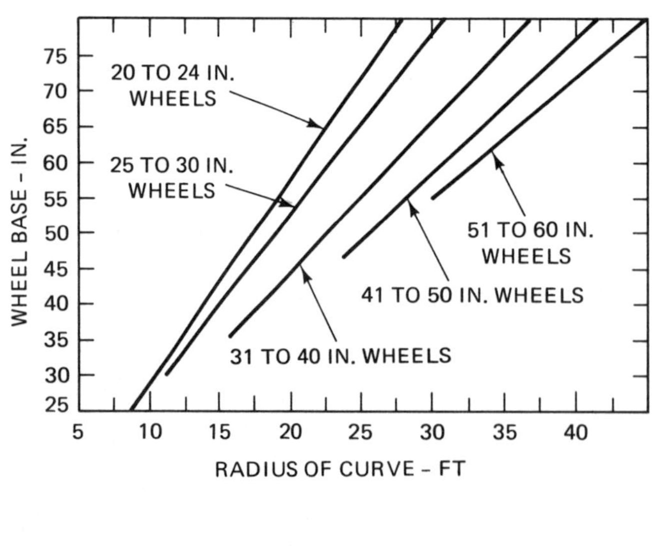

Figure 8-8 Minimum-radius curve. (From Peele, *Mining Engineering Handbook*, 3rd edition, Vol I, Wiley, 1945.)

radius for these conditions. The larger the radius, the less the track wear (lower maintenance) and the greater the speed (high productivity). More conservative designs call for track radii on mainline haulage between 300 and 500 ft.[8]

8.3.3 Methods of Layout Design and Field Use

The role of the track designer is to find (either by calculation or by descriptive graphics) the required radius or value of D that best blends the two merging entries. He or she must then lay out the information about this curve in such a way that the section foreman or survey party leader can easily guide the mining of this curve. Thus the layout draftsman needs to know the limitations and practices of constructing a curved heading.

Figure 8-9 shows the plan view of a typical curved heading or radius R. Several methods can be used to specify this curve so that it may be mined. The simplest and most common method is specifying the deflection angle and rib offsets, as shown in Fig. 8-10. The deflection angle, $\theta = 22.5°$, is the angle between the first tangent and the cord drawn from T.C. to a point, 1, on the centerline of the curve. The magnitude of the deflection angle θ from T.C. to point 1 is $\frac{1}{2}D$, the degree of curve when the cord T.C.–1 is 100 ft. For a cord less than 100 ft (typical in coal mining), the deflection

Figure 8-9 Initial layout of the first cord and the deflection angle.

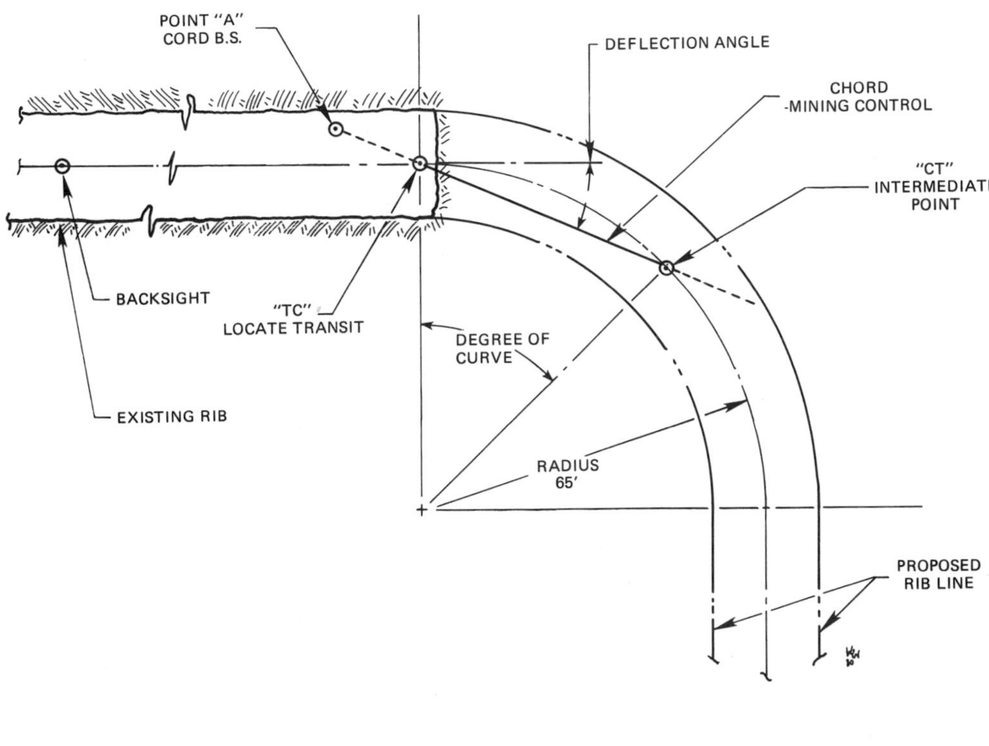

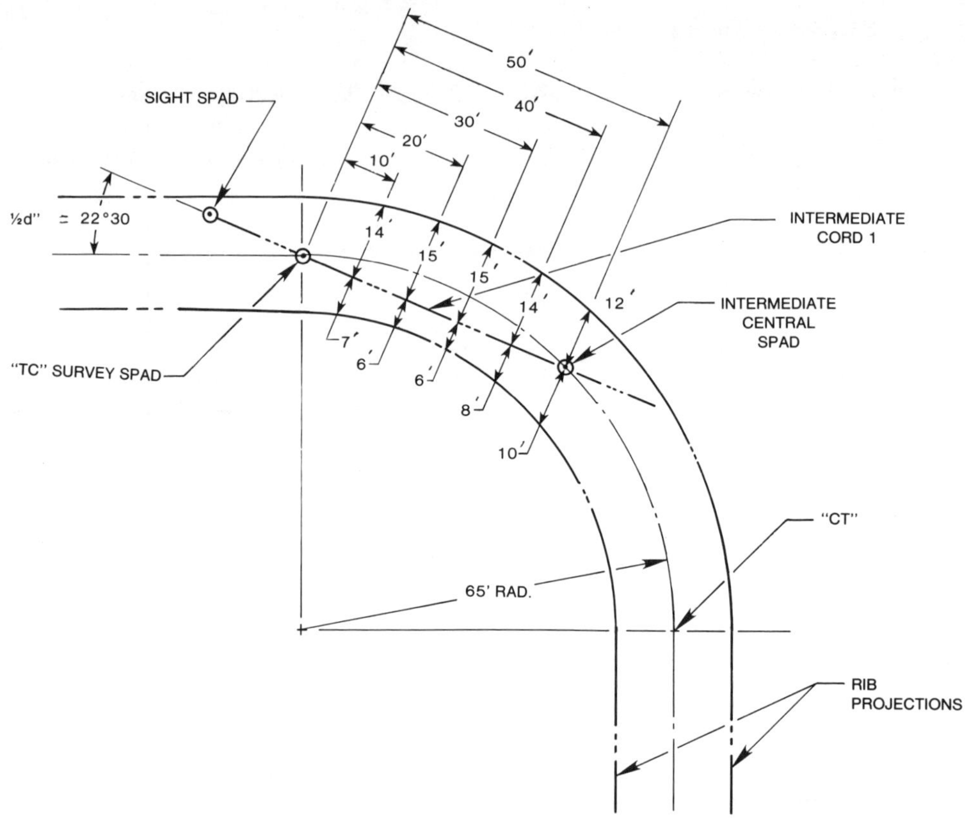

Figure 8-10 Layout of offsets defining a 65-ft radius.

angle is $\frac{1}{2}d$, the subangle. Similarly, the deflection angle that results between points 1 and 2 is $\frac{1}{2}d_{12}$. Mathematically then, the deflection angles are as follows:

$$\theta_{\text{T.C.-1}} = \tfrac{1}{2}d_{\text{T.C.-1}} \tag{8-3a}$$

$$\theta_{1\text{-}2} = \tfrac{1}{2}d_{1\text{-}2} \tag{8-3b}$$

$$\theta_{2\text{-}3} = \tfrac{1}{2}d_{2\text{-}3} \tag{8-3c}$$

It can also be stated that

$$\theta_{\text{T.C.-1}} = \tfrac{1}{2}d_{\text{T.C.-1}} \tag{8-4a}$$

$$\theta_{\text{T.C.-2}} = \text{T.C.-1} + \tfrac{1}{2}d_{1\text{-}2} \tag{8-4b}$$

$$\theta_{\text{T.C.-3}} = \text{T.C.-2} + \tfrac{1}{2}d_{2\text{-}3} \tag{8-4c}$$

etc.

In Fig. 8-10 the subangle $d_{\text{T.C.-1}}$ is 45°, thus the deflection angle $\theta_{\text{T.C.-1}}$ is 22.5° and $d_{\text{T.C.-2}}$ is 45°. Several points along the centerline can be specified by describing successive deflection angles and the length of the cord segment between them. Although any deflection angle and cord distance are equally convenient for the layout draftsman to use, these specifications should be made as simple as possible for the section foreman. It is advisable to segment the curve into equal and convenient cord lengths and deflection angles, avoiding fractions and decimals. As depicted in Fig. 7-9, most continuous miners advance by alternating 10-ft cuts left, then right, to facilitate proper advancement of face ventilation. A convenient cord length would then be a multiple of 10 ft. For this example, two 50-ft cords are used to span the curvature.

The deflection angle is fixed by a series of reference spads or controlling stations. Since forward controlling stations cannot be established as they are on the surface, a series of back stations are needed. The layout should show the exact location of these rear control spads needed to orient the deflection angle. These stations are established with a theodolite or transit; thus the simplicity of the numerical callout suggested for cord length layout is not as important for the deflection angle. Although the transit can be used to guide the miner, guide sticks are often suspended from the reference spads and the control of the miner will be approximated visually. In this way the transit and operator are freed for other work, and equipment maneuverability around the station site is simplified.

The actual mining of the rib is controlled by specifying the location of each rib relative to the distance along, and offset from, the cord. Note that this method of guiding the miner is identical to the offset side note method described in Chapter 4.

The procedures used underground are as follows (see Fig. 8-11):

1. Set transit at T.C.
2. Set vernier at O.
3. Set crosshairs on B.S.
4. Turn to establish cord back sight at A.
5. Measure off any convenient distance that permits a good visual perspective.
6. If the transit must be moved from the area, suspend guidesticks from spads at T.C. and A.
7. Measure the first rib dimension from the cord line and begin mining.
8. Control mining principally with the inside radius (offsets), as this is usually the most critical dimension.
9. Once position 1 is reached, locate a spad at 1 and check progress with transit and tape.
10. Position the transit under point 1 and repeat the foregoing procedures until point C.T. is reached.
11. Grade according to layout.

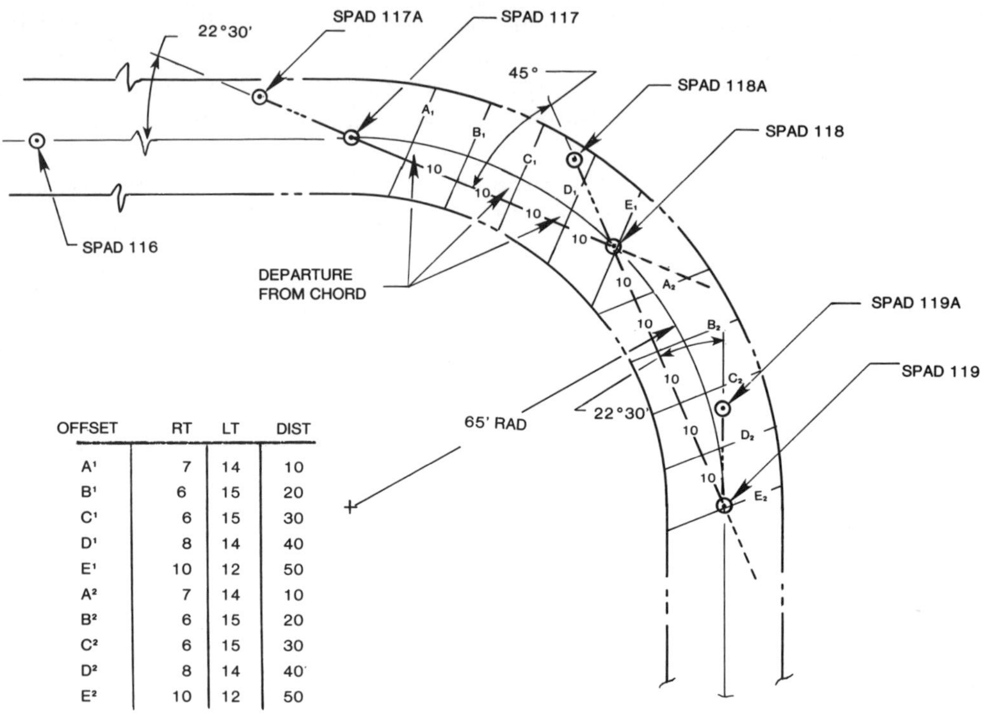

OFFSET	RT	LT	DIST
A^1	7	14	10
B^1	6	15	20
C^1	6	15	30
D^1	8	14	40
E^1	10	12	50
A^2	7	14	10
B^2	6	15	20
C^2	6	15	30
D^2	8	14	40
E^2	10	12	50

Figure 8-11 Completed rib construction layout.

The layout for guiding track curve construction should have the following characteristics: (1) show the position and orientation of the spads needed for construction control, (2) show deflection angles, and (3) show necessary rib offsets.

Curve grade and superelevation. The effects of curvature on a haulage trip are to increase the tractive effort and create a centrifugal force. Tractive effort is the combined resistance to motion offered by grades, mechanical bearing friction, rolling resistance slippage inertia (acceleration), wheel flange friction, and curves. On short curves the tractive effort contributed by the curve is small and often ignored; however, on longer curves the tractive effort contributed by the curve is counteracted by reducing the grade 0.04% for every degree of curve D (note that for a 100-ft cord, 1° curve = 5730 ft radius).

The centrifugal force created is countered by elevating the outer rail above the inner rail. The amount of elevation is directly proportional to the track gage and the square of the intended velocity, and inversely proportional to the radius of track curvature. Mathematically, the amount of superelevation is calculated from Eq. 8-5.

$$e = \frac{GV^2}{gR} \tag{8-5}$$

where e = superelevation (in.)
G = track gage (in.)
V^2 = trip speed (ft/sec)
g = gravitational constant (32.17 ft/sec^2)
R = track radius of curve (ft)

Example 2

Find the amount of superelevation required for a track with a radius of curve of 200 ft, a 36-in. gage, and a trip speed of 10 mph.

$$e = \frac{(36\text{ in.})10\frac{\text{mi}}{\text{hr}}\left(14.67\ \frac{\text{ft/sec}}{\text{mi/hr}}\right)^2}{32.17\text{ ft/sec}^2(200\text{ ft})}$$
$$= 1.2\text{ in.}$$

8.3.4 Turnouts

A *turnout* is a switching station and track leading from a main or another track for the purposes of:

1. Feeding a branch track used for general haulage or secondary haulage.
2. Feeding a siding, used for storage, loading/unloading, or for passing trips.
3. Feeding a parallel track (a crossover).

The principal parts of a turnout (Fig. 8-12) are (1) the switch, (2) the frog, and (3) the guard rail. The *switch* is that component which causes the trip to turn from one track system to another. The *frog* is a track structure used at the intersection of two rails, permitting the flange of the wheel to pass over the main track onto the turnout track. The *guard rail* is a section of rail laid parallel to the track rails which guides the flange of the wheel onto the proper side of the track. The guard rail is set opposite the frog and prevents the flange from bearing against the frog point or derailing.

Various types of switches are used. The most common is the split switch shown in Fig. 8-13. Figure 8-13 shows the switch set for mainline tracking. The ends of the switch (*AB* and *CD*) are tapered to allow proper tracking of the flange of the wheel in each respective position. The angle *DCE*, called the *switch angle*, is the angle formed between the gage lines of the stock rail and the switch rail. Figure 8-14(a) shows a simplified line

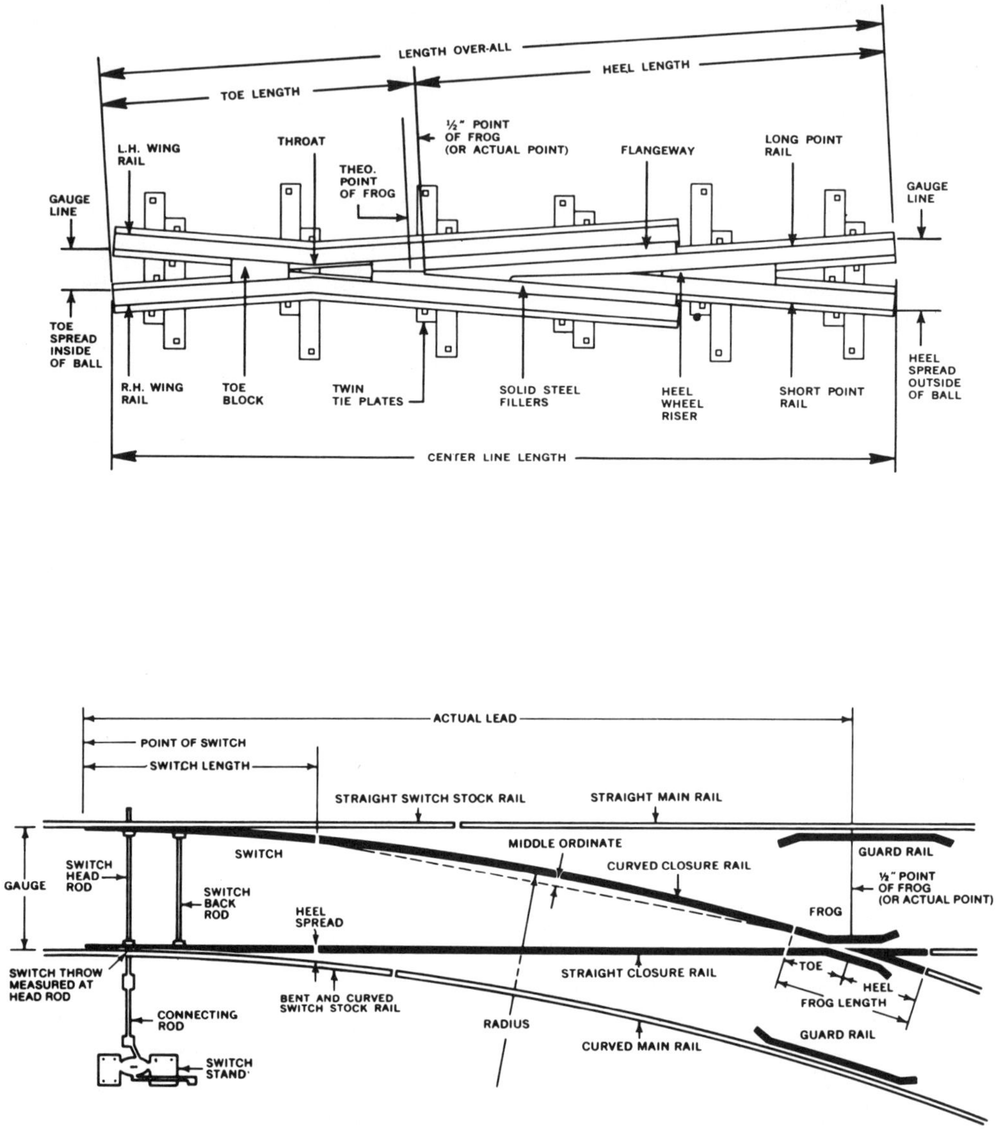

Figure 8-12 Turnout nomenclature. (Courtesy of L. B. Foster Co.)

drawing of a typical turnout depicting the switch angle and other critical dimensions which have been standardized by the American Mining Congress and the American Society of Civil Engineers. Tables 8-2 through 8-4 show the standardized dimensions for the switch and frog for the most common turnouts used underground.[9] Recent recommendations for mainline haulage turnouts call for sizes varying between number 5 and number 8.[8] The layout of a turnout usually follows the form shown in Fig. 8-14(b).

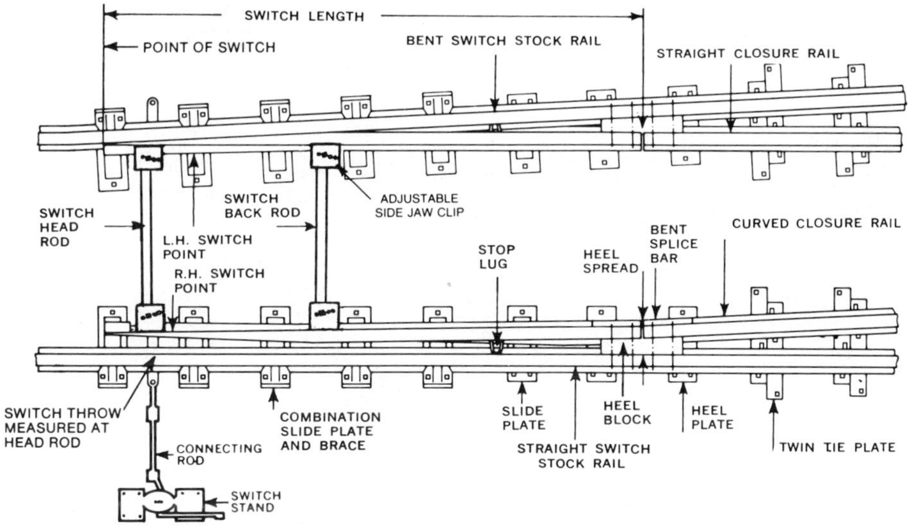

Figure 8-13 Split switch nomenclature. (Courtesy of L. B. Foster Co.)

Figure 8-14 (a) Simplified turnout. (Courtesy of L. B. Foster Co.) (b) Dimensional layout, No. 5 turnout switch, 42-in. gage, rail weight 75 lb, cast frog (two).

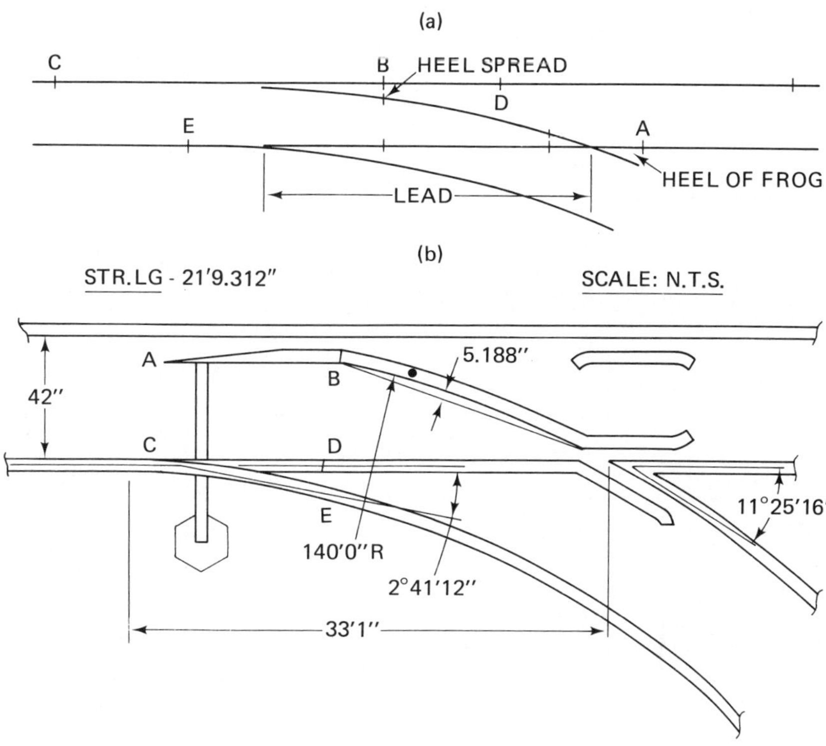

TABLE 8-2

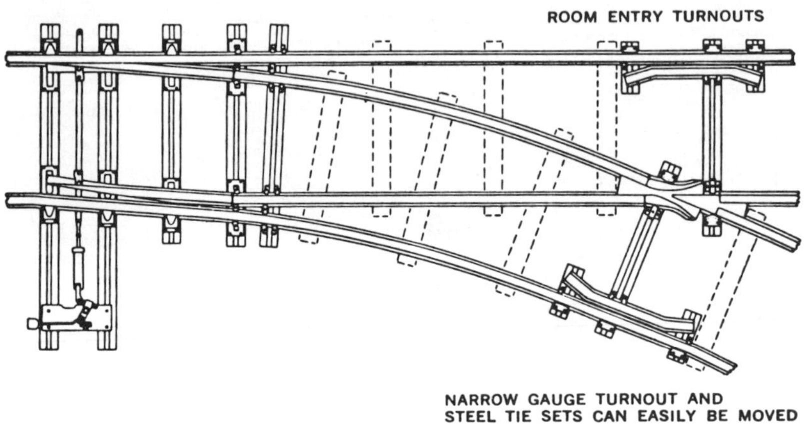

42″ GAUGE American Mining Congress has adopted this gauge as the standard for mines. This gauge is recommended where possible.

20# to 60# ASCE rail inclusive

No.	FROG Angle	FROG Type	FROG Rail Weight lb	FROG Point in.	FROG Actual Toe Length in.	SWITCH Length ft-in.	SWITCH Angle	SWITCH Point in.	SWITCH Heel Spread in.	ACTUAL LEAD ft-in.	CLOSURE RAILS CURVED Radius ft-in.	CLOSURE RAILS CURVED M.O. in.	CLOSURE RAILS CURVED Length ft-in.	CLOSURE RAILS STRAIGHT Length ft-in.
2	28° 04′ 21″	Cast	20 to 40	¼	6 11/16	3–6	6° 19′ 20″	⅜	5	12–7	22–0	5⅝	9–1½	8–6¼
		W.P.F.	20 to 30	¼	17	3–6	6° 19′ 20″	⅜	5	12–7	22–0	4¾	8–3¼	7–8
		W.P.F.	40	⅜	20	3–6	6° 19′ 20″	⅜	5	12–7	22–0	4½	8–0¼	7–5
2½	22° 37′ 12″	Cast	20 to 40	¼	8¼	5–0	4° 25′ 15″	⅜	5	16–11	35–0	5 13/16	11–8¼	11–2¾
		W.P.F.	20 to 30	¼	16½	5–0	4° 25′ 15″	⅜	5	16–11	35–0	5¼	10–11⅞	10–6½
		W.P.F.	40	⅜	20	5–0	4° 25′ 15″	⅜	5	16–11	35–0	4⅞	10–8½	10–3
3	18° 55′ 29″	Cast	20 to 60	¼	8⅜	5–0	4° 25′ 15″	⅜	5	19–0	52–0	5½	13–8½	13–3½
		W.P.F.	20 to 30	¼	16½	5–0	4° 25′ 15″	⅜	5	19–0	52–0	5	13–0¼	12–7½
		W.P.F.	40 to 60	⅜	24	5–0	4° 25′ 15″	⅜	5	19–0	52–0	4 11/16	12–5	12–0
3	18° 55′ 29″	Cast	20 to 60	¼	8⅜	6–0	3° 40′ 59″	⅜	5	20–6	51–0	5⅞	14–2¾	13–9½
		W.P.F.	20 to 30	¼	16½	6–0	3° 40′ 59″	⅜	5	20–6	51–0	5⅜	13–6	13–1½
		W.P.F.	40 to 60	⅜	24	6–0	3° 40′ 59″	⅜	5	20–6	51–0	4⅞	12–10¾	12–6
4	14° 15′ 00″	Cast	20 to 60	¼	11 1/16	5–0	4° 25′ 15″	⅜	5	23–0	96–0	4⅝	17–4½	17–1
		W.P.F.	40 to 60	⅜	27	5–0	4° 25′ 15″	⅜	5	23–0	96–0	4 1/16	16–0¾	15–9
4	14° 15′ 00″	Cast	40 to 60	¼	11 1/16	6–0	3° 40′ 59″	⅜	5	24–7	94–0	5⅛	17–11¼	17–7¾
		W.P.F.	40 to 60	⅜	27	6–0	3° 40′ 59″	⅜	5	24–7	94–0	4 7/16	16–7½	16–4
4	14° 15′ 00″	Cast	40 to 60	¼	11 1/16	7–6	2° 56′ 44″	⅜	5	26–10	91–0	5¾	18–8¼	18–4¾
		W.P.F.	40 to 60	⅜	27	7–6	2° 56′ 44″	⅜	5	26–10	91–0	5	17–4½	17–1
5	11° 25′ 16″	Cast	40 to 60	¼	13 13/16	7–6	2° 56′ 44″	⅜	5	31–0	148–0	5 3/16	22–7	22–4¼
		W.P.F.	40 to 60	⅜	30	7–6	2° 56′ 44″	⅜	5	31–0	148–0	4 11/16	21–2¾	21–0
6	9° 31′ 38″	Cast	40 to 60	¼	16 9/16	7–6	2° 56′ 44″	⅜	5	34–9	220–0	4⅝	26–0¾	25–10½
		W.P.F.	40 to 60	⅜	36	7–6	2° 56′ 44″	⅜	5	34–9	220–0	4⅛	24–5½	24–3
6	9° 31′ 38″	Cast	40 to 60	¼	16 9/16	10–0	2° 12′ 32″	⅜	5	38–10	210–0	5 7/16	27–7¾	27–5½
		W.P.F.	40 to 60	⅜	36	10–0	2° 12′ 32″	⅜	5	38–10	210–0	4¾	26–0¼	25–10

TABLE 8-3

30″ GAUGE American Mining Congress Standards

20# to 60# ASCE rail inclusive

No.	FROG Angle	FROG Type	FROG Rail Weight lb	FROG Point in.	FROG Actual Toe Length in.	FROG Length ft-in.	SWITCH Angle	SWITCH Point in.	SWITCH Heel Spread in.	ACTUAL LEAD ft-in.	CLOSURE RAILS CURVED Radius ft-in.	CLOSURE RAILS CURVED M.O. in.	CLOSURE RAILS CURVED Length ft-in.	CLOSURE RAILS STRAIGHT Length ft-in.
2	28° 04′ 21″	Cast	20 to 60	¼	6 11/16	3-6	6° 19′ 20″	⅜	5	9-5	13-0	3⅞	5-9½	5-4¼
		W.P.F.	20 to 30	¼	17	3-6	6° 19′ 20″	⅜	5	9-5	13-0	2 13/16	4-11	4-6
		W.P.F.	40 to 60	⅜	20	3-6	6° 19′ 20″	⅜	5	9-5	13-0	2½	4-8¼	4-3
2½	22° 37′ 12″	Cast	20 to 60	¼	8¼	5-0	4° 25′ 15″	⅜	5	12-10	21-0	4	7-5¾	7-1¾
		W.P.F.	20 to 30	¼	16½	5-0	4° 25′ 15″	⅜	5	12-10	21-0	3 5/16	6-9½	6-5½
		W.P.F.	40 to 60	⅜	20	5-0	4° 25′ 15″	⅜	5	12-10	21-0	3	6-6	6-2
3	18° 55′ 29″	Cast	20 to 60	¼	8⅜	5-0	4° 25′ 15″	⅜	5	14-4	32-0	3¾	8-11	8-7½
		W.P.F.	20 to 30	¼	16½	5-0	4° 25′ 15″	⅜	5	14-4	32-0	3 3/16	8-2¾	7-11½
		W.P.F.	40 to 60	⅜	24	5-0	4° 25′ 15″	⅜	5	14-4	32-0	2¾	7-7½	7-4
4	14° 15′ 00″	Cast	20 to 60	¼	11 1/16	5-0	4° 25′ 15″	⅜	5	17-0	60-0	3⅛	11-3½	11-1
		W.P.F.	40 to 60	⅜	27	5-0	4° 25′ 15″	⅜	5	17-0	60-0	2 7/16	9-11½	9-9
5	11° 25′ 16″	Cast	20 to 60	¼	13 13/16	7-6	2° 56′ 44″	⅜	5	23-0	93-0	3⅜	14-6¼	14-4¼
		W.P.F.	40 to 60	⅜	30	7-6	2° 56′ 44″	⅜	5	23-0	93-0	2 13/16	13-2	13-0
6	9° 31′ 38″	Cast	20 to 60	¼	16 9/16	7-6	2° 56′ 44″	⅜	5	25-7	139-0	3 1/16	16-10	16-8½
		W.P.F.	40 to 60	⅜	36	7-6	2° 56′ 44″	⅜	5	25-7	139-0	2½	15-2¾	15-1
6	9° 31′ 38″	Cast	20 to 60	¼	16 9/16	10-0	2° 12′ 32″	⅜	5	29-0	133-0	3 9/16	17-9	17-7½
		W.P.F.	40 to 60	⅜	36	10-0	2° 12′ 32″	⅜	5	29-0	133-0	3	16-1¾	16-0

TABLE 8-4

36″ GAUGE American Mining Congress Standards

20# to 60# ASCE rail inclusive

No.	FROG Angle	FROG Type	FROG Rail Weight lb	FROG Point in.	FROG Actual Toe Length in.	FROG Length ft-in.	SWITCH Angle	SWITCH Point in.	SWITCH Heel Spread in.	ACTUAL LEAD ft-in.	CLOSURE RAILS CURVED Radius ft-in.	CLOSURE RAILS CURVED M.O. in.	CLOSURE RAILS CURVED Length ft-in.	CLOSURE RAILS STRAIGHT Length ft-in.
2	28° 04′ 21″	Cast	20 to 60	¼	6 11/16	3-6	6° 19′ 20″	⅜	5	11-0	17-0	4⅞	7-5½	6-11¼
		W.P.F.	20 to 30	¼	17	3-6	6° 19′ 20″	⅜	5	11-0	17-0	3⅞	6-7¼	6-1
		W.P.F.	40 to 60	⅜	20	3-6	6° 19′ 20″	⅜	5	11-0	17-0	3 9/16	6-4¼	5-10
2½	22° 37′ 12″	Cast	20 to 40	¼	8¼	5-0	4° 25′ 15″	⅜	5	15-0	28-0	5 1/16	9-8½	9-3¾
		W.P.F.	20 to 30	¼	16½	5-0	4° 25′ 15″	⅜	5	15-0	28-0	4⅜	9-0¼	8-7½
		W.P.F.	40 to 60	⅜	20	5-0	4° 25′ 15″	⅜	5	15-0	28-0	4 1/16	8-8¾	8-4
3	18° 55′ 29″	Cast	20 to 60	¼	8⅜	5-0	4° 25′ 15″	⅜	5	16-9	42-0	4⅝	11-4¾	11-0¾
		W.P.F.	20 to 30	¼	16½	5-0	4° 25′ 15″	⅜	5	16-9	42-0	4⅛	10-8½	10-4½
		W.P.F.	40 to 60	⅜	24	5-0	4° 25′ 15″	⅜	5	16-9	42-0	3⅝	10-1	9-9
4	14° 15′ 00″	Cast	20 to 60	¼	11 1/16	5-0	4° 25′ 15″	⅜	5	20-0	78-0	3 15/16	14-4	14-1
		W.P.F.	40 to 60	⅜	27	5-0	4° 25′ 15″	⅜	5	20-0	78-0	3¼	13-0¼	12-9
5	11° 25′ 16″	Cast	20 to 60	¼	13 13/16	7-6	2° 56′ 44″	⅜	5	27-0	120-0	4 5/16	18-6¾	18-4¼
		W.P.F.	40 to 60	⅜	30	7-6	2° 56′ 44″	⅜	5	27-0	120-0	3¾	17-2½	17-0
6	9° 31′ 38″	Cast	20 to 60	¼	16 9/16	7-6	2° 56′ 44″	⅜	5	30-2	179-0	3⅞	21-5½	21-3½
		W.P.F.	40 to 60	⅜	36	7-6	2° 56′ 44″	⅜	5	30-2	179-0	3 5/16	19-10	19-8
6	9° 31′ 38″	Cast	20 to 60	¼	16 9/16	10-0	2° 12′ 32″	⅜	5	34-0	171-0	4 9/16	22-9½	22-7½
		W.P.F.	40 to 60	⅜	36	10-0	2° 12′ 32″	⅜	5	34-0	171-0	3 15/16	21-2	21-0

NOTE: W.P.F. denotes Welded Plate Frog

8.3.5 Transfer Stations

Transfer stations are used to unload mine cars, unload conveyor belts, provide surge capacity, and load mine cars or skip hoists. The most common transfer station is from secondary haulage to mainline track haulage. Several layout methods can be used to accomplish the necessary functions of unload, store, and load.

A common tipple site for automatic loading of mine cars is shown in Fig. 8-15. The site consists of a car retarder, a car spotter, a feed chute or tilt pan, a flight conveyor belt, a power unit (not depicted), and auxiliary trip monitors and controls (not depicted). The car spotter and retarder accept empties from the feeding rail, positions the transfer "dogs" (used to connect the spotter to the cars), and provides the necessary power to move the trip

Figure 8-15 Loading station (conveyor to track) layout.

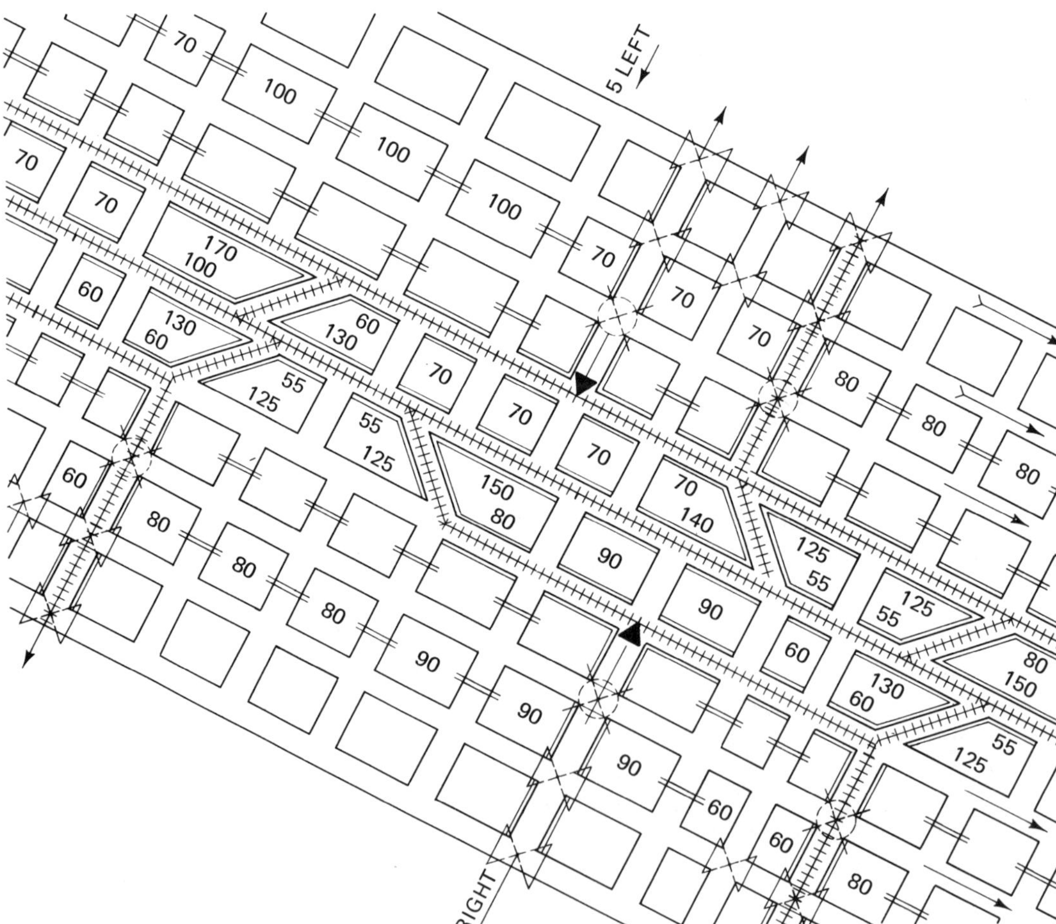

through the site. Since the car spotter controls the trip movement, the gathering locomotive is freed for other tasks. The tilt pan functions to feed the coal exiting the conveyor into the car gradually and to transfer the feeding from one car to the next. The flight conveyor transfers coal from the conveyor belt, or surge bunker, to the tilt pan. Flight conveyors equipped with transverse blades can transfer coal up 45° inclines (although angles greater than 25° are uncommon) and are therefore ideally suited to the confined spaces of the tipple site. The power unit is either hydraulically or electrically operated and is positioned in a crosscut on a separate split of air. Auxiliary controls function to monitor the level of coal within the car, to coordinate the tramming speed of the trip with the feed rate, to transfer the tilt pan or feed chute to empties automatically, and to provide manual override control. The haulage ventilation layout for several transfer stations is depicted in Fig. 8-16.

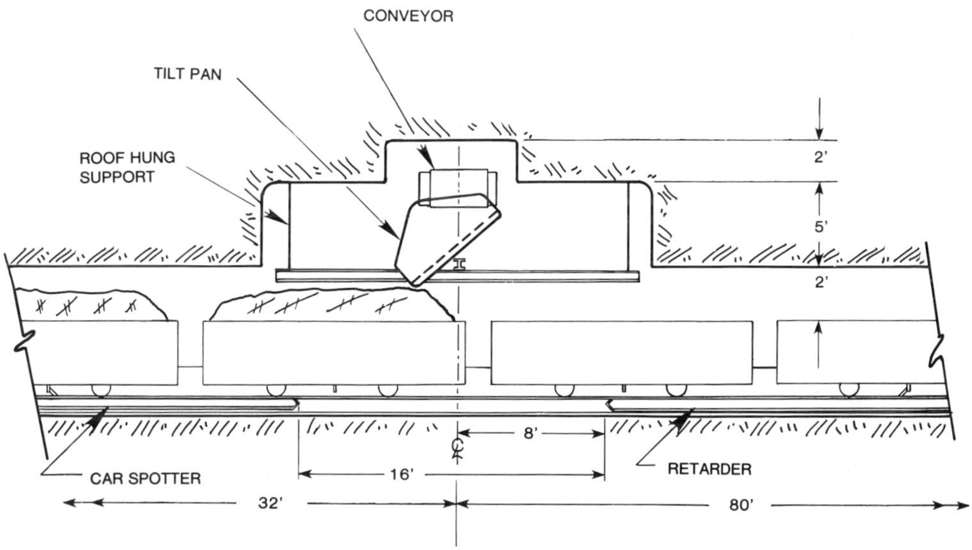

Figure 8-16 Transfer station. (After W. R. Stamler Corp.)

Skip hoist. Most mainline trips terminate at a skip shaft. Figure 8-17 shows a typical dual-compartment skip hoist layout using a rotary dumper to empty the mine cars. As with the loading station, a car spotter is used to position and move the cars through the tipple site. The individual cars are positioned and secured in the rotary dump. The entire assembly, including the small section of track, rotates causing the coal to fall into the skip pocket. The pocket then opens filling the skip. As the loaded skip ascends, the feed gate shifts to the second chute and another car is positioned. Once the second empty skip is in place, the car is rotated, filling the second skip pocket. The skip is filled by opening the skip pocket and the cycle continues, alternately filling the two skips.

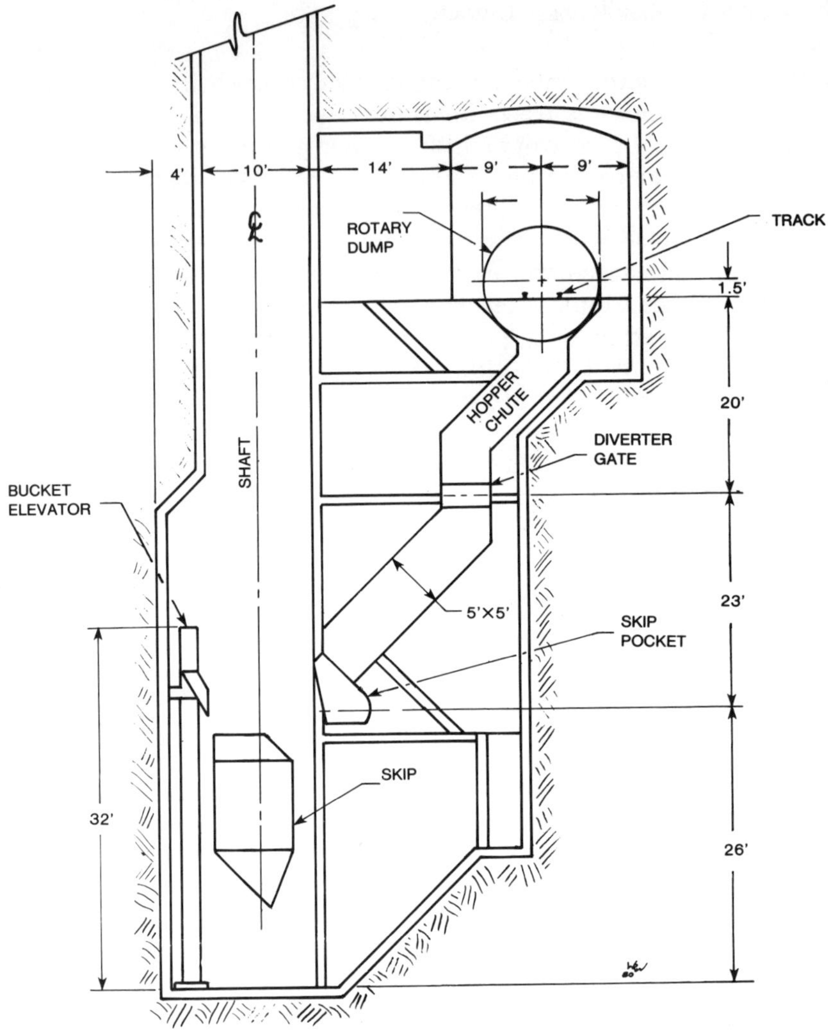

Figure 8–17 Typical layout of a duel skip hoist with rotary dumper.

8.3.6 Maintenance and Inspection

You will recall from Chapter 5 that inspection methods are classed as either observation or instrument assisted. The vast majority of inspections of the track haulage system are of an observational type and include checking the level and consistency of ballast, track alignment, gage spread on track and turnouts; checking for worn or deteriorated components; and observing missing spikes and bolts. Some instrument-assisted inspection procedures are used to maintain drive motors and other electrical equipment, but these are beyond the intended scope of this presentation.

Maintenance procedures include failure-cued maintenance, routine maintenance, and preventive maintenance. Failure-cued maintenance involves replacing broken track, correcting the causes and effects of trip derailment, and removing roof failures. Routine maintenance includes lubricating mechanical components, tightening loose components, ensuring ballast requirements, maintaining gage spread, cleaning components, and keeping the haulageway and shelter holes free of debris. Preventive replacement involves replacing track, switch gear, frogs, ties, and other worn equipment. The general requirements described in Chapter 5, covering the nature of these procedures should be reviewed.

8.4 CONVEYORS AND BELT LAYOUT

Numerous types of conveyors exist, including flight, drag chain, apron, pan, screw, shaker/vibrating, and belt. Of these, belt conveyors have the widest range of application in and around mines. Belts are growing in popularity as the best alternative for secondary track haulage, and in many cases as a more appropriate method for primary haulage, representing more than 20% of present mainline haulage. This growth in belt haulage can be attributed to the following advantages of belt conveyors:

1. Belts provide continuous haulage; thus as the number of continuous-mining methods, such as longwall and shortwall operations, increase, the number of belts providing the haulage function will also increase.
2. Belts require less head room than do tracked haulage methods, often with greater capacity.
3. Belts are limited to a grade of approximately 32% (the practical limit is closer to 23%), as opposed to the $4\frac{1}{2}$% grade of extended mine car haulage.
4. Vertical belt alignment is independent of regularity of bottom; thus extensive brushing is not required.
5. Belts can be supplied with devices for continuous weighing and with magnetic separators for automatic removal of scrap metal (roof bolts, etc.).
6. Belts can be more easily automated and monitored and are therefore more labor intensive than is track haulage.

As the number of "poorer" seams are opened, seam characteristics such as regularity, height of entry, and dip will dictate wider use of belt systems for haulage.

8.4.1 Belt System Components

The system of belt conveying shown in Fig. 8-18 can be divided into basic subsystems as follows:

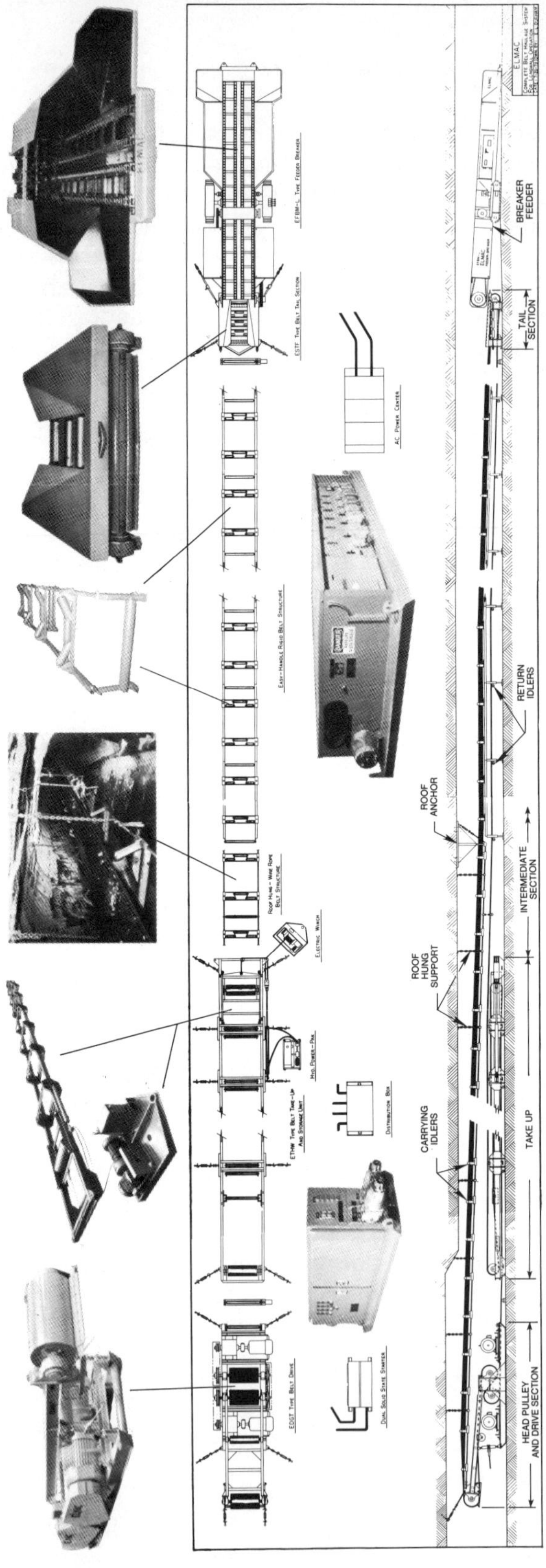

Fig. 8-18. Typical Conveyor Belt Layout (Courtesy of Elmac Corp.)

1. Head section and drive unit
2. Belt takeup section
3. Intermediate section (including frame and idler pulleys)
4. Tail section

The *belt* is composed of an inner core (which can be cotton duct, nylon, steel mesh, or steel cable) bonded together by an elastomer (a rubberlike material) and covered by either the same or a different type of elastomer. The belt should have the following attributes:

1. *Resistance to wear.* Wear from the coal or from the carrier system is the greatest maintenance problem associated with belt systems.
2. *Strength.* The belt must have the ability to withstand severe loading, which includes the lifting of loads up inclines.
3. *High resilience and toughness.* Both properties are related to the ability of the belt material to withstand high-impact loads without permanent distortion or surface damage.
4. *Tear resistance.* The rubberlike material of the belt will develop small cuts and tears over the normal course of operation. The material must be able to resist the propagation of these partings.
5. *Flexibility and conformability.* These characteristics apply to the ability of the belt to follow the drive, terminal, and trough pulleys without breaking or separating from the inner core material.

The *carrying idlers* support the belt, providing a smooth-rolling surface on the load-carrying side. Idlers should be spaced close enough to prevent the loaded belt from sagging between idlers. Underground, the spacing is nominally 5 ft. Excessive sag results in an impact type of load to the belt and subsequently shortens the useful life of the belt. For suspension belts the configuration of the idler assembly is usually three rollers across or offset, with the side rollers inclined 20 to 35° to the horizontal, thus forming a containment trough. The roller diameter usually varies from 4 to 5 in.

The *return idlers* support the belt on the unloaded return (bottom) side. Normally, these are straight rollers spanning the width of the belt. As with the carrying idlers, excessive spacing of return idlers will significantly shorten the life of the belt. Underground the nominal return idler span is 10 ft.

The *takeup assembly* is needed to maintain proper belt tension, thus preventing drive pulley slippage, excessive roller sag, or excessive tension on belt splices and terminal pulleys during operation. Most takeup assemblies have been gravity type (a naturally weighted type which depends on the load exerted by an adjustable counterweight to regulate the tension load on the belt), mechanical winch and cable type, or hydraulic type. Most underground belts today are hydraulic type, since this design provides more precise adjustment and occupies less space.

The *drive assembly* is located either at the head (offtake end) of the belt or as near the head as possible. (Otherwise, the belt would be pushing instead of pulling its load—an inefficient and improper design.) The drive pulleys should be as large as possible because the belt must flex around this radius without breaking. Various configurations of drive assemblies can be specified, from a single pulley to multiple drives, and manufacturers' literature should be consulted for the best arrangement for your needs.

8.4.2 Size Requirements

As previously noted, a belt conveyor may be of any length. Some overland conveyors are presently operating over 10 miles; however, the underground limit of economical efficiency is between 4 and 5 miles.[1] The practical length of the conveyor system depends somewhat on the width, speed, size, and weight of the material and the production capacity requirements.

The width of conveyor belts ranges from 14 to 72 in. The optimum width depends on (1) the largest single size of material expected (lump size should not exceed one-third the belt width), (2) the speed, and (3) the capacity. The most effective combination of the latter two factors is the narrowest belt operated at the greatest speed. This provides the lowest initial cost and best wear characteristics for the capacity output.

The linear speed of the belt depends on (1) the capacity; (2) the friability of the coal, that is, the amount of breakage allowable during transport and discharge; and (3) the inertial effects of the coal at the discharge chute. Surface belts operate at speeds up to 1200 ft/min; however, most underground operations range from 200 to 600 ft/min. It has been reported that doubling the belt speed increases belt wear by 25%.[10] This increased wear can be offset by either increasing the cover thickness or by replacing the belt with a more wear-resistant elastomer cover.

The capacity of the belt depends on three factors: (1) the speed of the belt, (2) the width of the belt, and (3) the weight (density) of the material. The effect of these factors is stated in the following formula:

$$C = \frac{(W + 180)W^2(S)(M)}{12{,}000{,}000} \tag{8-6}$$

where C = capacity (tons/hr)
W = width of belt (in.)
S = speed of belt (ft/min)
M = material bulk density (lb/ft^3)
12,000,000 = conversion constant

8.4.3 Operation

The most critical factor in proper belt operation is alignment. The *training* of the belt, as the alignment procedure is called, requires that the belt be installed to run true and well centered along both the load and return side of the conveyor. Belts that tend to ride up one side or off-center wear rapidly and need constant repair. The training of the belt should include the following procedures:

1. Check the straightness and alignment of the frame and idlers (carrier and return) with a transit or preferably with a laser.
2. Check the level of all idlers to ensure equal elevations.
3. Set all idlers perpendicular to the belt (use an idler square).
4. Check the straightness of belts and belt splices.
5. Lay out the loading method so that it is even and ideally in the same direction as belt movement. Avoid impacts that result from loading from too great a height. This may require the use of a breaker-feeder conveyor in front of the belt conveyor or a transfer belt arrangement such as that depicted in Fig. 8-19.

Figure 8-19 Transfer belt layout. (Courtesy of Elmac Corp.)

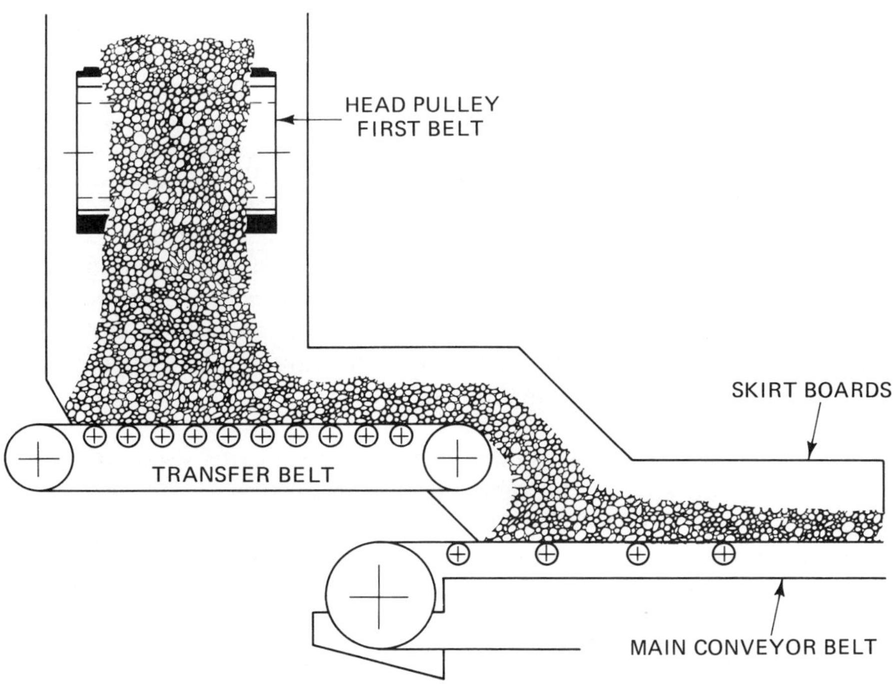

Other suggested factors affecting the proper operation of conveyor belt systems are:[11] (1) the head and tail assemblies should be firmly anchored at a minimum of four anchor points, as shown in Fig. 8-20; (2) the frame should be anchored such that minor side loading does not affect belt alignment; (3) belt tension should prevent drive pulley slippage but not be so tight as to overstress the belt material; (4) proper and periodic maintenance should be provided (Section 8.4.4), and (5) buildup of oil, grease, or water around the belt should be prevented.

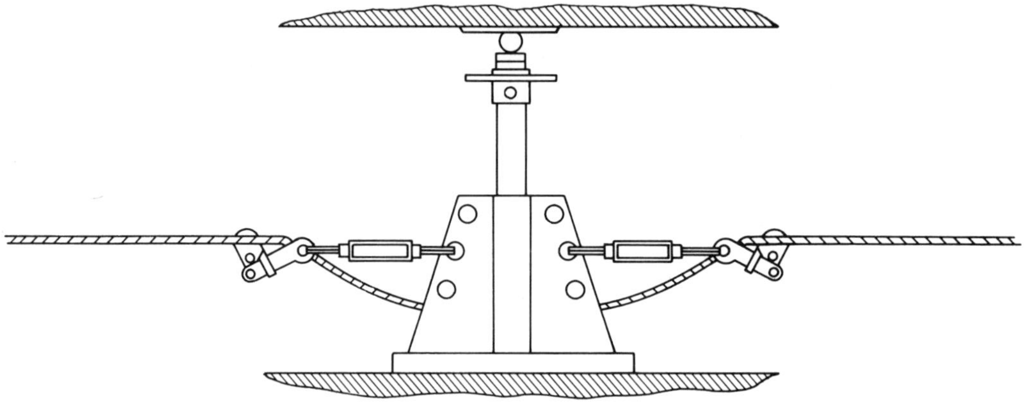

Figure 8-20 Typical anchor support for rope-mounted structure. (Courtesy of Elmac Corp.)

8.4.4 Layout

The layout of the belt entry is relatively simple, as only straight alignments can be used. The entry slope is characteristically less than the maximum possible grade of 32%. Concave transition curves are needed between horizontal beds and slope beds and slope entries and to follow major bedding irregularities. Convex transition curves are used to achieve horizontal discharge and are also used to account for bedding plane variations. Both types of vertical curves require large blending radii to minimize belt wear and assure proper tracking of the belt along idlers.

Support structure. Two types of structures used underground are the rope-mounted and the rigid side rail structures. The *rope-mounted structure* (Figs. 8-21 and 8-22) is the most common type because of its lower initial cost and ease of modification (e.g., advancement). The rope structure is supported either by floor-mounted jacks (Fig. 8-21) or by chain or wire rope hung from the entry top (Fig. 8-22).

The *rigid side rail structure* has carrying and return idlers secured to channel side rails. The advantages of the more conventional belt structure are ease of installation, alignment, and training of the belt. Also, since the

Figure 8-21 Floor-mounted rope conveyor structure. (Courtesy of Elmac Corp.)

Figure 8-22 Roof-mounted rope structure. (Courtesy of Elmac Corp.)

idlers are rigidly mounted, there is less idler movement and better maintenance. Like the rope-mounted structure, the rigid structure is mounted by floor jacks or hung from a roof chain. When headroom is ample, the maintenance advantages offered by the rigid structure often outweigh the cost and installation simplicity of the rope structure.

Belt alignment. Regardless of which structure is used, the most important consideration of belt haulage is the layout of perfect belt alignment. Two methods are employed to ensure this alignment: (1) careful layout and installation of engineering lines, and (2) the use of laser technology to align the structure.

Engineering lines needed for belt installation are the entry centerline and the side rope or rail alignment line. The entry centerline, thus the entry as well, must be perfectly straight in order to accommodate the installation of a conveyor belt. This requires careful attention by the section foreman and miner operator. The conveyor belt alignment line consists of a series of roof spads installed every 20 to 25 ft parallel to the entry centerline. The conveyor spad line is used to provide a plumbing alignment for (1) the initial installation of the rib side belt structure, (2) the subsequent inspection and maintenance of belt alignment, and (3) the reference guide for the installation of auxiliary equipment.

The conveyor spad line is offset from the entry centerline a distance that depends on the width of the entry and the desired tight-side clearance from the rib (at least 24 in.). One method for installing spads is to measure the offset distances from the entry centerline every 20 to 25 ft. Although relatively simple, this method cannot provide the accuracy needed for the proper installation and maintenance of even secondary haulage belts.

Secondary and mainline conveyor spad lines should be installed using a transit and taking as long a "shot" as possible to ensure greater accuracy. Several methods exist for doing this, the simplest being the *dual backsight coincidence* method shown in Fig. 8-23.[11] After the entry has been mined well past the proposed tail section position, the transit or theodolite is set up under the entry spad closest to and out from the proposed head section location. The rodman or chainman, standing at a spad in from the proposed tail section location, measures the necessary offset distance and installs a roof spad. Simultaneously, the transitman performs the same operation at his location. The angle from the entry centerline to the belt spad installed by the rodman is then turned and recorded. Next, the instrument is set up beneath the entry spad nearest the proposed tail section. A backsight is taken on the first entry spad and then the angle turned from the centerline to the belt spad installed by the transitman. The two angles will coincide if and only if the spad line is parallel to the entry centerline. If the angles do not coincide, appropriate adjustments are made until they do. After these outer belt spads have been positioned, the remainder of the belt spads can be installed at 20- to 25-ft intervals long the line between them (Fig. 8-23).

The alignment line is next used either to guide the setup of the floor jacks, or in the case of roof-hung structures, to guide the installation of roof anchor plates. If roof-hung mounts are to be used, the installation of the anchor plates is preceded by establishing an anchor plate alignment line parallel to both the entry centerline and the conveyor alignment line. Roof anchor plates are then bolted in parallel pairs along these two lines, which are separated by a distance equal to the gage of the rope structure. The center-to-center distance between anchor plate pairs is equal to the distance between the return idlers (see Fig. 8-22), which is nominally 10 ft.

An alternative to establishing a long series of belt spads is the use of the laser gun. The laser (Light Amplification by Simulated Emission of Radiation) projects a perfectly uniform, intense, and straight beam of light which

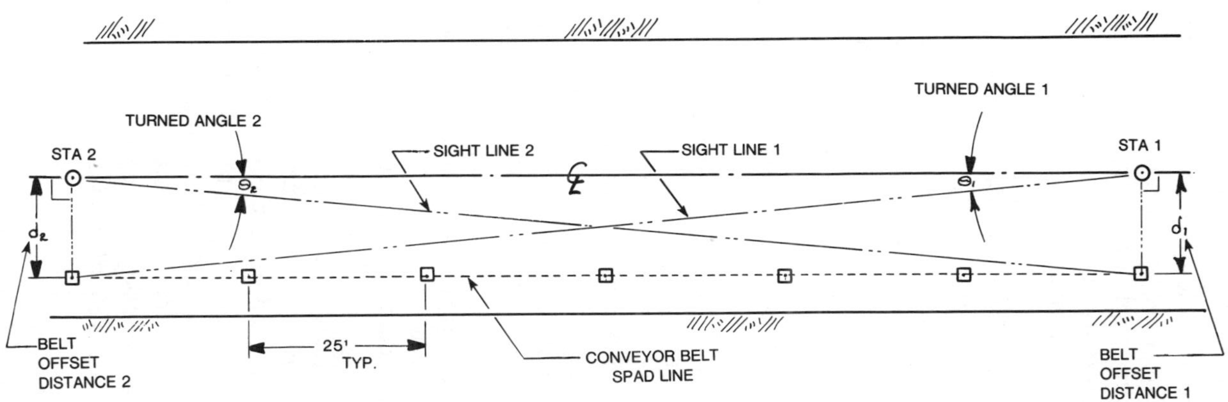

Figure 8-23 Dual-backsight coincidence method.

can be used to align the belt anchors, install the structure, and inspect and maintain this alignment after installation. The laser gun is advantageous because the setup is simple, quick, exact, not labor intensive, versatile (it can also be used to guide the entry development), and can be approved as permissible.[12] The laser gun guides the precise installation of the rib side roof anchors, and simple offset measurements can be taken to establish the second-row alignments. For long conveyor installation or for frequent advancement of the tailpiece, laser gun technology is without equal for maintaining alignment.

For initial or new installations, the tail section, takeup, and auxiliary equipment are taken in from the proposed head section location and distributed where needed along the entry. The head section foundation is then poured (preferable for more permanent belt installations) or a cribbing foundation built. The installation of the head section is critical to the subsequent alignment of the structure and eventually to the maintainability of the belt. The head section requires leveling and installation of at least four anchoring points to prevent lateral or transverse movement of the head pulley.

Next, the takeup section is installed. The belting is threaded over and through the head and takeup pulleys and attached with the wire rope to the tail section. The tail section is then dragged down the entry. Once the components are distributed along the entry, the structure can be assembled and aligned. The conveyor structure must be anchored at various points along the beltline. The anchoring stations (Fig. 8-24) are located at each advancement point, nominally every 250 ft.[11] This distance often corresponds to the length of tailpiece advancement.

Following the installation of the belt structure and prior to its operation, the necessary auxiliary equipment must be installed. This equipment includes electrical components, safety equipment, and additional belt support equipment.

The electrical support equipment for conveyor operation includes belt transformer (or power center), motor controller, and power cables from the motor controller to the drive motor. The belt power center and motor controller supplying the induction motor are installed as close to the belt as possible to avoid excessive voltage drop and power losses. Federal regulations[13] require that the belt power transformer be located on a separate split of air on the same side of the belt as the drive motor.

The safety equipment that must be installed to support the belt system[13] includes water lines running parallel to the belt, equipped with firehose tees at 300-ft intervals as well as head and tail assemblies. At various locations along the entry, 500 ft of firehose must be stored. The minimum delivery capacity of the water system is 50 gal/min at a 50-psi nozzle pressure. Two portable fire extinguishers must be provided at each permanent electrical installation. Automatically actuated fire suppression equipment

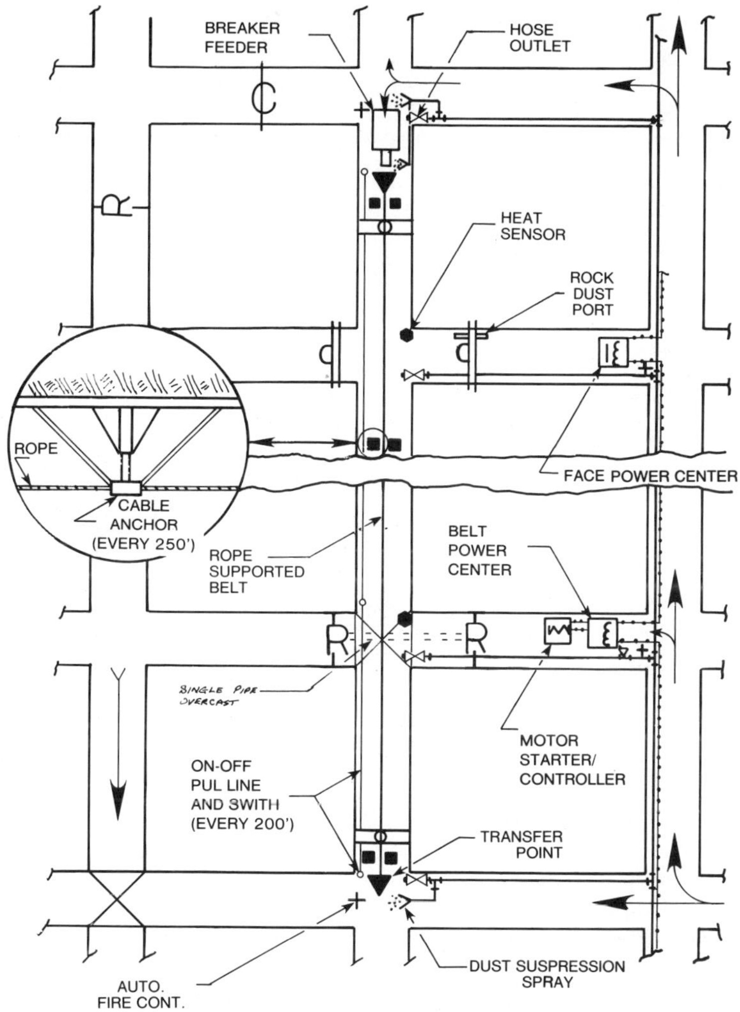

Figure 8-24 Conveyor belt layout.

(deluge-type water spray, foam generating systems, or dry powder chemical) must be installed at all mainline and secondary belt drive locations. Figure 8-25 shows a schematic diagram for one such system. Fire sensors which are automatically actuated must be installed along the length of the belt to aid in locating fire positions. On-off controls must be extended the entire length of the belt. A pull cord with switches positioned every 200 ft is the most common method for meeting this requirement. Finally, the belt must be ventilated on a separate split of air. Most of the safety equipment is installed on the tight side of the belt, suspended from the top or rib.

Other types of equipment that must be installed include hydraulic power pack discharge chutes, transfer belt or chute belt cleaners, and limit switches.

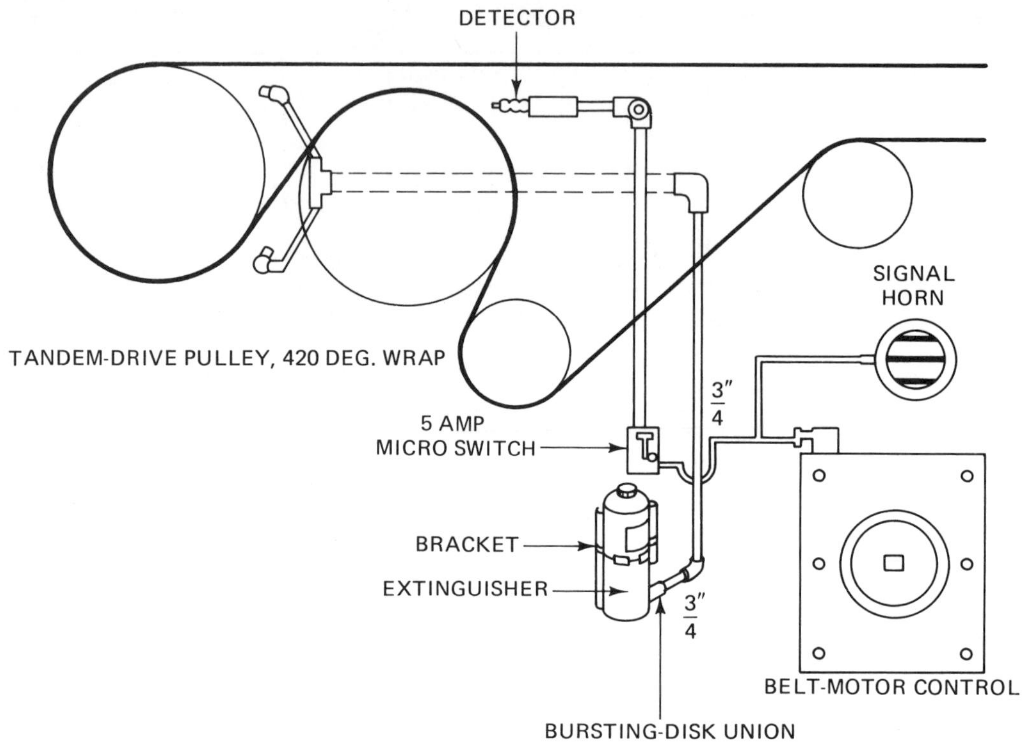

Figure 8–25 Fire suppression and control layout for a conveyor belt drive. (Courtesy of Elmac Corp.)

8.4.5 Reliability/Maintainability

The most significant single factor affecting the reliability and maintainability of conveyors is the initial installation alignment. This requirement is followed in importance by the adequacy of the inspection/maintenance program, particularly well-designed and enforced routine maintenance.

The belt conveyor, once properly installed, is usually quite predictable in its performance. The electrical power, measured by monitoring the current at a constant voltage, is directly proportional to the frictional load or drag of the belt. This frictional drag on an unloaded belt at constant tension and speed is, in turn, directly proportional to the bearing drag offered by the idlers. Thus a form of the BITE method (see Chapter 5) of inspecting the condition of the system is easily achieved by merely monitoring and recording the daily amperage required by the system. Gradual increases in current draw indicate a constant and an expected degradation in the status of the system, which is connected by routine maintenance. Sudden changes are usually more critical, possibly indicating idler failure and requiring failure-cued maintenance strategies.

8.5 FLUID ENTRAINMENT SYSTEMS

The most recent innovative developments in mainline haulage systems are various fluid entrainment systems. Various entrainment media have been used, but the two most economical and available are water and air. Water slurry systems have been used to carry coal and other products aboveground and are beginning to be adapted to commercial underground operations. Continental Oil Co., using the Robinson Run Mine 95-A facilities of its subsidiary Consolidation Coal Co., initiated the field development and testing of one such adaptation. The relative success of this prototype installation resulted in the installation of a similar but larger-capacity system at Consolidated's Loveridge Mine.

This demonstration system, which is paralled by a track system for reliability, is the first full-scale demonstration of the potential advantages offered by entrainment haulage: (1) reduction of dust and spillage in haulage entries; (2) elimination of separate splits or the need to locate haulage only on intake entries; (3) lowering of space requirements; (4) automation of the system, that is, less labor intensive; (5) conformation of the haulage system to the requirements of newer hydraulic extraction methods; and (6) elimination of transfer or tipple stations, therefore less handling. Initial demonstrations began in August 1980 and additional tests are currently under way.

Air or pneumatic transport systems have also been used in aboveground applications such as grain transport. One of the first attempts to investigate the adaptability of pneumatic transport for coal hoisting was conducted in the 1950s at the Morgantown Energy Research Center, then operated by the U.S. Bureau of Mines.[14] The system was operated using both a pressurized and a vacuum mode, with the latter offering the least number of problems (e.g., pipe plugging). The principal advantage that pneumatic transport has over slurry is that the material can be handled dry, thereby avoiding the cost of extensive dewatering.

8.5.1 Components

Figure 8-26 depicts one possible schematic layout for an underground slurry transport installation. Coal is conveyed from the working sections to a loading station surge bunker or hopper. A movable boom flight conveyor can selectively charge either the slurry system or a track system (used principally for transport of workers, equipment, and supplies). Chain conveyors feed roller crushers, which break the lump coal into entrainment sizes. Screens can be located at each successive charging chute to separate clean coal from refuse, thereby decreasing the cost of hauling and preparing refuse.

Clean coal is delivered to a mixing sump, where water and coal are agitated to attain as homogeneous a slurry mixture as possible. The slurry can

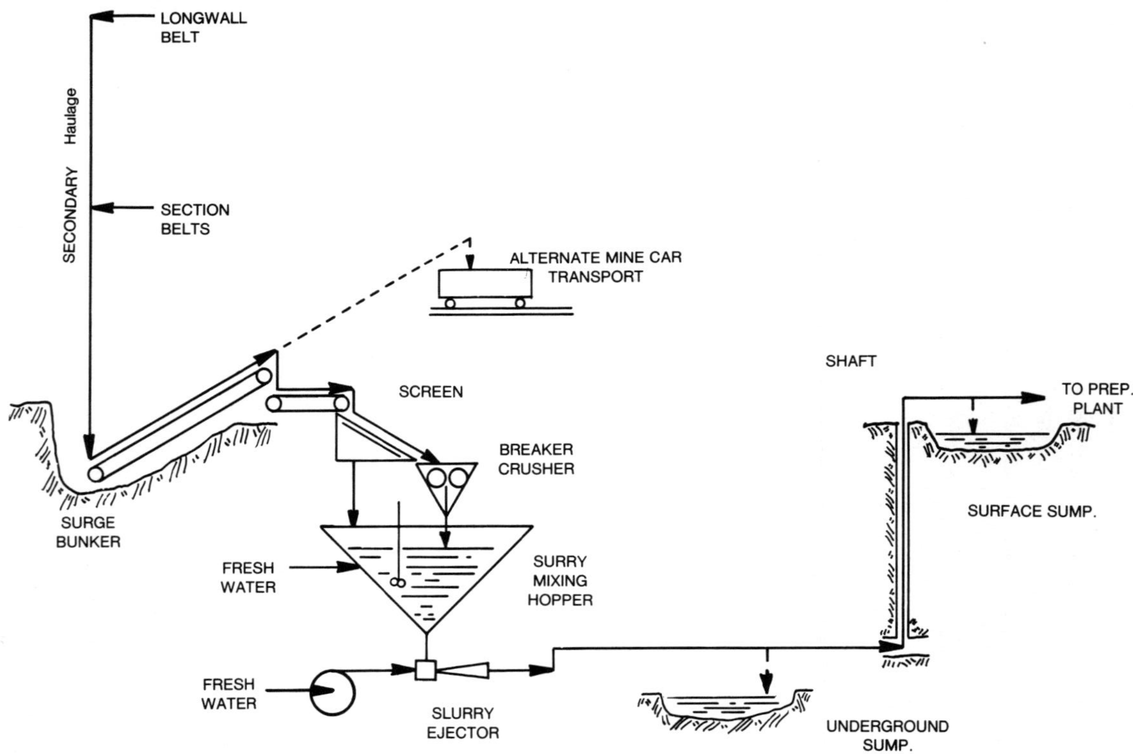

Figure 8-26 Slurry transport for underground haulage.

then be injected into the pipeline using (1) direct pump, suction side, intake; (2) lock hopper arrangement; (3) entrainment injection; or (4) pressurized coal feeders. Once the slurry has been injected into the pipeline, the fluid velocity must be such that the coal particles can be kept in homogeneous suspension.

The mixture is pumped to successive slurry ponds, both underground and on the surface, and eventually to the preparation plant. The arrangement and location of the ponds depends on the reliability of the system and the effect on the system in the event of downtime failure.

8.5.2 Current Status

Many development and design problems currently inhibit the use of underground slurry transport. The most pressing problems are those related to development of reliable components, particularly pipeline coal feeding equipment. The feeder design problems center upon creating a reliable pressurized seal arrangement that will permit the input of solid particles at atmospheric pressure to some higher-pressure vessel or pipe. The solution to this problem is offered to you as an exercise.

REFERENCES

1. Thomas M. Nasiatka, "Continuous Face to Plant Coal Haulage Systems," in Nicholas P. Chironis, ed., *Coal Age Operating Handbook of Underground Mining*, McGraw-Hill Book Company, New York, 1977, p. 151.
2. "Steep Gradient Locomotive," *Coal Age Magazine*, December 1977, pp. 144-145.
3. Hurolo Davis, "Wide Conveyors Haul Overburden Out of Deep Pits," *Coal Age Magazine*, November 1977, pp. 96-99.
4. C. Frank Allen, *Railroad Curves and Earthwork*, 7th ed., McGraw-Hill Book Company, New York, 1931.
5. Robert S. Lewis, *Elements of Mining*, 3rd ed., John Wiley & Sons, Inc., New York, 1964.
6. J. W. Brantner, "Mine Haulage Locomotive Calculations," in Ivan A. Given, ed., *SME Mining Engineering Handbook*, Vol. 1, Society of Mining Engineers, AIME, New York, 1973.
7. Edwin C. Holden, "Underground Transport," in Robert Peele and John Church, eds., *Mining Engineer's Handbook*, Vol. 1, 3rd ed., John Wiley & Sons, Inc., New York, 1945.
8. "Mechanical, Hydraulic, Pneumatic Options in Deep-Mine Transport," in Ref. 1, p. 144.
9. *Trackwork: Frogs, Switches, Turnouts, Light Trackwork and Fastrak*, L. B. Foster Co., Catalog, Pittsburgh, Pa., 1979.
10. E. R. Traxler, "Get the Most from Your Conveyors," *Coal Age Magazine*, March 1978, p. 80.

11. *Installation Maintenance and Lubrication of Underground Belt Conveyors*, ELMAC Corp., Huntington, W.Va., 1974.
12. "Laser Beams Provide Accurate Advance for Underground Mining," in Ref. 1, pp. 12-15.
13. Federal Mine Safety and Health Act of 1977, Public Law 91-173 as amended by Public Law 95-164, U.S. Government Printing Office, Washington, D.C., 1977.
14. J. D. Spencer, T. J. Joyce, and J. H. Faber, eds., *Pneumatic Transportation of Solids*, Proceedings, Institute of Gas Technology IC8314, Bureau of Mines Symposium, Morgantown, W.Va., October 19-20, 1965.

PROBLEMS

1. A seam outcrops at point A. Point B is located 3 miles N82°W of A and 1584 ft below it. Point C is located $1\frac{1}{2}$ miles S43°W of A and 1480 ft below it. What is the practical bearing angle range for a mainline track haulage system?
2. A curve with a radius of 70 ft would require what track gage?
3. Design and specify an entry curve of $R = 45$ ft between two perpendicular entries each 16 ft wide. What is the maximum track gage?
4. Design and specify a curve with a $R = 25$ ft to blend a 20-ft track entry, N45°E, with a 16-ft track entry, N15°W. What superelevation would be needed for a 5-mph trip with a 42-in. gage?
5. Each shift, 650 tons of coal is produced underground. At a bulk density of 42 lb/ft^3, what is the minimum belt width if the belt speed is 200 ft/min?

9

Water Control System Layout

9.0 INTRODUCTION

The flow of water into a mine can present problems with effects ranging from minor nuisances, such as slowing normal operations, to catastrophic problems that cause the closing of the mine. The type of problem caused by water intrusion varies with the location affected.

1. Water within face operations can:
   (a) Create general discomfort for workers (especially in "low-coal" mines), resulting in discontent with the work environment.
   (b) Produce a boggy bottom, which interferes with proper equipment tramming, people walking, and haulage traction.
   (c) Increase maintenance problems due to sticky and gumming intrusions into bearings, gears, and so on.
   (d) Increase electrical short-circuit and grounding hazards, thus decreasing safety.
   (e) Corrode metal thereby decreasing equipment life and overall reliability.
   (f) Inhibit deterioration of explosive charges, slowing production and increasing danger to the workers.
   (g) Accumulate in the face so rapidly that work in that section must be abandoned.
   (h) Flood the working section and drown workers when a flooded abandoned mine or high-yield aquifer is inadvertently opened.

2. Water out from the face can:
   (a) Add to the humidity of mine air, which when routed to the face increases worker discomfort and decreases work output.
   (b) Prevent proper maintenance of the track when a boggy or soft bottom is encountered.
   (c) Corrode track, conveyor rope, and other metallic equipment, decreasing life and reliability.
   (d) Hasten the rotting of timbers (ties, roof supports, etc.), increasing maintenance cost, decreasing safety, and adding to the ventilation (CO_2) load.
   (e) Alternate dryness and wetness of entry walls, producing corresponding contractions and expansions in certain types of roof strata (particularly roof shales), resulting in premature roof failures.
   (f) Accumulate in low spots to the point that headings can be blocked to haulage or even ventilation.
   (g) Cause freezing in the winter months, inhibiting proper transportation (conveyor head pulleys and idlers may freeze) or creating safety hazards such as ice in escapeways.
3. Water affecting surface operations can:
   (a) Inhibit "dry screening" of coal at the preparation plant, decreasing plant efficiency and adding to the cost of the operation.
   (b) Affect water table levels.
   (c) Affect the quantity and rate of subsidence.
   (d) Present environmental pollution hazards since most mine water is acidic or highly laden with particulate matter.
   (e) Present surrounding downstream communities with flood danger.

The effects of mine water can be summarized as decreasing productivity, decreasing safety to both mining and nonmining personnel, decreasing mine and equipment reliability, decreasing mine and equipment maintainability, and generally decreasing the overall values of operations. The magnitude of these effects must be compared with the difficulty and cost of correcting them. In most eastern U.S. mining operations the presence of water is too great to ignore, since bituminous mining averages 5 to 6 tons of water pumped for each ton of coal.[1] Because of geohydrology in the East, mining without some inflow is often impossible. Thus the major concerns for mine designers are how to treat and control the water once it is in the mine. However, in certain western U.S. regions the hydrology is quite different, in that precipitation is less and the geologic structure affords shallow and limited storage. Various industries, such as agriculture, energy conversion, and mining, compete for the scarce water resources. Preservation of the water status by premining stabilization is therefore the prime consideration of mine designers.

9.1 FUNCTIONAL ANALYSIS

Controlling water in and around mining operations entails several subsystems, each accomplishing distinct tasks. The layout of the components that comprise these subsystems depends on knowledge of (1) the functions being performed, (2) the principles of accomplishing those functions and, (3) the practices used to implement the layout. The water control system is founded upon accomplishing the following:

1. Controlling water
2. Ensuring safety
3. Maximizing reliability
4. Improving maintainability
5. Optimizing value

9.1.1 Control Water

The *control water* function implies that water inflow to the mine can (1) be partially prevented or avoided; (2) can be routed, collected, and distributed away from critical areas; and (3) can be treated and disposed. Each of these operations (Fig. 9-1) forms a distinct subsystem with methods and activities used to achieve the needed results.

Prevention and avoidance. The two most effective and least costly measures that can be taken to control mine drainage are (1) preventing water from entering the mine, and (2) the avoidance of water in mining areas where inflow is likely. Both prevention and avoidance require a thorough understanding of the sources of mine water gained in the premining investigations of the hydrology of the region.

Hydrology is the study of water: its occurrence, circulation, distribution, physical and chemical properties, and its reaction with the environment. The hydrologic cycle is divided into three principal phases: (1) precipitation, (2) evaporation, and (3) runoff (including both surface and groundwater).[2,3] Within each phase occurs water transport, temporary storage, and change of strata. The *runoff* phase is shown in Fig. 9-2. Water enters the soil and infiltrates the surface due to the combined effects of gravity and capillary action of the soil. Gravity acts only vertically on the water, whereas capillary action acts both vertically and laterally to divert the water. Part of the water is stored within the soil profile; the remainder percolates into the groundwater system. It is the nature, location, and discharge characteristic of the groundwater system that is of importance to water control designers.

Groundwater is defined as the saturated zone of water storage below the surface of the earth occurring in both consolidated and unconsolidated

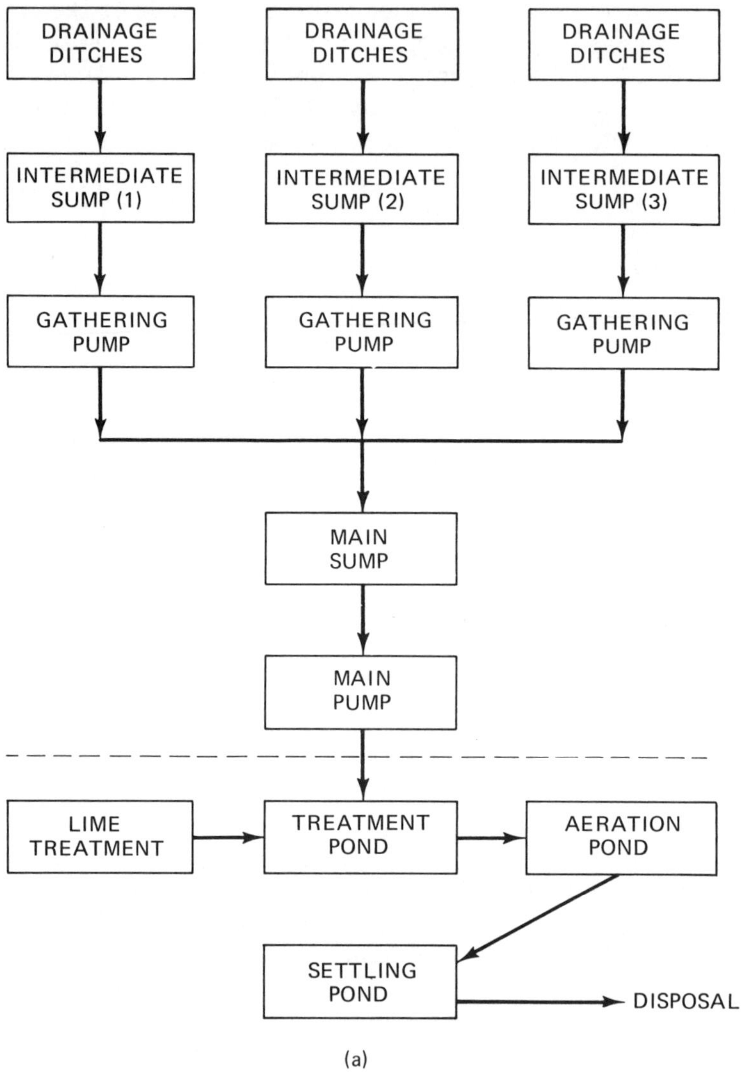

Figure 9-1 (a) Drainage flowsheet. (b) Phases and relationships needed to control mine water.

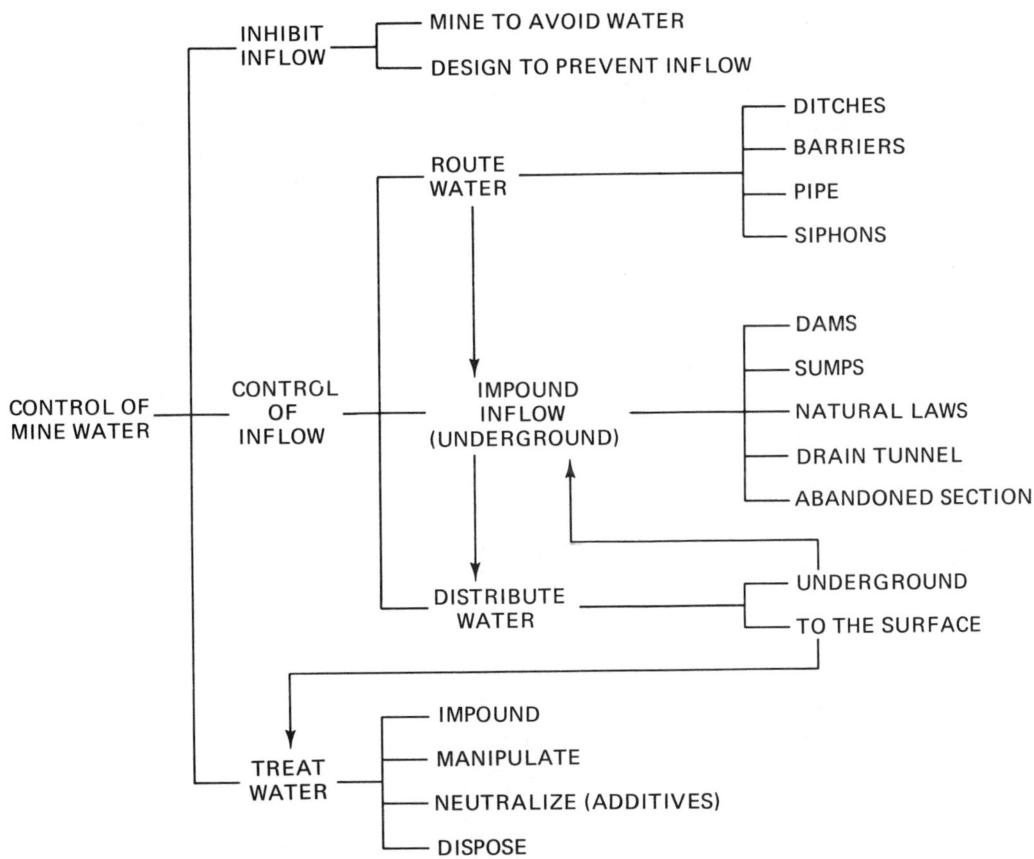

(b)

Figure 9-1 (continued)

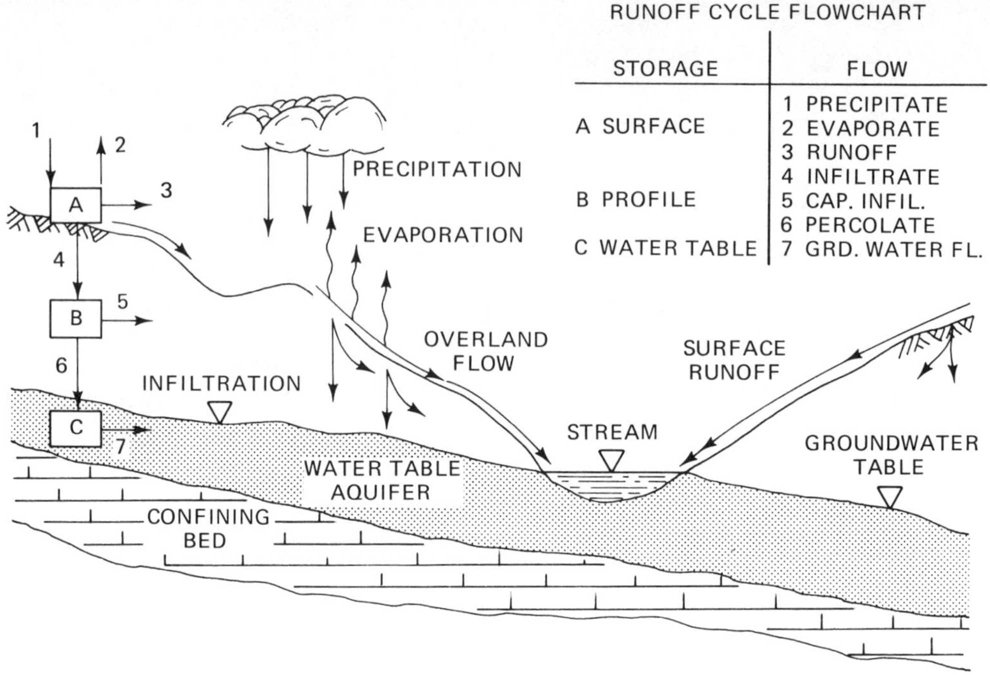

Figure 9–2 Profile view illustration of geohydrology concepts.

formations. Aquifers are the most important class of groundwater formations, with sedimentary materials constituting the largest fraction of that class. An *aquifer* is a water-bearing stratum of porous and permeable rock which can yield water at a significant rate.

The *porosity* of a material formation is the ratio of the volume of openings to the total volume of a formation. The porosity of a formation is a measure of how much water can be stored and is an indirect indication of the potential yield of water from that formation. Thus more-porous sandstone formations are generally a higher-yield aquifer than either unconsolidated limestone or shales. The finer the grains of the material and the closer these grains are compacted, the less will be the yield from a given formation. Thus loosely compacted sandstones yield more than tightly cemented sandstones.

Sandstone, limestone, and shale beds are all commonly encountered in mining and form an important class of aquifers called *sedimentary aquifers.* Together with normal percolation, sedimentary aquifers are the major sources of inflow. Additional sources of water include:

1. Normal surface water runoff and that resulting from flood conditions entering through outcrops, bore holes, and portals.
2. Abandoned and flooded mines which are broken into.
3. Mine methods causing caving, which destroys the competent, impervious rock formations above the mine, allowing percolation drainage from water tables, lakes, rivers, and so on, into the mine.

Prevention of inflow by mining in such a way as to *avoid* inflow is premised upon the premining data and availability of hydrographic maps. Hydrographic maps show the locations of bodies of water, rivers, streams (above and below ground level), the groundwater data (table density), the character and locations of aquifer (water-bearing stratum usually positioned between nonporous strata), and the geologic structures and formations (as they may affect the location and concentration of water). Many of these data are available from the USGS, which is charged with the task of gathering and publishing information about water resources. For each state the survey issues a folder and principal map to describe the hydrology investigations under way or recently concluded.[4] These studies should be consulted before developing prevention and avoidance plans.

Avoidance activities entail positioning portals in areas where direct surface runoff is unlikely; not mining to the outcrop; ensuring that the dust suppression equipment is turned off when not being used; not mining near abandoned mines or underground impoundments; avoiding sandstone aquifers; leaving head coal when pumping shales constitute the immediate top; pressure grouting an area prior to sinking a shaft; and leaving substantial pillars under lakes, streams, and rivers.

Prevention entails pressure grouting and cemetation of aquifers and pumping strata accidentally intersected during mining; locating ditches or dams around portals to divert groundwater runoff; capping or grouting bore holes; locating water collection rings at intermediate elevations on shafts; installing watertight collars on shafts; using barrier pillars to isolate wet sections from the rest of the mine; driving entries uphill in wet mines to maintain a dry face, while in dry mines entries should be developed downhill since retreat pillaring may cause groundwater inflow; and preleveling or draining the water table in advance of mining.

The effectiveness of avoiding or preventing inflow depends on the accuracy of the hydrographic data and maps, the location and type of aquifers, the type of mining, and so forth. Most preventive tactics that are taken are only partially successful, and the designer should assume that some water will be present in the mine.

Routing, collection, and distribution. Most mines will have some amount of water. The methods used by the designer to control the effects of that water constitute the second and generally most extensive subsystem. Three common practices are (1) routing the water with drain ditches, barriers, pipes, and siphons; (2) collecting the water with dams, sumps, or naturally occurring low spots; and (3) distributing the water away from critical areas or out of the mine, usually with pumps and piping systems.

Water flows in a direction perpendicular to the strike or bed contours and at a rate proportional to the slope. The nature of drainage is governed by the grade and uniformity of entry. Drain ditches used in haulageways and permanent entries are the easiest and least expensive method of drainage control. For particularly troublesome entries, drain pipe or tile may be required; however, the expense of such equipment is often prohibitive.

Sumps are underground impoundments of mine water. Sumps are either cut or shot into the mine floor and range in volume from a few cubic yards to entire sections. The size of each sump depends on the rate of flow into it, the desired duty cycle of the pumps that empty it, and the expense of installing and operating the sump. The purpose of a sump is to collect water in large enough volume that it may then be pumped either to a larger main sump or to various treatment impoundments located on the surface. (An exception to this is clean, neutral water which can be used to supply the water needs of the dust suppression and fire control systems.) Sumps are located at strategic locations in the mine, such as the bottoms of shafts, the ends of main drifts, and at naturally occurring lower profiles on the seam.

The distribution of water is achieved by arranging a sump, pump, and piping circuit. The most common circuit is the main sump, which is the largest and most permanent sump installation. It is located as close to a surface outlet as possible. Because of the large volume of water handled, centrifugal pumps are the most commonly used, although vertical turbine pumps are also used. For extremely deep mines, centrifugal pumps are positioned in series to supply the needed head requirements. These requirements and fluid flow fundamentals are discussed in detail in Section 9.2.

Water treatment. Most water collected in the mine and pumped to surface impoundments is acidic and high in suspended solids. The acidity of mine water is the primary problem because it can severely affect the ecology and aesthetics of an area. Acidic mine water is a chemical product resulting from the reaction of water with various sulfur compounds, principally iron pyrite. Equation 9-1 is the symbolic representation of the basic chemical reaction producing mine acid.[5]

$$FeS_2 + 7O_2 + 2H_2O \rightleftharpoons 2FeSO_4 + 2H_2SO_4 \tag{9-1}$$

where FeS_2 = iron pyrite
O_2 = oxygen (from the air)
H_2O = water
$FeSO_4$ = ferrous sulfate
H_2SO_4 = hydrosulfurite acid

The sulfuric acid must be treated prior to disposal or release to the surrounding region. The process of treatment is called *neutralization*. Most neutralization processes adjust the pH of mine water above 6.0 (pH is a relative measure of the acidic or alkaline nature of a fluid, with acidic fluids having pH values less than 7.0 and alkaline fluids have pH values greater than 7.0. Water is neutral, having a pH of 7.0.) A secondary result of neutralization is the formation of insoluble compounds. These must be removed before discharge to the surroundings. (Usually, these compounds are allowed to settle in a holding pond.)

The selection of the most appropriate neutralization process is a complicated comparison of the factors that present the most value for a particular mine acid condition. A discussion of these factors is beyond the scope of this book.

9.1.2 Ensure System Safety

The safety considerations in the design of the drainage system must first be for personnel who interact with the system and then for the equipment and property. In all but the most massive and catastrophic types of failures, such as mining into a flooded adjacent mine, the failure of the drainage system rarely threatens mine personnel. Similarly, equipment and property safety hazards directly tied to failures of the drainage system are few and can usually be avoided by designing sump capacities large enough to account for greater inflow over extended downtime. (Note that this is an assessment of the drainage system as outlined in Section 9.1.1 and must be considered as being unrelated to the dangers or hazards of not having the system, as noted in the introduction.)

The inherent safety of the drainage system can be increased by implementing some of the safety schemes presented in Chapter 5:

1. *Isolation schemes.* Locate large sumps away from commonly traveled areas and use barrier pillars to isolate wet sections of the mine.
2. *Alerted contact schemes.* Exploratory drilling must be undertaken within 200 ft of suspected accumulations of water. Most drilling follows the pattern depicted in Fig. 9-3.

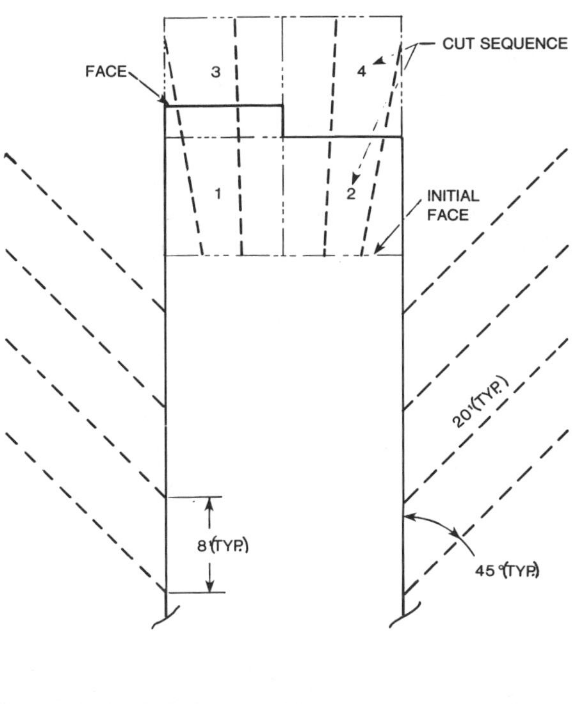

Figure 9-3 Methods of drilling in advance of mining.

9.1.3 Maximize Reliability

Like most mechanical systems, the reliability of the drainage system depends on the modes of system failure and the factors presented in Chapter 5. The majority of drainage system failures affecting the downtime percentage of the drainage system are pump or pump-motor failures. The most important modes of these failures include:

Aging failures. Corrosion by acidic mine water is a common failure, which can be avoided by use of corrosion-resistant materials such as plastics, Cr–Ni stainless steels, or nickel-compounded cast irons.

Degradation failures. Particle and cavitation erosion are both experienced underground. The effects of the former are reduced by placing strainers on the pump intake lines, selecting wear-resistant materials (such as nickel compounds), and limiting flow velocities. The cavitation results from poor motor–pump matching and too fast an operation speed. Both conditions are obviously avoidable.

Improper use, abuse or maintenance. This mode is the most prevalent and easily avoided. Most pump problems occur on the intake (suction) side of the pump and are related to improper installation. The nature of the installation is discussed in detail later.

The reliability factors that most increase the life expectancy of the drainage system are the following:

Select reliable components. The proper specification of pipe, pump, and auxiliary components comprising the drainage system is basic to improving the reliability index of all systems. Material selection is especially critical because of the acidic conditions of the mine water.

Use redundant circuits. The most critical component in the system and the one with the lowest reliability index is the main sump pump. Depending on the capacity of the main sump, the rate of inflow into the mine, and the consequences to productivity if failure were to occur, a redundant main sump pump arrangement may prove advantageous.

Provide proper installation. The proper installation of components is second only to the proper initial selection of those components in maximizing the reliability of the system. The pump layout is the most important factor governing proper installation, followed by piping installation and then by the layout of support equipment.

Provide regular inspection and maintenance. As with all equipment, periodic inspection and routine maintenance are minimum requirements for system performance. Bearings need to be lubricated, packing adjusted or replenished, strainers backflushed and cleaned, and so forth.

These are the four most significant factors affecting the drainage system. The other factors are important but to a lesser degree. For example, simplicity was not listed because a drainage system is inherently simple.

9.1.4 Improve Maintainability

Maintainability as a function has been divided into:

1. The probability of restoring to service a failed system and the forecasted time needed for restoration.
2. The ability to advance, retreat, or expand the functions of the system.
3. The ability to inspect and maintain the operational status of a system.

Service restoration. The restoration of service depends on the cause of the failure, the location of the failure, and the severity of failure. Most drainage system failures are pump failures. Accordingly, most (centrifugal) pump troubles occur on the suction side of the pump and may be as high as 85%.[6] This estimate is for a typical aboveground installation. Underground the percentage is higher, since the suction-side factors contributing to failure (e.g., particulate suspension, acid water, sump stability) are so unpredictable. These factors are avoidable through careful suction side layout, which is discussed in later sections.

The worst pump failure scenario would be one in which the pump casing fails, necessitating pump replacement. Because of the simplicity of the system and the modular fitting of the components, the amount of time required for replacement varies from several hours to several shifts. The only system failures that would present an exception to this time estimate would be a major roof fall or a mine flood that blocked restoration efforts. Restoration is then dependent on access and the severity of spillover damage.

System modification. The ability to advance, retreat, or modify the water control system is the simplest of any of the functional systems. The system is truly modular in construction, combining pipe, pumps, and fittings together to adapt easily to entry requirements. Three joints are commonly used: (1) welded joints, (2) screw ends with couplings, and (3) quick-disconnect couplings. *Welded joints* are used on mainlines, where modifications are few and high bending or axial loads may be experienced. *Screwed ends and couplings* increase ease of assembly or disassembly of the pipeline. The screwed ends and couplings are the most common joints and are only slightly less reliable than welded ends. *Quick-disconnect couplings* permit rapid assembly, disassembly, and modification to the line underground at a minimum labor cost. The quick-disconnect joint should be used only on those gathering operations where the ability to advance the system rapidly is more critical than maximum reliability.

Inspection and maintenance. Together, periodic inspection and routine maintenance rank as the second most significant tasks in improving system reliability. Observational inspections include examining packing and casing areas for leaks; checking for excessive vibration or noise, indicative of bearing malfunctions; and measuring oil and grease levels. Instrument inspections include temperature measurement of both pump and motor, pump discharge pressure, speed of rotation, flow rate, and power consumption. Of these, the simplest and most important measurement is discharge pressure. Too high a discharge pressure may indicate blocked flow, closed valves, or excessive speed; too low a pressure may indicate blocked sump supply, loss of suction, drawing of air, worn impeller, or too slow operation speed.

The simplicity and predictability of the drainage system components can make preventive replacement strategies quite successful. Preventive replacement can be used for seals, seats, shafts, and impellers but is more frequently applied to replacement of the entire unit. However, for preventive replacement to be effective, the cost of the preventive action must be less than if failure were allowed to occur. Since the drainage system is less than crucial to the mining operation, the cost of replacing a pump before failure may equal the failure-cued replacement cost. Thus preventive replacement may not be beneficial.

9.1.5 Optimize Value

Determination of the comparative value of different layouts and methods assume that equal primary benefits are derived. For example, assume that water inflow is a problem in setting up a longwall panel. Any layout alternative must provide a relatively water-free environment as the primary benefit. Following this initial requisite, each alternative must be compared based on additional benefits, limitations, and relative costs. This particular problem requires a knowledge of fluid mechanics and underground practices.

9.2 FLUID MECHANICS

The principles that govern the flow of water are identical to those discussed previously governing the flow of air. The two most important concepts, continuity of mass and conservation of energy, are summarized by the following equation:

$$M_{in} - M_{out} = M_{accumulated} \quad (7\text{-}4)$$

or

$$(VA)_{in} - (VA)_{out} = (\text{volume})_{acc} \quad (7\text{-}7)$$

and

$$\frac{P_1}{\gamma} + \frac{V_1^2}{2g_c} + Z_1 - \frac{P_2}{\gamma} + \frac{V_2^2}{2g_c} + Z_2 = \text{head loss} \quad (7\text{-}14)$$

or

$$\frac{P_1}{\gamma} + \frac{V_1^2}{2g_c} + Z_1 + H_{pump} = \frac{P_2}{\gamma} + \frac{V_2^2}{2g_c} + Z_2 + H_{loss} \quad (7\text{-}20)$$

where H_{pump} is the additional head resulting from the input of a pump located between points 1 and 2.

9.2.1 Siphon Application

The *siphon* is a nonmechanical method of lifting water above containment obstructions and discharging at a point below the initial fluid level. Theoretically, a perfect vacuum, 0 psia, can lift a column of water 34 ft at sea level. Figure 9-4(a) (page 316) depicts this condition. Accordingly, Eq. 7-20 can be written as follows:

$$\frac{P_1 - P_2}{\gamma} = H_2 - H_1$$

$$\left[14.7 - 0\left(\frac{\text{lb}}{\text{in.}^2}\right)\right] 2.307\left(\frac{\text{ft}_{H_2O}}{\text{psi}}\right) = \Delta H_2$$

$$33.91 \text{ ft} = \Delta H_2$$

The actual lifting capacity of a siphon is less since it depends on the atmospheric pressure, the velocity of the flow, and the resistance losses through the line. In general, no more than 70% of theoretical lift is possible. Thus, if 34 ft is theoretically attainable, 24 ft would be the practical maximum lift. The practical limit varies with atmospheric pressure. The ratio between actual atmospheric pressure and sea level and corresponding lifting heights are shown below.

Elevation	P/P_{atm}	Practical lifting height (ft)
Sea level	1	24
5,000 ft	0.832	19.97
10,000 ft	0.6877	16.50

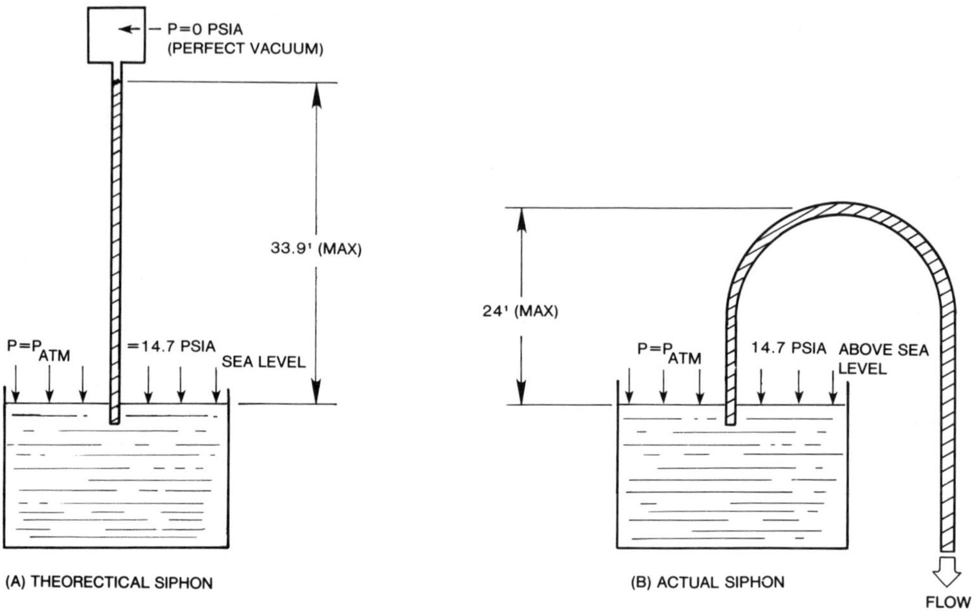

Figure 9-4 Siphoning as a means of water control.

The theoretical velocity of flow depends on this practical lifting difference. The energy equation can now be written

$$\frac{P_{\text{atm}}}{\gamma} + \frac{(V_{\text{res}})^2}{2g_c} = \frac{P_{\text{bend}}}{\gamma} + \frac{(V_{\text{bend}})^2}{2g_c} + \Delta Z + H_L \qquad (9\text{-}2)$$

$$\frac{P_{\text{atm}}}{\gamma} - \frac{P_{\text{bend}}}{\gamma} - \Delta Z - H_L = \frac{(V_{\text{bend}})^2}{2g_c}$$

At sea level these values can be substituted:

$$\sqrt{2g_c(34\text{ ft} - 24\text{ ft} - H_L)} = V_{\text{bend}}$$

Under ideal flow conditions ($H_L = 0$),

$$V_{\text{bend}} = 25.4\text{ ft/sec} = V_{\text{pipe}}$$

Siphons can be constructed to control fluid level in reservoirs "automatically." Figure 9-5 depicts one such design. Water level rises in the reservoir until the siphon is filled and a steady-state condition is obtained. As inflow to the sump decreases, the siphon causes the level of water to decrease until air intake into the siphon breaks the vacuum, thereby stopping the flow.

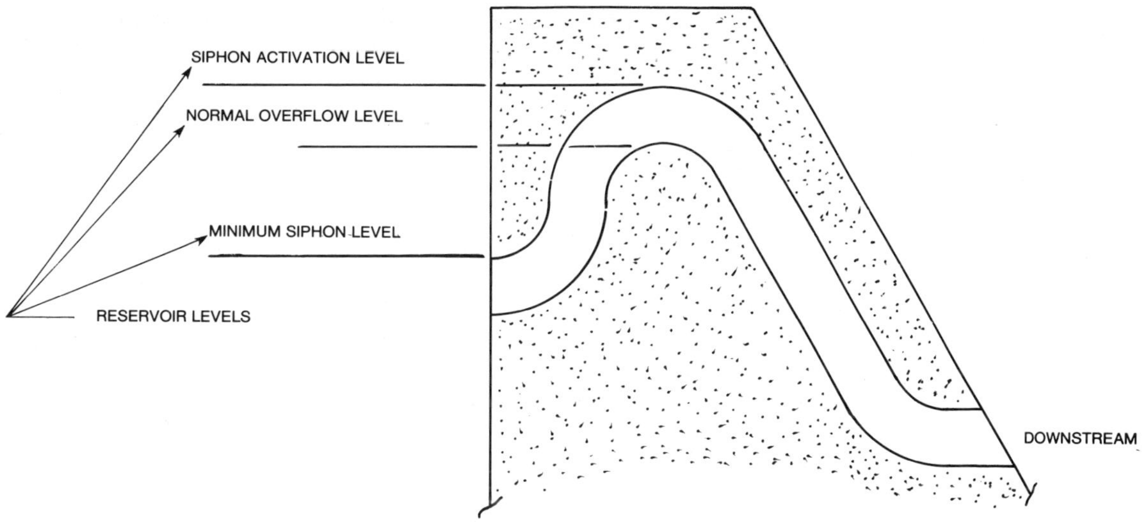

Figure 9-5 "Automatic leveling" of reservoir using a siphon.

9.2.2 Pipeline Flow

By far the most common method for controlling the presence of water underground is the use of pumps and pipelines to convey water from one sump to another. The flow of water is principally governed by the continuity equation (Eq. 7-4) and the energy equation (Eq. 7-20).

The head loss term in Eq. 7-20 is the summation of frictional losses resulting from the resistance of the pipe to flow and minor losses contributed by pipe fittings, valves, and bends. The amount of friction head loss varies according to the type and condition of the pipe, the velocity of the fluid, the diameter of the pipe, and the length of the pipe. Mathematically, these can be stated

$$h_f = f\left(\frac{L}{D}\right)\frac{V^2}{2g_c} \tag{9-3}$$

where h_f = frictional head loss (ft)
f = frictional coefficient of pipe (nondimensional)
L = length of pipe (ft)
D = inside pipe diameter (ft)
V = average velocity of flow (ft/sec)
g_c = gravitational constant (32.17 ft/sec^2)

The friction coefficient is an empirical value that depends on the relative roughness of the pipe and the Reynolds number. The relative roughness, ϵ/D, is a ratio of the absolute wall roughness, ϵ (ft), to the inside diameter of the pipe, D (ft). Steel pipe has an absolute roughness of 0.00015; cast iron, 0.00085; galvanized iron, 0.0005; and plastic pipe, less than 0.0001, or "smooth."

The *Reynolds number* is a ratio of the inertial forces to the viscous forces and is calculated using Eq. 9-4:

$$\mathrm{Re} = \frac{VD}{\nu} \tag{9-4}$$

where Re = Reynolds number (dimensionless)
V = velocity (ft/sec)
D = diameter (ft)
ν = absolute viscosity (ft/sec^2)
(1.40×10^{-5} ft/sec = ν_{H_2O} at 60° F)

The empirical relationship between the Reynolds number, relative roughness, and the resulting frictional factor is shown in the Moody diagram, Fig. 9-6. For Re $\leqslant$ 2000, the flow is laminar; for Re $\geqslant$ 2500, the flow is turbulent. Most pipeline flows are turbulent and the resulting frictional factor for mining conditions is between 0.015 and 0.03.

Example 1

Determine the head loss resulting from the flow of 1000 gal/min of mine water through 750 ft of 10-in.-diameter Schedule 40 cast-iron pipe.

Solution

1. Determine the velocity of flow from Eq. 7-7:

$$V = \frac{Q}{A} = \frac{\left(1000\frac{\text{gal}}{\text{min}}\right)\left(0.1337\frac{\text{ft}^3}{\text{gal}}\right)\left(\frac{1}{60}\frac{\text{min}}{\text{sec}}\right)}{\frac{\pi}{4}\left(\frac{10.02^*}{12}\right)^2}$$

2. Determine the Reynolds number:

$$\mathrm{Re} = \frac{4.07(10.02/12)}{1.40 \times 10^{-5}} = 2.43 \times 10^5$$

(which is highly turbulent flow).

3. Determine the relative roughness:

$$\frac{\epsilon}{d} = \frac{0.00085}{10.02/12} = 0.001$$

4. From the Moody diagram, find the Reynolds number on the abscissa. Move straight up until intersecting the 0.001 relative roughness curve. Moving to the left parallel with the abscissa, read the corresponding frictional coefficient, f = 0.021.

*Table 9-1: 10-in. Schedule 40 pipe, inside diameter V = 4.07 ft/sec.

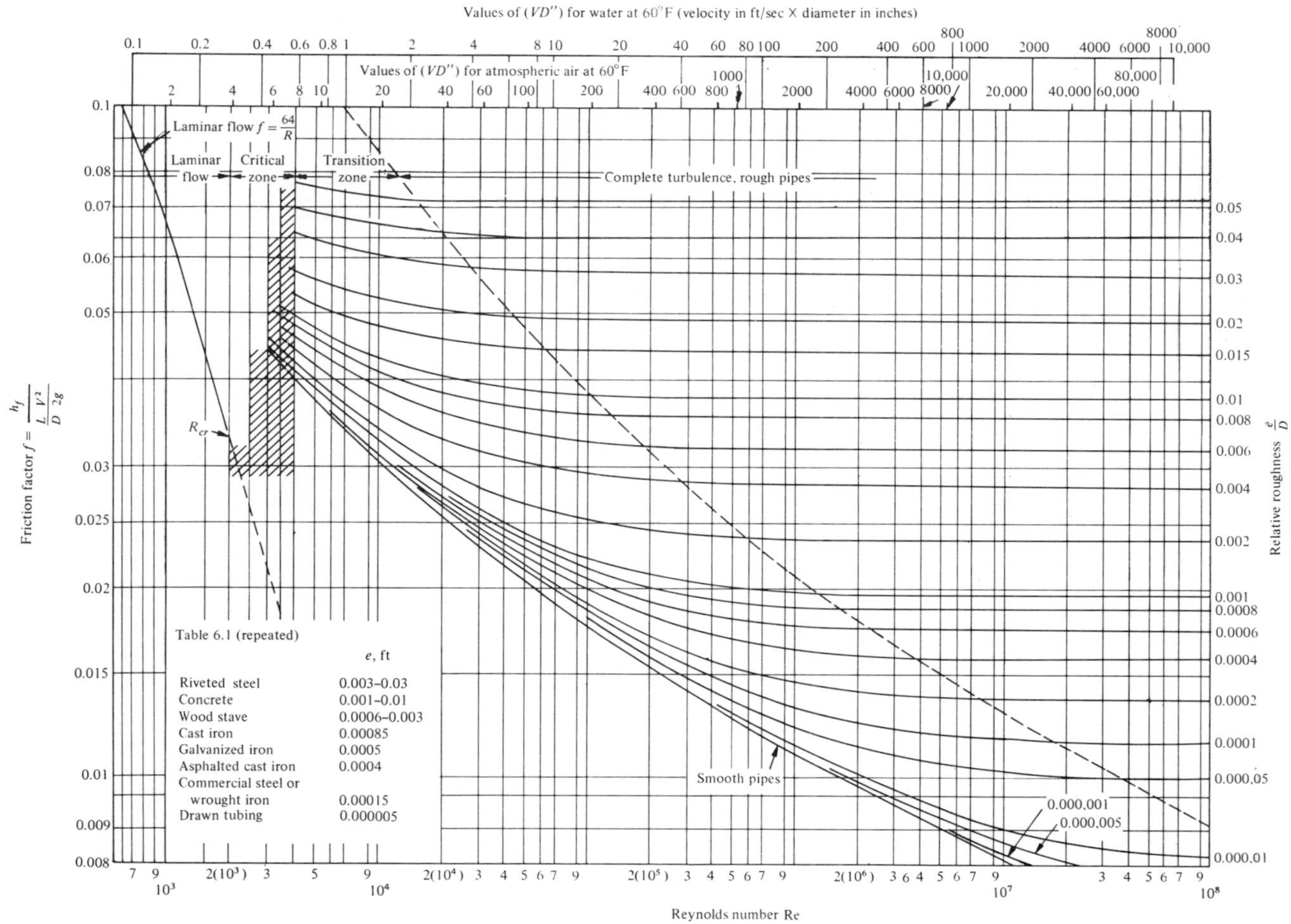

Figure 9-6 Friction factors for any kind and size of pipe. (Courtesy of the Hydraulic Institute, New York, 1961.)

TABLE 9-1 STANDARD PIPE DIMENSIONS

Nominal size (in.)	Pipe OD (in.)	Inside diameter (in.)									
		Sched. 10	Sched. 20	Sched. 30	Sched. 40	Sched. 60	Sched. 80	Sched. 100	Sched. 120	Sched. 140	Sched. 160
$\frac{1}{8}$	0.405				0.269		0.215				
$\frac{1}{4}$	0.540				0.364		0.302				
$\frac{3}{8}$	0.675				0.493		0.423				
$\frac{1}{2}$	0.840				0.622		0.546				0.466
$\frac{3}{4}$	1.050				0.824		0.742				0.614
1	1.315				1.049		0.957				0.815
$1\frac{1}{4}$	1.660				1.380		1.278				1.160
$1\frac{1}{2}$	1.900				1.610		1.500				1.338
2	2.375				2.067		1.939				1.689
$2\frac{1}{2}$	2.875				2.469		2.323				2.125
3	3.500				3.068		2.900				2.624
$3\frac{1}{2}$	4.000				3.548		3.364				
4	4.500				4.026		3.826		3.624		3.438
5	5.563				5.047		4.813		4.563		4.313
6	6.625				6.065		5.761		5.501		5.189
8	8.625		8.125	8.071	7.981	7.813	7.625	7.439	7.189	7.001	6.813
10	10.750		10.250	10.136	10.020	9.750	9.564	9.314	9.064	8.750	8.500
12	12.750		12.250	12.090	11.934	11.626	11.376	11.064	10.750	10.500	10.126

5. Substitute these values into Eq. 9.3 and calculate the head loss:

$$h_f = 0.021 \left(\frac{750}{10.02/12}\right) \frac{(4.07)^2}{2(32.17)}$$

$$= 4.856 \text{ ft}$$

Minor losses such as the shock losses in the ventilation system can be computed as part of the frictional losses if the appropriate equivalent length (L/D) is known. For piping systems some of the equivalent lengths for common fittings and valves are well known, as shown in Table 9-2.[7] The total head loss can be calculated according to Eq. 9-5.

$$H_L = f[(L/D)_p + (L/D)_v] \frac{V^2}{2g_c} \tag{9-5}$$

TABLE 9-2 EQUIVALENT LENGTH OF NEW STRAIGHT PIPE FOR VALVES AND FITTINGS (TURBULENT FLOW ONLY)

Fittings			Pipe Size																				
			$\frac{1}{4}$	$\frac{3}{8}$	$\frac{1}{2}$	$\frac{3}{4}$	1	$1\frac{1}{4}$	$1\frac{1}{2}$	2	$2\frac{1}{2}$	3	4	5	6	8	10	12	14	16	18	20	21
Regular 90° ell.	Screwed	Steel	2.3	3.1	3.6	4.4	5.2	6.6	7.4	8.5	9.3	11	13										
		Castiron										9.0	11										
	Flanged	Steel			0.92	1.2	1.6	2.1	2.4	3.1	3.6	4.4	5.9	7.3	8.9	12	14	17	18	21	23	25	30
		Castiron										3.6	4.8		7.2	9.8	12	15	17	19	22	24	28
Long radius 90° ell.	Screwed	Steel	1.5	2.0	2.2	2.3	2.7	3.2	3.4	3.6	3.6	4.0	4.6										
		Castiron										3.3	3.7										
	Flanged	Steel			1.1	1.3	1.6	2.0	2.3	2.7	2.9	3.4	4.2	5.0	5.7	7.0	8.0	9.0	9.4	10	11	12	14
		Castiron										2.8	3.4		4.7	5.7	6.8	7.8	8.6	9.6	11	11	13
Regular 45° ell.	Screwed	Steel	0.34	0.52	0.71	0.92	1.3	1.7	2.1	2.7	3.2	4.0	5.5										
		Castiron										3.3	4.5										
	Flanged	Steel			0.45	0.59	0.81	1.1	1.3	1.7	2.0	2.6	3.5	4.5	5.6	7.7	9.0	11	13	15	16	18	22
		Castiron										2.1	2.9		4.5	6.3	8.1	9.7	12	13	15	17	20
Teeline flow	Screwed	Steel	0.79	1.2	1.7	2.4	3.2	4.6	5.6	7.7	9.3	12	17										
		Castiron										9.9	14										
	Flanged	Steel			0.69	0.82	1.0	1.3	1.5	1.8	1.9	2.2	2.8	3.3	3.8	4.7	5.2	6.0	6.4	7.2	7.6	8.2	9.6
		Castiron										1.9	2.2		3.1	3.9	4.6	5.2	5.9	6.5	7.2	7.7	8.8
Teebranch flow	Screwed	Steel	2.4	3.5	4.2	5.3	6.6	8.7	9.9	12	13	17	21										
		Castiron										14	17										
	Flanged	Steel			2.0	2.6	3.3	4.4	5.2	6.6	7.5	9.4	12	15	18	24	30	34	37	43	47	52	62
		Castiron										7.7	10		15	20	25	30	35	39	44	49	57
180° return bend	Screwed	Steel	2.3	3.1	3.6	4.4	5.2	6.6	7.4	8.5	9.3	11	13										
		Castiron										9.0	11										
	Regular flanged	Steel			0.92	1.2	1.6	2.1	2.4	3.1	3.6	4.4	5.9	7.3	8.9	12	14	17	18	21	23	25	30
		Castiron										3.6	4.8		7.2	9.8	12	15	17	19	22	24	28
	Long radius flanged	Steel			1.1	1.3	1.6	2.0	2.3	2.7	2.9	3.4	4.2	5.0	5.7	7.0	8.0	9.0	9.4	10	11	12	14
		Castiron										2.8	3.4		4.7	5.7	6.8	7.8	8.6	9.6	11	11	13
Globe valve	Screwed	Steel	21	22	22	24	29	37	42	54	62	79	110										
		Castiron										65	86										
	Flanged	Steel			38	40	45	54	59	70	77	94	120	150	190	260	310	390					
		Castiron										77	99		150	210	270	330					
Gate valve	Screwed	Steel	0.32	0.45	0.56	0.67	0.84	1.1	1.2	1.5	1.7	1.9	2.5										
		Castiron										1.6	2.0										
	Flanged	Steel								2.6	2.7	2.8	2.9	3.1	3.2	3.2	3.2	3.2	3.2	3.2	3.2	3.2	3.2
		Castiron										2.3	2.4		2.6	2.7	2.8	2.9	2.9	3.0	3.0	3.0	3.0
Angle valve	Screwed	Steel	12.8	15	15	15	17	18	18	18	18	18	18										
		Castiron										15	15										
	Flanged	Steel			15	15	17	18	18	21	22	28	38	50	63	90	120	140	160	190	210	240	300
		Castiron										23	31		52	74	98	120	150	170	200	230	280
Swing check valve	Screwed	Steel	7.2	7.3	8.0	8.8	11	13	15	19	22	27	38										
		Castiron										22	31										
	Flanged	Steel			3.8	5.3	7.2	10	12	17	21	27	38	50	63	90	120	140					
		Castiron										22	31		52	74	98	120					
Coupling or union	Screwed	Steel	0.14	0.18	0.21	0.24	0.29	0.36	0.39	0.45	0.47	0.53	0.65										
		Castiron										0.44	0.52										
	Bell mouth inlet	Steel	0.04	0.07	0.10	0.13	0.18	0.26	0.31	0.43	0.52	0.67	0.95	1.3	1.6	2.3	2.9	3.5	4.0	4.7	5.3	6.1	7.6
		Castiron										0.55	0.77		1.3	1.9	2.4	3.0	3.6	4.3	5.0	5.7	7.0
	Square mouth inlet	Steel	0.44	0.68	0.96	1.3	1.8	2.6	3.1	4.3	5.2	6.7	9.5	13	16	23	29	35	40	47	53	61	76
		Castiron										5.5	7.7		13	19	24	30	36	43	50	57	70
	Reentrant pipe	Steel	0.88	1.4	1.9	2.6	3.6	5.1	6.2	8.5	10	13	19	25	32	45	58	70	80	95	110	120	150
		Castiron										11	15		26	37	49	61	73	86	100	110	140
	Sudden enlargement	$h = \frac{(V_1 - V_2)^3}{2g}$ feet of liquid; if $V_2 = 0$, $h = \frac{V_1^2}{2g}$ feet of liquid																					

Source: Reproduced with permission from *Pipe Friction Manual*, 3rd ed., Hydraulic Institute, New York, 1961.

where $(L/D)_p$ = equivalent length contributed by the pipe
$(L/D)_v$ = equivalent length contributed by fittings and valves

Most mining applications attempt to keep the range of the ratio of total head loss to total head between 0.1 and 0.05. The proper design and sizing of the pipe is critical to effecting pumping. Increases in the quantity of flow result in squared increases in the head loss and a corresponding change in H_L/H_T ratio.

Power requirements. The power required to move water over some distance is proportional to the flow rate, the specific weight of the fluid, and the total head as expressed in Eq. 9-6a.

$$\text{hp} = \frac{QH_T}{550} \tag{9-6a}$$

where Q = volumetric flow rate (ft^3/sec)
H_T = total head (ft)
γ = specific weight of fluid (lb/ft^3)
550 = conversion constant [(ft-lb/sec)/HP]

Recall that the total head for a fluid system was expressed as

$$H_T = \frac{\Delta P}{\gamma} + \frac{\Delta V^2}{2g_c} + \Delta Z + H_L \tag{7-20}$$

and the specific weight of water is

$$\gamma_{\text{water}} = 62.4 \text{ lb/ft}^3.$$

Now, if water is being pumped from one *open* sump to another, that is, intake and discharge pressures are equal, Eq. 9-6a can be written

$$\text{hp} = \frac{Q\,[(\Delta V^2/2g_c) + \Delta Z + H_L]}{8.817} \tag{9-6b}$$

In most cases the percentage of velocity head compared to the total head is small, usually less than $\frac{1}{2}$%. Thus Eq. 9-6a is simplified further by dropping the velocity head term, to yield

$$\text{hp} = \frac{Q(\Delta Z + H_L)}{8.817} \tag{9-6c}$$

Then assuming a head loss of 5% of the total head, the equation can be further reduced to

$$\text{hp} = \frac{Q(1.05\,\Delta Z)}{8.817}$$

or

$$\text{hp} = \frac{Q\,\Delta Z}{8.4} \tag{9-6d}$$

where 8.4 is the conversion constant [(ft^4/sec)/hp]

Equation 9-6d can be used with confidence for cases in which large vertical heads are encountered; however, for horizontal systems Eq. 9-6b is more appropriate.

Example 2

Determine the ideal pump horsepower needed to pump 500 gal/min of mine water 600 ft up a vertical shaft.

Solution A. From Eq. 9-6d

$$\text{hp} = \frac{500\,\dfrac{\text{gal}}{\text{min}}\left(0.1337\dfrac{\text{ft}^3}{\text{gal}}\right)\left(\dfrac{1}{60}\dfrac{\text{min}}{\text{sec}}\right)(600\text{ ft})}{8.4}$$

$$= 79.58\text{ hp}$$

Solution B. To solve this problem using Eq. 9-6b, we must make some assumptions: (1) assume (initially) that the head loss is 5% of the total head; (2) assume that the frictional coefficient of the pipe, f, is 0.02; and (3) assume (initially) that the total head is the vertical head plus the frictional head.

Approximate the size of the pipe:

$$\text{total head} = \text{vertical head} + \text{frictional head} = 1.05\,(600) = 630\text{ ft}$$

The maximum frictional head is therefore 30 ft:

$$H_f = f\,\frac{L}{D}\,\frac{V^2}{2g_c} \tag{9-3}$$

Substitute

$$V = \frac{Q}{A} = \frac{Q}{(\pi/4)D^2}$$

into Eq. 9-3 to yield Eq. 9-7.

$$H_f = \frac{8fLQ^2}{\pi^2 g_c D^5} \tag{9-7}$$

Substitute values for the terms:

$$30 = \frac{8(0.02)(600\text{ ft})[500(0.1337)\tfrac{1}{60}]^2}{\pi^2(32.17)D^5}$$

where $D^5 = 0.0125$

$D = 0.416$ ft

$= 5.0$ in. (approximate pipe size)

From Table 9-1, 5-in. Schedule 40 pipe is 5.047 in. in ID.

Using this size, the actual head loss is calculated as illustrated in Example 1, using

$$V = 8.096$$

$$\text{Re} = 2.41 \times 10^5$$

$$\frac{\epsilon}{d} = 0.002$$

From the Moody diagram:

$$f = 0.025$$

$$H_L = 0.025\left(\frac{600}{5.047/12}\right)\frac{(8.096)^2}{2g_c}$$

$$= 36.3 \text{ ft}$$

Thus the calculated horsepower using Eq. 9-6b is

$$\text{hp} = \frac{(1.114)\left[\dfrac{(8.096)^2}{2(32.17)}\right] + 600 + 36.3}{8.817}$$

$$= 80.54 \text{ hp}$$

The velocity head, 1.02 ft, contributes only 0.16% to the horsepower computations! The remainder is contributed by the head loss resulting from the necessary pipe sizing. Obviously, using Eq. 9-6d is much faster and as accurate as Eq. 9-6b for vertical installation.

Efficiency. No pumping operation is 100% efficient. The series of equations 9-6 must be modified to reflect this inefficiency according to Eq. 9-8.

$$\text{hp}_{\text{input}} = \frac{QH_T}{550\eta} \tag{9-8}$$

where η is the decimal equivalent of the ratio of output power to input power. The value of η depends on several factors but primarily upon the type of pump being used. Positive-displacement pumps have an efficiency that ranges from 70 to 95%, and nonpositive displacement pumps range from 50 to 80%.

Example 3

For Example 2, assume a centrifugal pump is to be installed with an efficiency of 70%. What input horsepower is required?

$$\text{hp}_{\text{input}} = \frac{80.54}{0.7}$$

$$= 115.1 \text{ hp}$$

The amount of electrical energy consumed (input to the electrical motor) will be this amount divided by the drive train and motor efficiencies.

9.2.3 Modifications

In actual operation the foregoing calculations provide a working baseline from which the actual specifications are calculated. One of the most significant modifications or design considerations will be the condition of the pipe over time. Head loss will usually increase with time. In some cases this is due to the increasing relative roughness of the pipe causing an increase in f. This increase in f can be as much as 500% in 20 years for 4- to 8-in. pipes.[7] Another factor is the buildup of deposits inside the pipe, which increases the velocity and thus causes a squared increase in head loss (see Eq. 9-3). In general, a doubling of the head loss over the operation life is common. To maintain head loss at less than 10% of the total head, a design guideline of 5% should be used for initial calculations.

Another factor that results in modification to the system component selection is the pump efficiency degradation over time. For example, wear, erosion, and corrosion of the pump impeller result in a significant decrease in the pumping efficiency. A decline of 5 to 10% is usually enough to justify rebuilding the pump (see Problem 4).

9.3 COMPONENTS

The components used in a mine drainage system can be divided into three general classes:

1. Piping hardware, including hoses, lines, and fittings.
2. Valving hardware, including operational and safety valves used to control the flow.
3. Pump-related hardware, including the pumps and motors used to increase the total head of the system.

9.3.1 Piping Hardware

The type of pipe or hose used for a given application depends on the permanency of the situation, the volume and head requirements, and the erosive and corrosive quality of the mine water. Piping costs usually represent from 50 to 80% of the installation cost. Various types of pipe materials are available. The best material is, of course, that which provides the function at the lowest cost. For low-head applications, plastic pipe of less than 200 psi is a far better value than metal pipe. Compared to metals, plastic (which covers an extremely large grouping of materials) provides better corrosion resistance, better erosion resistance, less head loss, faster and easier installation, and less initial cost (usually). Disadvantages of plastic include limitations of temperature, impact strength, and internal pressure (in larger-diameter pipes).

The internal pressure limitations must be qualified. For example, the hydrastatic design stress of A. O. Smith's Silver Thread gloss reinforced vinyl polyester pipe is 6800 psi (hoop stress) at 75°,[9] whereas the usual design limit for centrifugally cast iron is 6000 psi.[8] In general, maximum continuous working pressures are confined to 150 psi (346 ft) at the working temperature, although some are as high as 550 psi in smaller sizes. When greater pressures and larger sizes are needed, such as in shafts or bore hole discharge arrangements, metal pipe is usually needed.

The underground piping system is subject to various hazards and loads not normally encountered aboveground. In haulageways the piping should be buried or otherwise protected from possible damage resulting from occasional derailment. In shafts the pipe should be anchored at two points along every 20-ft joint. Long pipelines or those installations with quick-acting valves require air chambers to be installed along the route to lessen the impact forces resulting from too quick a shutoff. The high fluid pressures resulting from a mass of water being suddenly stopped, known as *water hammer*, can cause the pipe to rupture. Air chambers permit the shock to be partially absorbed and dissipated by compression of the gas.

9.3.2 Valves

The categories of valves shown in Fig. 9-7 are divided according to the purpose they serve, including on–off service, regulation of flow, safety service, and flow diversion.

On–off valves (e.g., gate, ball, and plug valves) are normally used fully open or fully closed. The use of these valves as pressure or flow regulation valves is rarely recommended because of high noise and excessive erosion rates. Gate valves have been the traditional on–off valve. Ball and plug valves, however, are usually best for extremely dirty service since the seat and seal are both protected. Full-ported ball valves provide a lowhead loss, an ability to "rod out" the line, a price competitive with gate valves, and most important, an ability to automate the drainage system easily. (Only a 90° rotation is required of ball valves, whereas gate valves usually require more than 360° to fully open and close the valves.)

Regulation valves (e.g., globe, needle, pinch) are used to change the percentage of flow opening, thereby changing the pressure drop across the valve and velocity. Because of their straight-through nature pinch valves are usually best for extremely erosive flows; however, the maximum operational pressure is limited to less than 150 psi. As a result globe valves with hardened and renewable seats are used when flow regulation is needed. (For infrequent regulation or low flow velocities, ball valves can be used for regulation service.)

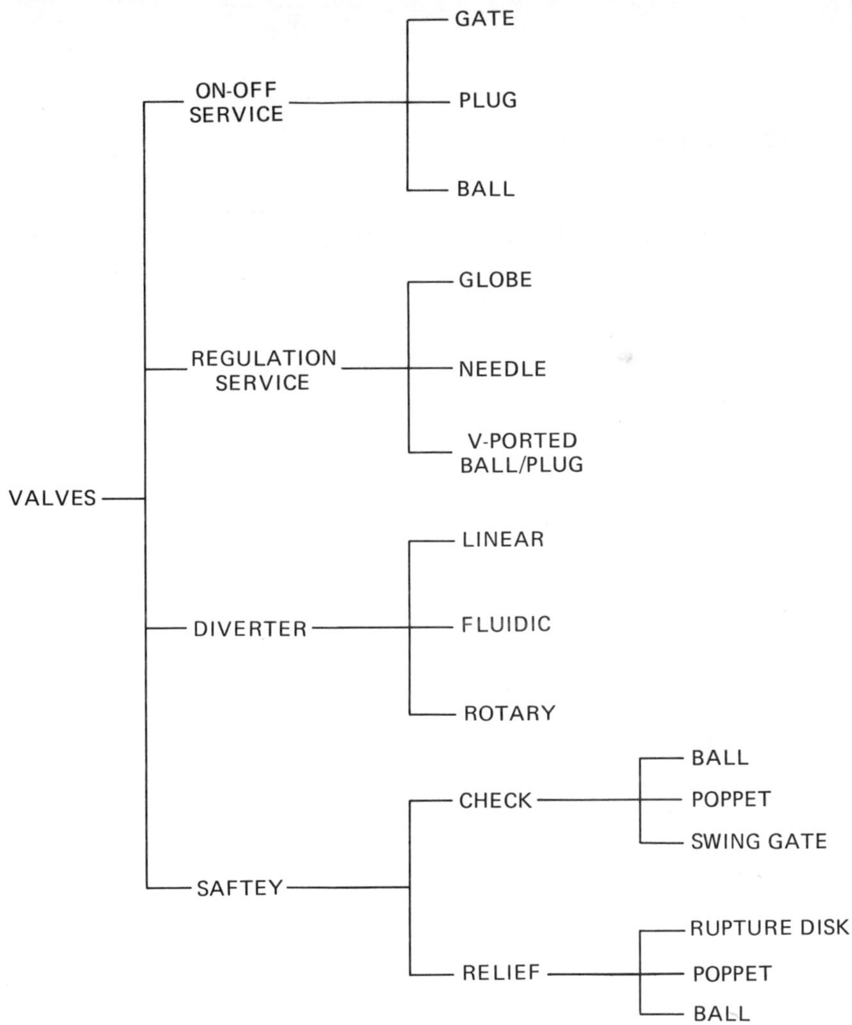

Figure 9-7 Classification of valves by operation.

Safety valves used for the drainage system include check valves (e.g., swing check, ball check, poppet) and safety relief valves (e.g., poppet). Check valves are used to ensure a single direction of flow. In the event of a reversal or loss in pressure, the valve closes, preventing backflow. For dirty service the ball check provides the most reliable service since the ball is free to rotate and therefore wears uniformly. Relief valves are usually closed by a predetermined spring load. When the upward force caused by static fluid pressure on the port area reaches or exceeds the downward force of the spring on the ball, the valve opens, relieving the internal pressure. When the force caused by the pressure is less than the force of the spring, the valve closes automatically.

Diversion valves permit two or more parallel lines to operate simultaneously. Normally, the piping used to effect drainage is of the series type and diversion valves are not used.

9.3.3 Pumping Hardware

Figure 9-8 shows the wide range of pumps available for obtaining the head requirements of the drainage system. The family of pumps are most frequently divided into positive and nonpositive flow displacement groups. A *positive displacement pump* inputs a given volume of fluid to the system regardless of downstream conditions. A *nonpositive displacement pump* inputs a quantity of fluid to the system inversely proportional to the square root of pressure.

Figures 9-9(a) and (b) show the principal performance difference between the two categories. Figure 9-9(a) shows that for a positive displacement pump, as the pressure downstream increases, the flow rate remains the same. Figure 9-9(b) shows the pressure-flow rate curve for a typical nonpositive displacement pump. As the downstream pressure increases, the pump output flow rate decreases.

Both positive and nonpositive displacement pumps are used underground. Centrifugal pumps are the major type of nonpositive displacement pump, and rotary design pumps are the most frequently used type of positive displacement pump. The characteristics of each type are summarized in Table 9-3. Centrifugal pumps are most appropriate for pumping larger volumes (50 gal/min and greater) of fluid at moderate pressures (less than 500

Figure 9–8 Classification of major pumps.

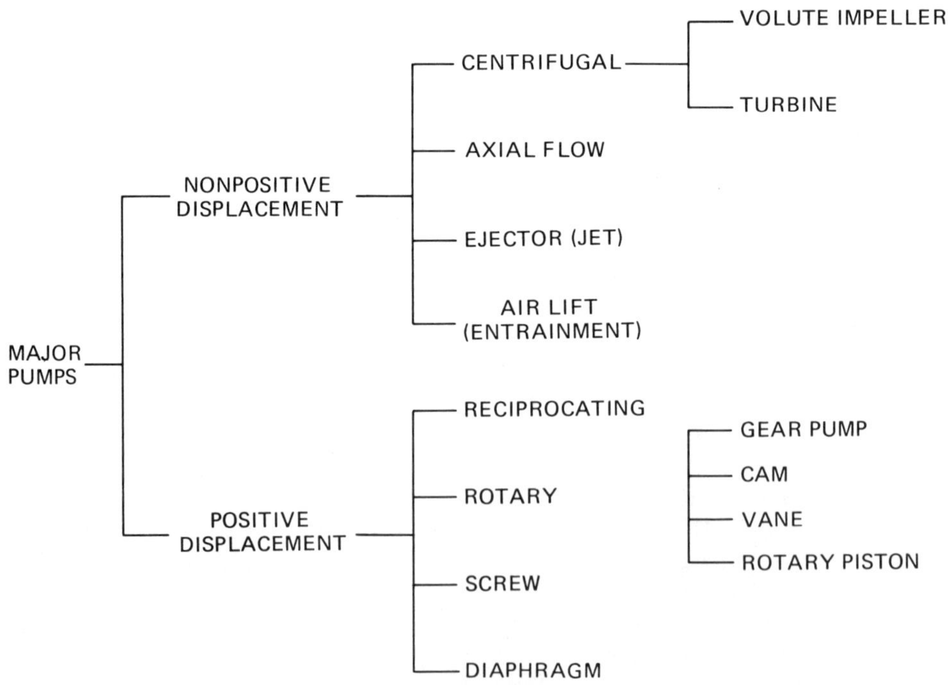

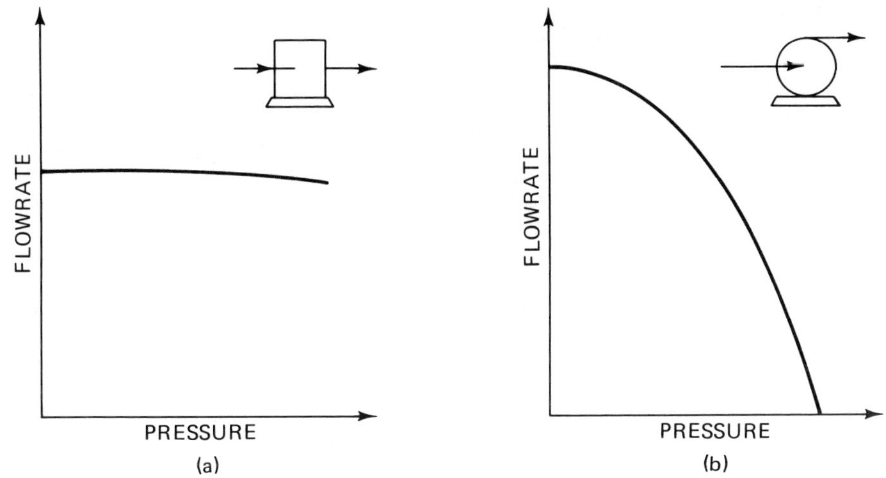

Figure 9-9 (a) Characteristic curve for a positive displacement pump (b) Characteristic curve for a non-positive displacement pump.

TABLE 9-3 CHARACTERISTIC COMPARISON OF CENTRIFUGAL AND ROTARY PUMPS

Characteristic	Centrifugal pumps	Rotary pumps
Flow rate	$Q = f(P_2 S_1 D)$	$Q = f(S, A)$
	Q = flow rate	Q = flow rate
Factors	P = upstream pressure	
	S = shaft speed (fixed)	S = shaft speed (variable)
	D = impeller diameter (fixed)	A = pump size
Range	0 to 20,000 gal/min	0 to 5000 gal/min
Pressure	$P = f(Q, N, S^2)$	$P = f(S, N)$
	P = upstream pressure	P = upstream pressure
	Q = flow rate	
Factors	S = shaft speed [Note: when S is constant $P = f(D^2)$]	S = shaft speed
	D = impeller diameter	
	N = number of stages	N = type of design and stress limits
Range	0 to 200 psi/stage	0 to 1000 psi
Efficiency	$\eta = f(N, T)$	$\eta = f(N, T)$
	η = efficiency	η = efficiency
Factors	N = design and size	N = design (slip)
	T = time, wear, erosion	T = time, wear, erosion
Range	50% to 85%	80% to 95%
Horsepower (input)	hp = $f(S^3, Q, P, \eta)$	hp = $f(S^3, Q, \eta)$
	hp = horsepower	hp = horsepower

TABLE 9-3 (Continued)

Characteristic	Centrifugal pumps	Rotary pumps
Factors	S = shaft speed Q = flow rate [f(P)] P = upstream pressure [$f(Q)$] η = efficiency	S = shaft speed Q = flow rate [$f(S)$] η = efficiency
Other		
Initial cost	\$ = $f(M, H, Q)$ M = materials of construction H = total head Q = flow rate usually higher than rotary	\$ = $f(M, H, Q)$ (same)
Maintenance cost	\$ = $f(S, F)$ S = shaft speed F = fluid characteristic usually lower than rotary	\$ = $f(S, F)$

psi), as is needed for main sump discharge. Rotary pumps, usually lighter, more portable, and less costly, are used to pump water from remote (face) locations on the main sump.

Pump drives. The drive power for pumps is usually furnished by electrical motors. Ac squirrel cage induction motors (SCIM) are the most common, least expensive, and most reliable type of drive motors. The speed and torque characteristics of the dc class of motors are rarely needed. Additionally, the higher initial cost and greater frequency of maintenance make the dc motor clearly the second choice. An exception is in older mines in which only dc power is available. In these cases the dc shunt motor is most commonly used for pumping service.

9.3.4 Selection Criteria

Performance. The effective operation of the system depends first and foremost on the proper selection of the components. The first criterion is performance. *Performance* is the ability to move the quantity of water at the necessary head. The design flow rate, Q, depends on the size of the main sump, the desired duty cycle of the pump, and the rate of inflow into the main and sump. For continuous, steady-state operation the flow rate of the pump must just equal the inflow. This is rarely done since inflow changes with hydrographic activity. The duty cycle is usually intermittent, being automatically controlled by the fluid-level monitors in the sump. In these cases the minimum flow rate of the system must be greater than the maximum rate of inflow. The maximum system flow rate is often two, three, or four times the maximum rate of inflow. For a duty cycle of 1-hr pump operation every 4 hr, the flow rate must be four times the inflow rate.

The total head is principally the sum of the height potential and the head loss. The height is usually known and a design criterion for head loss was suggested as 5% of the total head. Given this criterion and the flow rate, Q, the diameter of the pipe can be selected, as demonstrated in Example 2, Solution B.

Another design criterion is based on *pump efficiency*. Figure 9-10 shows a typical set of performance curves for a centrifugal pump with a $2\frac{1}{2}$-in. discharge port. At maximum pump efficiency, 80%, the flow rate is 160 gal/min at a 230-ft head. The corresponding velocity in a $2\frac{1}{2}$-in. Schedule 40 pipe is 10.72 ft/sec. A range of pump output velocities from 8 to 12 ft/sec is characteristic of centrifugal pumps operating at peak efficiency. The best pump velocity may or may not correspond to the velocity resulting from a 5% head loss criterion. An "optimum" design is achieved when the two criteria produce equal-velocity outcomes.

Reliability and maintainability. Selection factors most affecting reliability and maintainability are operation speed, mounting type, materials of construction, and design. Different operation speeds for the same head and

Figure 9-10 Motor mounts.

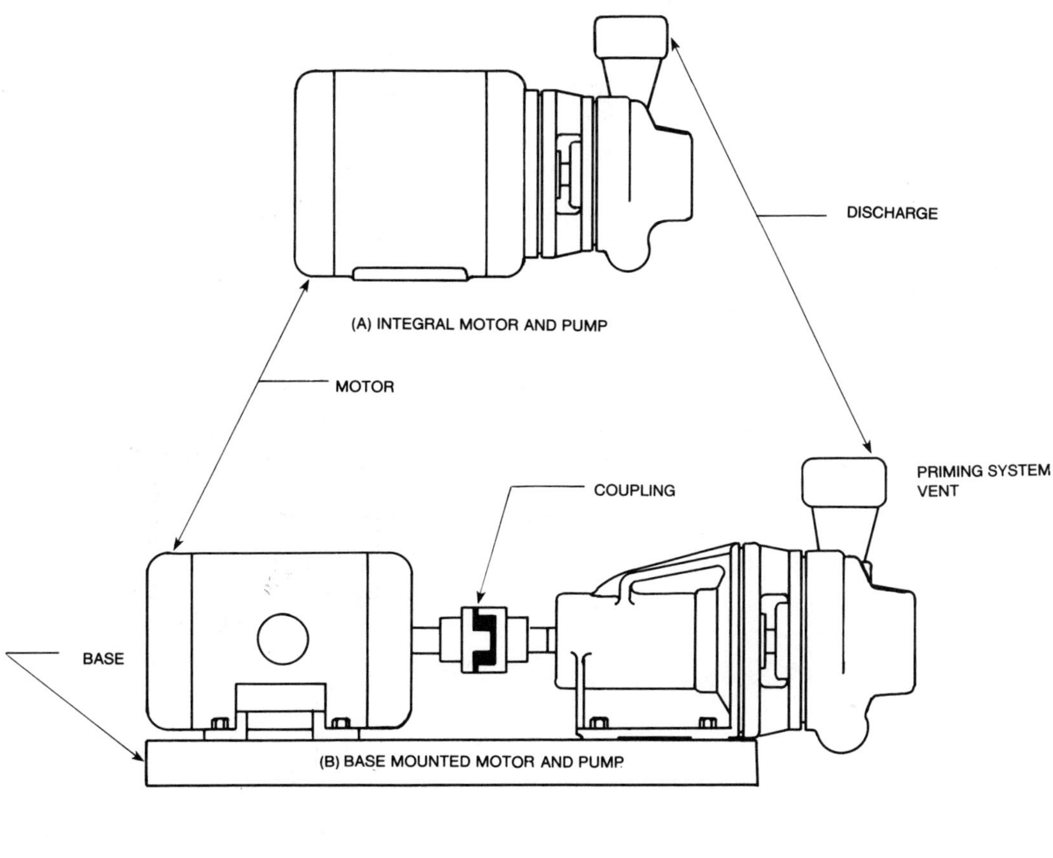

capacity characteristic are available from different pump designs. The expected life and maintenance cost are directly proportional to speed. Thus a pump rotating at 900 rpm will usually last twice as long as one operating at 1800 rpm. (Although slower-speed pumps last longer, they also cost significantly more; thus the expected benefit-to-cost ratio must be considered.)

The two types of motor-pump mounts depicted in Fig. 9-11 are (1) the *close-coupled mount* and (2) the *free base mount*. The close-coupled unit requires no site alignment and requires much less maintenance than the free base unit. Additionally, the close-coupled unit is less expensive, more compact, and easier to install. Failure of either the pump or the motor requires a change out of the entire assembly, which is a costly disadvantage. Another limitation is that close-coupled units are usually produced for typical sizes and applications; however, most mining service is highly atypical.

Various materials are used in pump construction. Erosion and corrosion caused by acidic mine water require special material considerations. Traditionally, the 300 series of stainless steels have been chosen as impeller and case materials; however, high cost, poor erosion resistance, and the availability of new coatings and composite materials have altered purchasing practices. High-alumina ceramics, glass-reinforced plastics, and silicon carbide

Figure 9-11 Flowchart and layout of pump house.

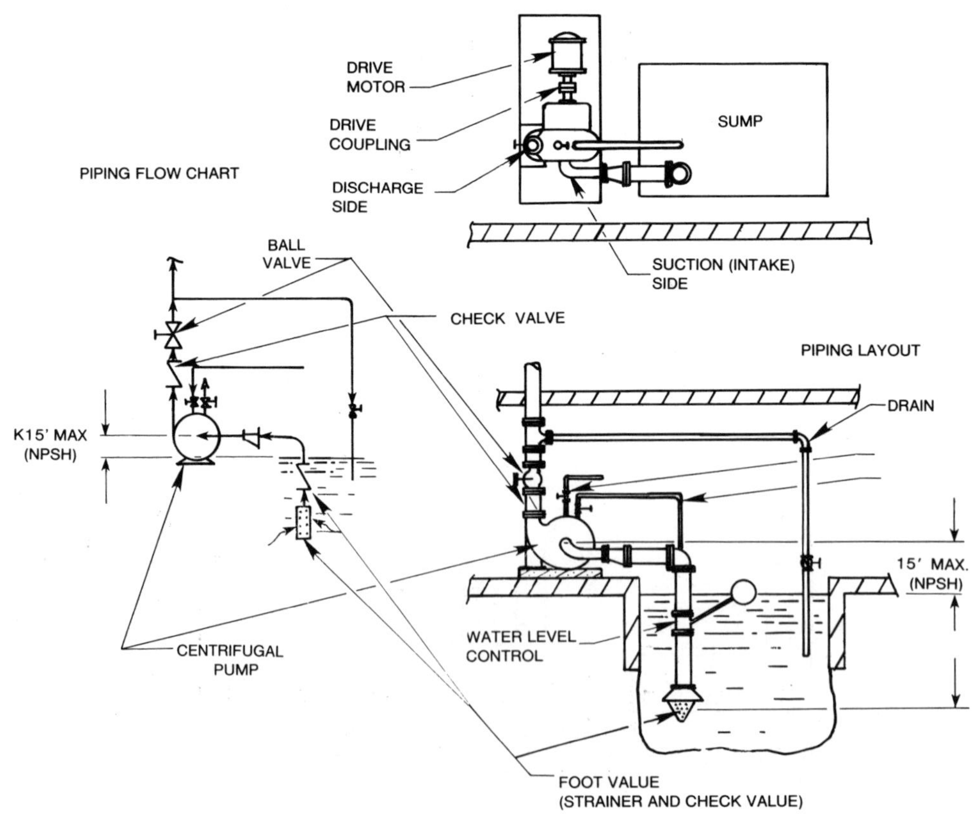

spray coatings are examples of the newer materials improving the reliability and maintainability of pump components. Manufacturers should be consulted concerning the use of alternative materials for your particular case.

Reliability and maintainability depend on good design more than any other factor. Several indicators of good design were discussed in Chapter 5. For centrifugal pumps you should particularly look for multiple bearing support for the impeller shaft, deep packing glands, exclusion seals on shafts preventing coal dust infiltration, seal cages using either water or grease traps to seal the impeller shaft, and above all, complete maintenance specification schedules and procedures.

Value. Once the pump performance reliability and maintainability characteristics are established, the last selection criterion is the value or relative cost of various alternatives. Piping for most installations represents the greatest item of cost. The initial sizing of the pipe using the 5% head loss design criterion should be considered only an approximation of the optimum size. The optimum pipe size depends on (1) the initial and per year cost of pipe, and (2) the cost of power. Small-diameter pipe has a lower initial cost; however, because of the greater head loss it may eventually cost much more than larger-diameter pipe. The yearly cost of various sizes over the life of the system is best illustrated by example.

Example 4

Assume that various alternatives in pipe size are being considered for Example 2, in which 500 gal/min is pumped 600 ft. Initially, we calculated a 5-in. pipe size to give a 5% head loss. Is this the most cost-effective size?

Solution. Assume:

1. Life of installation 20 years, at which time the pipe will require replacement,
2. Interest rate, yearly maintenance cost, and 20-yr replacement percentage to average 22% per year,
3. The cost of power to be 5.5 cents/kW-hr.
4. The efficiency of the pump to be 65% and the pump operates a total of 10 hr/day every day (3653 hr/year).

Procedure:

1. The fixed overhead costs per size will be the product of the initial cost times 22%.
2. The yearly operating costs will be the cost of power needed to overcome head loss. The kW-hr value is calculated as

$$\frac{\text{kW-hr}}{\text{yr}} = \frac{Q(\text{head loss/size})(0.746)(\text{hr/year})}{8.817} \tag{9-9}$$

where 0.746 = conversion constant (kW/H_p)

hrs = hours of operation during each year

Substitution yields

$$\frac{\text{kW-hr}}{\text{yr}} = \frac{(1.114)(\text{head loss})(0.746)(3653)}{(8.817)(0.65)}$$

$$= 529.791(\text{head loss})$$

The cost is

$$C = (529.791)(0.055)(\text{head loss}) \tag{9-10}$$
$$= \text{cost of lost power/yr}$$
$$= 29.138(\text{head loss})$$

3. Construct a table as shown.

Nominal pipe size(in.)	Initial cost	22% of initial cost	Head loss (ft)	Cost/yr	Total cost
3	$ 2,086	460	441	$12,850	$13,310
4	3,056	672	108	3,147	3,819
5	4,431	975	36.3	1,058	2,033
6	5,419	1,192	12.6	367	1,559
8	7,808	1,718	3.7	108	1,826
10	10,758	2,367	1.3	38	2,405

Analysis: Although the 5-in. pipe provides a 5% head loss, both the 6- and 8-in.-diameters appear to provide the lowest yearly cost. (A more detailed analysis would require a pump-motor sizing for the 5-, 6-, and 8-in. designs and the resulting costs determined similarly.)

9.4 MAIN SUMP LAYOUT

The most important subsystem in the drainage network is the main sump layout and installation. The installation is controlled by the piping layout, which depicts line sizes, critical dimensions, component parts and materials, and needed construction details. Although the designer has some leeway in the layout, much of the design is governed by the pumping requirements.

9.4.1 Suction-Side Layout

As noted earlier, roughly 85% of the problems of pump installations can be attributed to defective suction side or pump inlet layout. If a pump is to perform to its rated potential, the layout must be capable of supplying the quantity of water at some pressure at the intake side of the pump. The pump inlet pressure, measured in psia and called the *net positive suction head* (NPSH), is responsible for pushing fluid into the impeller. When the available NPSH is less than the required NPSH, the pump is "starved" for fluid, the vapor pressure of the water is reached, and cavitation occurs. This results in a rapid performance decline and eventual failure.

The ability of the pump to lift water is inversely proportional to the discharge flow rate. Stated another way, the greater the flow rate capacity, the more NPSH that must be available to the pump. The theoretical maximum NPSH available would be 34 ft, just as for the siphon. Similarly, the more realistic maximum NPSH limit at 5000 ft would be 19.97 ft. This value would decrease as the suction pipe became longer and smaller in diameter (thereby increasing head loss). Historically, 15 ft has been established as the maximum suction height. Fifteen feet is only a rough installation guideline, depending principally upon the altitude (available atmospheric pressure), the head loss of the suction pipe (length, diameter, and number of fittings), and the pump discharge. A pump may require 5 ft NPSH at 100 gal/min but at 235 gal/min will require at least 15 ft NPSH. The greater the required NPSH, the less will be the lift capabilities of the pump and the lower must be the pump installation.

Underground suction side layouts shown in Fig. 9-12 require that strainers be installed to eliminate excessive particle contamination. The strainer internal area should be four to five times the internal pipe area to avoid excessive suction side head loss. Similarly, strainers should be cleaned periodically to avoid declining NPSH availability.

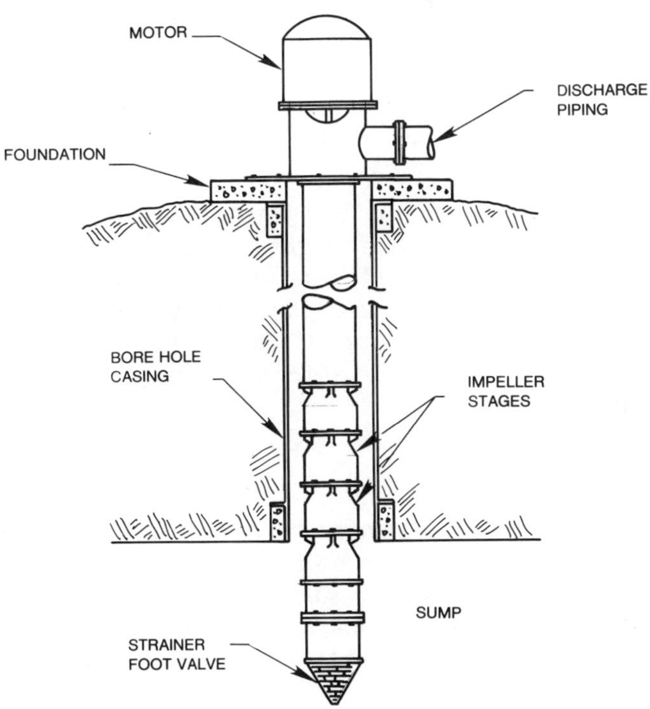

Figure 9-12 Vertical turbine pump.

Centrifugal pumps are not self-priming; that is, they cannot lift water when air is in the line. Foot valves are installed, often as part of the strainer assembly, to prevent the suction side water from backflowing to the sump after the pump stops. Care should also be taken to ensure the absence of air pockets in the suction side piping. When installed, the valve stems should be pointed horizontally or slightly downward to prevent air from being trapped in the suction side. For centrifugal pumps, auxiliary priming is needed when lift is required.

To summarize the characteristics of the suction side piping:

1. Design the suction side pipe to be as short as possible, have as few bends as possible and be one size larger than the pump inlet port diameter.
2. Design the available NPSH at the site equal to that NPSH required of the pump and as small as possible. The usual maximum is 15 ft.
3. Install foot valve and strainer, and add 15% head loss calculations as a penalty.
4. Avoid any pipe design or fitting layout resulting in trapped air.
5. Provide complete support for the suction side piping independent of pump.
6. Place the intake end of pipe at least 3 ft below the free surface of the reservoir.

9.4.2 Discharge-Side Layout

The size of the discharge line can be several sizes greater than the pump size but should never be less. A ball check and on-off valve should be installed just downstream of the pump, with the check valve adjacent to the pump. The check valve prevents backflow from the line through the pump and eliminates the stress on the pump caused by the weight of water in the line.

The discharge piping should be supported independently of the pump. If water hammer is possible, air cushions or surge tanks should be installed to avoid excessively stressing the pipe or pump. If welded pipe joints are used, unions need to be installed on either side of the pump to permit rapid pump disassembly and change out.

9.4.3 Pump-Motor Layout

The location of the main pump house is governed by the location of the main sump. The main sump is located at the bottom of shafts, naturally occurring low spots or in close proximity with surface treatment facilities. Sumps take the form of water held back by dams, reservoirs cut or shot into the entry bottom, or lower abandoned workings that are allowed to fill with water. Dams are generally the best type of reservoir design for providing maximum NPSH since the pump inlet may be located below the free surface of the water. Other sump designs necessitate a lifting capacity and lower available NPSH.

Regardless of the sump design, the pump should be located as close to the sump discharge as possible. Enough space should be provided to meet the inspection and maintainability needs. When possible, roof-mounted I-beams should be installed above the foundation to permit assembly of portable crane or block and tackle. The foundation is ideally 4- to 6-in.-thick concrete base with an area large enough to provide a floor bearing load of $\frac{1}{2}$ to 1 ton/ft^2. Permanent pump houses are of block construction, surrounding and protecting the motor pump instation. These stations must be located on separate air splits.[10]

The installation of the pump requires leveling between inlet and discharge piping and leveling between the pump and its drive motor. Alignment between the motor and pump is critical to ensuring packing reliability and minimizing wear. Flexible couplings are installed between the motor drive and the pump impeller shaft to prevent excessive vibration transmission between the two components but should not be used to alleviate poor alignment. Once alignment is completed, grouting is added around pump, motor, and piping supports to ensure alignment.

9.4.4 Alternative Installation

Vertical turbine submersible pumps provide an alternative method for pumping water vertically or for draining otherwise inaccessible mines. Figure 9-12 shows a submersible turbine pump installation. The advantages of vertical turbine pumps are that centrifugal pump problems associated with the suction side piping are practically eliminated, elaborate sump and pump station layouts are not needed, and accessibility to the region is attained through vertical bore holes from 6 to 24 in. in diameter. Since the vertical turbine pump is an adaptation of the centrifugal concept, efficiencies and performance are roughly the same. Initial cost, however, is higher and a detailed cost evaluation should be performed before a final selection is made.

REFERENCES

1. John K. Berry, "Drainage," in Samuel M. Cassidy, ed., *Elements of Practical Coal Mining*, Society of Mining Engineering, AIME, New York, 1973.
2. James M. Murray, "Groundwater Hydrology," in Donald M. Gray, ed., *Handbook on the Principles of Hydrology*, National Research Council of Canada and Water Information Center, Inc., Port Washington, N.Y., 1970.
3. Henry M. Morris and James M. Wiggert, *Applied Hydraulics in Engineering*, 2nd ed., The Ronald Press Co., New York, 1972.
4. M. M. Thompson, *Maps for America*, U.S. Dept. of the Interior, U.S. Geological Survey, U.S. Government Printing Office, Washington, D.C., 1979.

5. R. L. Loofbourow, "Ground Water and Ground Water Control," in Arthur B. Cummins and Ivan A. Given, eds., *SME Mining Engineering Handbook*, Vol. 2, Society of Mining Engineers, AIME, New York, 1973.
6. Tyler G. Hicks, *Pump Operation and Maintenance*, McGraw-Hill Book Company, New York, 1958.
7. Clarence B. Anderson, "Mechanics of Fluids," in Theodore Baumeister and Lionel S. Marks, eds., *Standard Handbook for Mechanical Engineers*, 7th ed., McGraw-Hill Book Company, New York, 1967.
8. Reno C. King, "Pipe and Pipe Fitting," in Ref. 7.
9. *Engineering and Design Guide Fiberglass Reinforced Piping Systems, Manual No. 9600*, A. O. Smith Inland, Inc., Little Rock, Ark., 1979.
10. *Federal Mine Safety and Health Act of 1977*, Public Law 91-173 as amended by Public Law 95-164, U.S. Government Printing Office, Washington, D.C., 1977.

PROBLEMS

1. Determine the head loss resulting from pumping 1250 gal/min through 900 ft if 8 in. Schedule 40 steel pipe is used.
2. What size pipe would give a 5% head loss pumping 600 gal/min 450 ft to the surface? How many horsepower are required?
3. The cost of rebuuilding the pump in Example 2 is \$213. If the cost of electrical power is 6.8 cents kW-hr and the pump must operate continuously, when will the pump rebuild cost equal the power lost costs, assuming that the pump efficiency falls from 70% to 60%?
4. Determine the head loss for the system depicted below. Should any modification to this design be made?

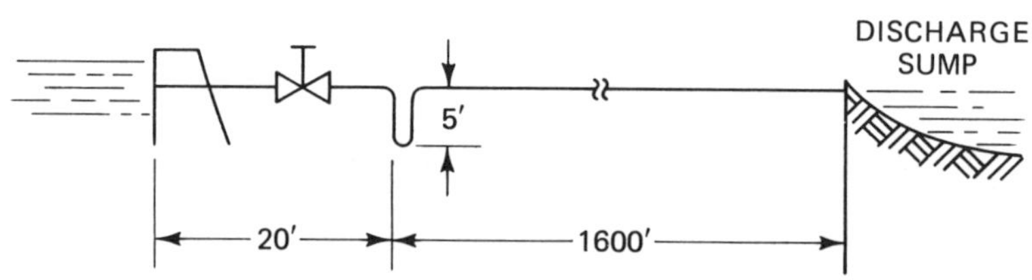

SPECIFICATIONS

1. FLOW RATE - 40 gpm
2. $2\frac{1}{2}''$, SCH. 40, CARBON STEEL PIPE

5. Determine the ideal pipe (material and size) needed to pump 65 gal/min 620 ft to the surface over a 15-yr period. Assume that the power cost is 6.2 cents/kW-hr. (Obtain additional costs from local suppliers.)
6. Prepare an installation layout for a pump sump to provide a 2:1 duty cycle for the pump in Problem 4.

10

Electrical Power System Layout

10.0 INTRODUCTION

The early half of the twentieth century saw the introduction of electricity in underground mining as a comparatively safe means for powering pumps, cutting machines, and locomotives. Today most underground mines use electricity as their principal power supply for all phases and functions of the mining cycle. (The only exception is the use of diesel-powered haulage equipment in "hard rock" mines and in some coal mines.) This dependence upon electricity as an energy source will probably continue for some time since electrical power is easily controlled, easily distributed, safe (even in the adversive environments found in mines), commonly available, quite versatile in its application (conversion), highly reliable, and relatively economical.

The requirements of the electrical network underground differ substantially from those of aboveground industrial users because:

1. Underground mines require both ac and dc power supplies. Dc motors are still the motor of choice for driving loads requiring speed control, reversibility, electrical braking capability, and high starting torques, as exemplified by traction motors.
2. The mine electrical distribution system is constantly advancing or retreating together with mining activities. Thus the more reliable permanent connections and installation procedures cannot be used.
3. The distribution system is subject to an ever-present threat of ground movement in the form of rib bursts and roof falls, which can interrupt service at points in from the fault.

4. The space underground is limited to the size of entry and since both electrical feeders (power cables), and haulage equipment must use intake air courses, an ever-present danger exists where workers and equipment can contact the electrical power system.
5. The typical underground environment is dusty, dirty, muddy, and wet (with highly corrosive and conductive mine water acid), all of which adversely affects the performance of the electrical system. Also, the presence of explosive gases such as methane and coal dust necessitates the total enclosure of all electrical equipment, thus complicating cooling and maintenance of the equipment. Finally, the low illumination, excessive noise, and confining space present a greater psychological stress and therefore a greater danger of human error causing electrical system faults and subsequent life-threatening situations.
6. Finally, the mining cycle is such that the system periodically experiences severe startup and surge loads, which tax the capabilities of the system and tend to shorten the longevity of the electrical system.

The remainder of the chapter expands on the foregoing requirements of underground mining as they affect the layout and distribution of the electrical network.

10.1 FUNCTIONAL ANALYSIS: OVERVIEW OF THE POWER SYSTEM

As with the other functional operations discussed in previous chapters, the electrical power network is made up of subsystems each performing specific purposes or functions. To understand the principles and practices of electrical power layout properly, you must first understand the functions an electrical system must deliver. Each electrical power system, regardless of whether it is above or below ground, is premised on achieving the following five functions:

1. Providing power
2. Ensuring safety
3. Maximizing reliability
4. Improving maintainability
5. Optimizing value

Although each of these functions must be met, they are often at odds with one another. The selection of the primary function is usually a simple task. However, when discussing electrical power, which can be distributed underground at 13.2 kV or greater, many designers will often select ensuring safety as the primary function. Regardless of the apparent hazards presented by the electrical system, the primary function is to provide power. It is a mistake to believe otherwise, since to do so may result in an inadequate, unreliable, and therefore potentially dangerous system design. This is not to say that all possible safety precautions, devices, and interlocks should not be utilized, nor does it imply that less money should be appropriated for the safety function than for the primary function. It implies, rather, that an adequate provision of power when and where needed is the *first* consideration of the electrical system designer.

10.1.1 Provide Power

Providing the necessary electrical power when and where needed requires the coordination of several subsystems of the power network:

1. The power delivery subsystem.
2. The mine power control subsystem(s).
3. The mine power distribution subsystem.
4. The power utilization subsystem(s).

Figure 10-1 depicts the proximal interactions and relationships among these four subsystems.

Figure 10-1 Overview of the electrical power distribution system.

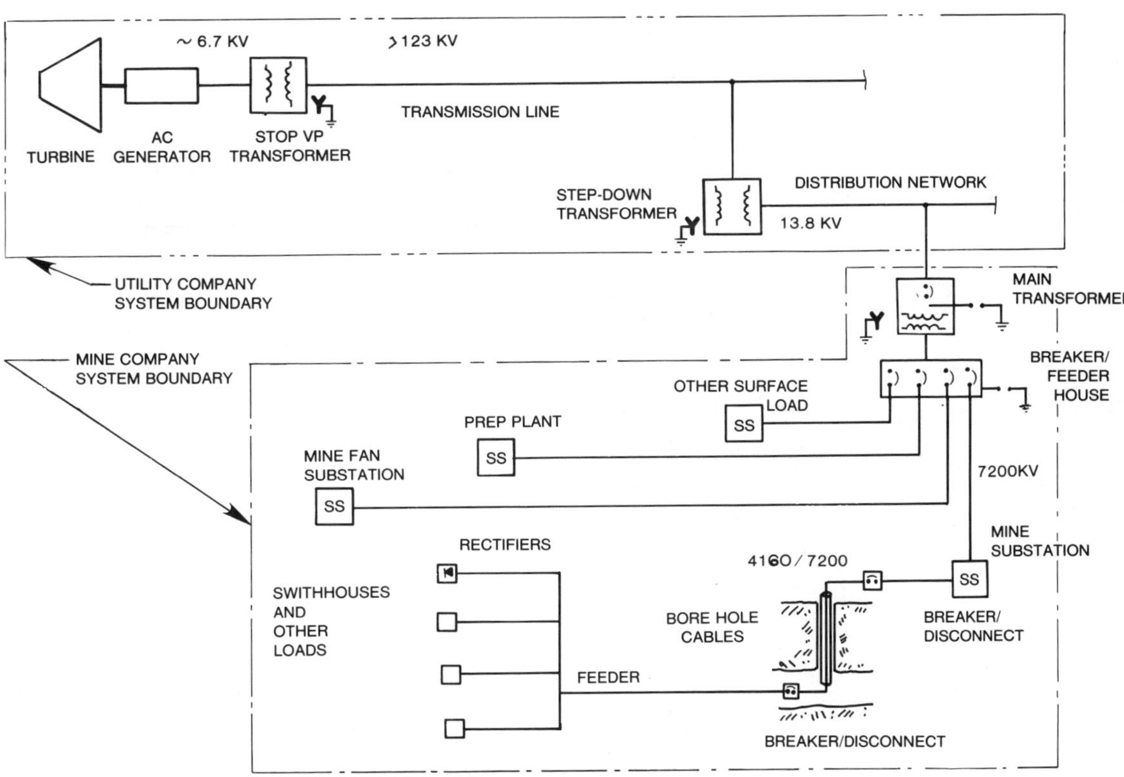

Power delivery subsystem. Most electrical power used by mines today is purchased from utility companies. Steam turbines or other suitable prime movers turn large synchronous generators, which produce alternating current at a frequency of 60 Hz (common in North America). The voltage of the generated power is then increased using step-up transformers often exceeding 69 kV (69,000 V).

Distribution Losses. The reasons for stepping up the voltage to such high levels are related to the economics of power distribution. Power is defined as energy per unit time. The load or demand for electrical power is the product of current, voltage, and the power factor (Eq. 10-1).

$$P = IE(\cos \theta) \tag{10-1}$$

where P = power demand [watts (dc) or kVA (ac)]
I = current (amperes)
E = voltage (volts)
$\cos \theta$ = power factor (for dc circuit $\cos \theta = 1$; thus $P = IE$)

The voltage drop across any given resistance is defined as the product of the current and resistance (Eq. 10-2).

$$V_{\text{drop}} = IR \tag{10-2}$$

where V_{drop} = voltage drop (volts)
I = current flow across resistance (amperes)
R = resistance (ohms)

(Note that the voltage drop in a true ac circuit would be the product of current, I, and impedance z.)

When energy is used, distributed, converted, or manipulated, some amount of that energy is lost (or otherwise unavailable for use), usually by being converted into heat. The "lost energy" in an electrical distribution system depends on the product of current squared and the resistance through which it must flow (Eq. 10-3).

$$P_{\text{lost}} = I^2 R \tag{10-3}$$

where P_{lost} = lost power (heat) (watts or kVA)
I = current (amperes)
R = resistance of circuit (ohms)

The power that must be input to a distribution system therefore depends on the power needs at the other end (IE) plus the power that will be lost during distribution. This is summarized in Eqs. 10-4a and 10-4b.

$$P_{\text{in}} = P_{\text{out}} + P_{\text{lost}} \tag{10-4a}$$

$$IE_{\text{in}} = P_{\text{in}} = I_{\text{out}}E_{\text{out}} + I^2{}_{\text{dist}}R \tag{10-4b}$$

since

$$I_{\text{in}} = I_{\text{out}} = I_{\text{dist}}$$

If power is delivered at low voltages, the current must correspondingly be higher than if power is delivered at a higher voltage. The higher the value of current, the greater will be the I^2R losses within the system and the more energy (and cost) will be needed to supply the same load. Therefore, utilities and mines that must distribute power over long distances employ the highest possible distribution voltage and then step down the voltage at the point of utilization.

Other factors, such as voltage regulation, line impedance, line-to-line capacitance, and line cost, contribute to the desirability of high distribution voltages. Complete analyses of line selection and design are offered by Johnson.[1]

Mine power control subsystem.[2] The power supplied by the utility company is delivered to the mine site at a voltage level that can range from 4160 V to 123 kV (or greater). Pragmatically, the actual site voltage depends on the voltage availability in the region of the mine rather than the needs of the mine (manipulation of the mine's power system is far simpler than modifying the utility's network). The incoming power is routed through circuit breakers (used to protect the "line side" distribution network) to a primary transformer which steps down the voltage to a range of 480 V to 13.8 kV but most commonly to 7200 V. The smaller voltages would correspond to smaller drift mine operations, where utilization of the power is relatively close to the transformation site. The higher voltages would correspond to larger operations, which require long on-site distribution and multiple utilization voltages. In general, larger mines with correspondingly larger loads purchase the transformer since monthly power costs will be substantially less, while smaller mines or those that will not be in operation very long usually rent the primary transformers from the utility company.

From the primary transformer, the power is routed to a feeder-breaker switchyard, which provides a distribution bus with multiple circuits to allow distribution to various surface and underground load centers. Each circuit is protected by circuit breakers, fuses, and lightning (surge) arrestors to prevent electrical system faults from adversely affecting the system or those persons interacting with the system. The number of individual circuits depends on the number of major load centers. Various types of loads requiring separate circuits include the mine fan(s), which by law must be powered on a circuit separate from any other, the preparation plant, other surface facilities, (such as maintenance operations, bath house, and offices), and the underground operations.

Depending on the size of the operation, separate transformers called *unit substations* may be required for each major load center. The function of the individual substations is to provide optimum voltage control for the load served. The location and number of substations delivering power underground depend on the size of the distribution network, the nature of the load demand, and the degree of selectivity, reliability, and safety desired. These factors are considered in greater detail later.

Since more than one substation is often used to supply power and since the low conductivity of the surrounding strata provides poor grounding, the ground for the electrical power system is usually established at the main substation and carried underground with the power feeder cables. Circuit breakers, lightning and voltage surge arrestors, and visible disconnect switch gear are mandated by law as integral components of the power control system at each location where power is fed underground.

Mine power distribution system. The underground power distribution system, referred to generally as the *primary distribution system*, can be laid out using a wide variety of possible arrangements, depending on the degree of safety, reliability, and maintainability desired. The various arrangements are discussed in more detail in Section 10.2, However, to facilitate a functional analysis, a simplified series or "radial" system will be presented.

In the simplest electrical distribution system (Fig. 10-2), power is delivered underground through bore hole cables or along feeders placed within a drift or slope entry to an underground switch house. A switch house can be a simple high-voltage visible disconnect switch servicing a single outgoing circuit or a more complicated switching station with an internal bus servicing several circuits simultaneously. From this initial switch house, power is fed through high-voltage cables to sectionalizing switch houses, which function to divide the power circuit.

Figure 10-2 Simple radial layout of the underground electrical power system.

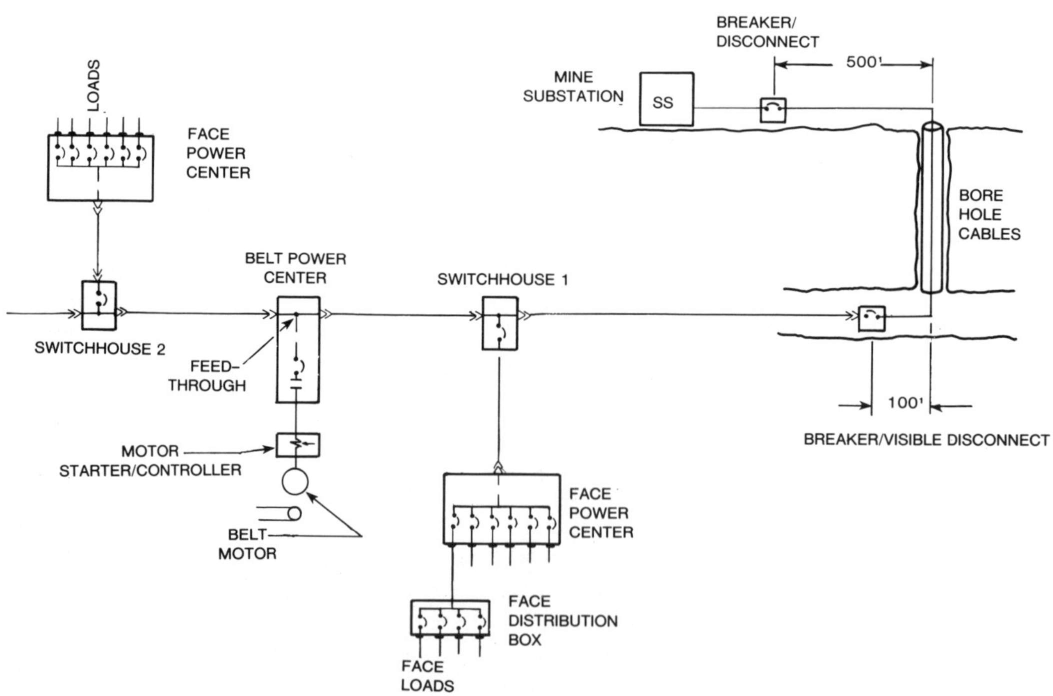

Power is then routed to a belt power center, where the circuit is again divided, with part of the power delivered to working areas and part delivered to the primary winding of an interval step-down transformer. The power from the secondary winding of the belt power center transformer is routed to the belt motor starter-controller, then to the belt drive motors.

Power delivered to locations by the belt power center is routed to a face power center. The functions of the face power center are (1) to step down the voltage to appropriate utilization levels; (2) to provide the necessary internal bus, protection devices, and outgoing circuits to supply face equipment; and (3) in some cases to rectify the transformed voltage to provide dc power when and where needed.

Power utilization subsystem. The last major partitioning of the provide power function is the utilization power delivery system. Power is distributed from the load centers, rectifiers, and transformers directly to the utilization equipment, separate motor controllers, or distribution centers, depending on the machinery and section being serviced. For large loads, such as belt drives, underground tipples, and pump houses, power is usually directed to separate motor controllers and then through permanently mounted cables. The regulated power is fed to the drive motors. For track haulage, the rectified power is directed through the dc circuit breaker to the trolley line. Finally, power needed for face operations is supplied to the face equipment through trailing cables attached on large take-up spools mounted on the equipment. To help reduce the frequency of advancing the face power center as the face advances, distribution centers are often employed. These centers act as an umbilical extension of the trailing cable, providing several circuits closer to the face operation.

This discussion of the radial layout serves to review the major functions of the electrical power system. The functions of ensuring safety, extending reliability, improving maintainability, and optimizing value will effect modifications of the radial as well as other layouts. These are discussed next.

10.1.2 Ensure Safety

The electrical system is perhaps the most monitored and regulated system of all underground functional systems. Strict safety regulation arises in part from the nature of electrical power and electrical hazards. Since electrical energy cannot be "observed" without the aid of instruments, neither its presence nor possible safety hazards appear obvious to an unassisted observer. Thus an unsafe condition can occur without any outwardly physical indications, and merely awaits a final misdirected link, such as unguarded human contact, to cause a catastrophic "accident."

The elements of safety (see Section 6.2.1) combine to inhibit the various links that can contribute to contacts between the agent (A) and the electrical system (FS). Unlike some of the systems discussed previously, the complex nature and potentially hazardous status of the electrical system require the use of all of the safety schemes:

1. Isolation schemes
2. Controlled-link schemes
3. Protected agent schemes
4. Protected equipment schemes
5. Fail-safe design schemes
6. Deenergized contact schemes
7. Alerted contact schemes
8. Environmental exclusion

Isolation schemes. The strategy of isolation attempts to avoid A–FS contact by maintaining a physical space between the two. For the electrical system this is achieved through (1) the layout of equipment away from concentrations of personnel or the traffic of equipment, and (2) installing certain components in inaccessible locations. An example of the first would be positioning power centers, disconnects, and rectifiers at the rear of crosscuts, thereby avoiding most personnel and equipment traffic. An example of the latter is installing power feeders or permanent cables in trenches, usually recessed in the roof and secured by cable hangers. In both cases system components are physically "isolated" from agent contact.

Controlled-link schemes. Standard utilization voltages underground include 480, 550, and 960 V. Under "ideal" conditions a human being can be electrocuted by voltages as low as 27 V.[3] Since motors that supply the energy for pumps, belts, and mobile equipment must be controlled by a human operator, some low-voltage circuit and mechanical link must exist between the operator and the power circuit. This link is usually called the *system control circuit* or simply "controls." The control circuit consists of relays (low-voltage control switches), magnetic coils, and related mechanical mechanisms, combined to activate high-voltage power contactors, which in turn regulate the parameters (voltage, current, etc.) of the power circuit.

Protection barrier schemes. A protection barrier or protected contact scheme is such that either the human being is covered, enclosed, or otherwise physically restricted in his contacts with the electrical system, or the electrical system is enclosed, preventing potentially dangerous contact. The first scheme, exemplified by gloves, insulated suits, and insulated work pads, is used when the human being must interact directly with energized or possibly energized components such as might be required during maintenance, installation, or inspection operations. The latter, exemplified by fences enclosing substations, trolley wire drops, insulation surrounding wire, and so forth, is a strategy used when a part of the electrical system could be subjected to harsh treatment, abuse, or unauthorized tampering.

Monitored deenergizing schemes. The safety strategy that monitors the A–FS interface and deenergizes the electrical system when a potentially hazardous condition exists is called the monitored deenergized scheme. This scheme is used quite often on the electrical system, since malfunctions of the electrical system, such as electrical shorts, overloads, and ground faults, can be easily detected and the system deenergized before endangering personnel or proximal equipment.

The best example of a monitored deenergizing scheme is the ground fault monitoring scheme. Many differing ground faults can occur: for instance, the ground wire may be severed or inadvertently disconnected. To "monitor" the continuity of the ground wire, an additional wire, called a *pilot wire*, is taken underground with the ground wire. A small current is passed across a relay winding through the pilot wire and returned along the ground wire. The relay controls the main power circuit breaker. If the ground wire is broken, the current that is energizing the relay fails to flow through the wire. This causes the relay to open, and this opening in turn causes the main power contactors to open, thereby deenergizing the power system. Several different deenergized schemes are in existence[4-6] and you should study these references in detail to appreciate their logic fully.

Fail-safe schemes. Since electrical power cannot be directly observed, the effect of a deenergized system energizing suddenly and without indication would be potentially hazardous to both equipment and personnel interfacing with the system. Thus any time the system is shut down or fails, all relays and contactors open. Before current can again flow, all controls have to be started intentionally by supervising personnel. This fail-safe scheme is basic to all electrical systems.

Alerted contact schemes. The alert scheme neither prevents contact nor deenergizes the equipment, but rather attempts to notify personnel of a potentially dangerous situation. Alert schemes may be quite simple and passive, such as signs warning of "high voltage," or the schemes may be more complex, incorporating sound or light warning systems, such as those activated by tampering with equipment enclosures.

Environmental exclusion schemes. The underground environment into which the electrical system is placed is potentially explosive (due to methane gas and coal dust) and conductive to ground faulting since the environment is wet (usually acidic). Because of such adversive environmental conditions, safety schemes must include schemes for either preventing the environment from dangerously contacting the functional system (exclusion) or, if contact does occur, preventing the "hazard" from propagating externally (containment). The primary examples of environmental exclusion and containment are the electrical permissibility requirements for face equipment. With permissibility, face equipment must be constructed to withstand an internal explosion without permitting the explosion to propagate to the surroundings.[7]

Each of the foregoing schemes or combination of schemes is used to ensure the safety of personnel, equipment, and property that interacts with or is a part of the electrical system. Many of the safety-related devices, layout, and procedures are specified by the mining agencies of the state and federal governments. It should be noted that although certain situations are specified by various governing agencies, a safe work place cannot be legislated and requires the concerted cooperation of designers, management, and hourly personnel, who ultimately benefit from a safe electrical system.

10.1.3 Maximize Reliability

Reliability has been defined previously as the probability that a given system will perform its intended function over a predetermined period of time. For the electrical power system, we are interested principally in maintaining the continuity of electrical service. The costs of unreliable systems were noted in Section 5.1.3 as were some of the basic factors related to improving any system's reliability. The influence of these reliability factors is affected by the possible modes of system failure. Electrical components are really quite reliable, with component indexes typically greater than 0.99.[8] The majority of failure modes experienced underground are more often than not mechanical in nature rather than electrical and include mechanical wear, damage from other equipment, roof falls, transportation damage, and so forth.[9] Because of the extreme environmental conditions experienced underground, the reliability of the electrical equipment may be 1% or less of that normally experienced in typical industrial situations.[10] In addition to the mechanical and environmental failures, electrical failure modes that affect the reliability of the system include the following.

Defect failure. Equipment defect failures occur less frequently with electrical equipment than with other equipment previously discussed because fewer mating mechanical parts are involved. Those defect failures that do occur usually result from improper application of materials, resulting in dielectric breakdowns, corona cutting of insulation, and the like.

Aging failures. The principal aging failures associated with electrical components involve the degradation of insulator properties, such as in ozone embrittlement or in the surface oxidation of (principally aluminum) conductors, particularly at cable junctions or splices.

Degradation failure. Degradation such as wear, erosion (electrical and mechanical), and so forth is rarely found in electrical equipment except for those contactors, relays, and other electrical devices subject to relative motion between mating parts or extremely high current densities.

Improper use, abuse, or maintenance. Failures resulting from deliberate or accidental human maltreatment contribute the greatest number of electrical system failures. These include such failures as trailing cables which have been inadvertently run over, improperly hung power cables, the use of power cables for welding by driving spikes into the cables to obtain "instant" power, and so forth.

Accumulative microscopic failures. This class of failures normally involves the marginal load-related stresses (e.g., the skin effect, in which the current density concentrates nearer the surface, adding stress to the insulator), internal ionization, corona discharge, and so forth. These microscopic failures accumulate, causing insulator breakdown and eventually system failure.

Cumulative macroscopic failures. Macroscopic failures generally occur in components that move or must be periodically transported (during advancing or retreating operations). Macroscopic failures also occur when the safe current density is exceeded, resulting in insulator breakdown which over several trials causes the system to fail.

Spillover failures. Spillover failures include failures resulting from proximal relationship with another failed system or component. For instance, a transformer malfunctions, causing excessive voltage to be applied to the secondary winding; this in turn causes a tripping of a fused circuit. These also result from other functional systems, causing failure of the electrical system; for example, roof falls which sever the power cable are classified as proximal failure.

Reliability of the electrical system can be improved using the following layout principles.

Simplify the design. As stated in Chapter 6, as the number of independent components in series in any given system increases, the reliability decreases. Independent components can be as obvious as a power center transformer or as subtle as a connection or splice. Remember, each time the system is treated or manipulated, the reliability of that operation reduces the total reliability of the system, based upon Eq. 6-1.

Select reliable components. The designer needs to understand the design advantages and limitations of each piece of equipment specified. Obviously, the more reliable the components, the more reliable will be the total system. Refer to Eq. 6-1 for a mathematical representation of the effects of individual component reliability.

Utilize redundant circuits. For components that have low reliability indexes or are critical to the proper functioning of the system, the designer should select redundant circuits. Redundant circuits are also needed for components that can benefit from derating, and for those circuits that require system deenergizing to be maintained properly and safely. Remember however, that redundancy carries with it the penalty of added cost, control complexity, and size. These advantages and limitations are discussed later in comparing various layout schemes.

Employ derating. The effective life of certain components, such as power cable, can be extended through operating at levels below the rated operation level; however, derating is expensive and with proper design should not be necessary for most electrical components.

Control the operating environment. As much as is practical, the electrical components should be kept clean, dry, and as free from dust as possible. In selecting components you should select those designs that have specific seals designed for excluding the environment.

Use human factors engineering in the layout. As much as possible, locate equipment, junctions, disconnects, and controls away from untrained personnel. For instance, "tamperproof" cable couplers, circuit breakers, and so forth that require a special "key" to open are available and may be necessary to maintain system reliability. Note that many safety-related devices presently required on electrical systems often need to be designed with the premise that a person is trying to defeat the system safeguards.

Provide proper installation. Observe the manufacturer's primary or secondary installation procedures. Do not attempt installation methods other than those proposed without first seeking the advice of the manufacturer.

Provide the proper training and instruction in equipment operation. Never assume that you or your workers know how something operates or should operate. Although direct experience is often described as the "best teacher," it generally is not. Experience is too selective in the scope of lessons it teaches, and in the case of electricity, the lessons taught by experience alone carry severe penalties for ignorance and mistakes.

Inspect and maintain the equipment regularly. Proper inspection, routine maintenance, and prompt repair will provide extended system longevity and continuity of service. This factor forms a distinct function that is discussed in more detail in Section 10.1.4.

Each of the nine factors discussed above contributes to ensuring system reliability. The first, second, third, fifth, sixth, seventh, and ninth factors most affect the reliability of eventual electrical layout. The actual reliabilities for various layouts are discussed in Section 10.2.1.

10.1.4 Assure Maintainability

Maintainability of the electrical system involves:

1. Time estimates and the probability of restoring service of a failed portion of the system.
2. The ability to advance, retreat, or expand the functions of the electrical system.
3. The ability to and requirements for inspecting and maintaining the electrical system.

The dominant factors influencing each of the categories above were discussed generally in Section 6.1.4 and are described here briefly as they affect the status of electrical system.

Service restoration. The functional restoration of a failed system depends on (1) type, location, and severity of the failure; and (2) the layout of the system. The first factor determines (1) the type of needed repairs, and (2) the repair, replace, or removal decisions and the personnel required to effect the restoration. The second factor determines (1) the accessibility of the failure, (2) the ability to deenergize this portion of the system while not deenergizing related elements (based principally on the use of redundant layout strategies), and (3) the simplicity of the system in the failure locale.

One element that is essential to electrical system restoration is the ability to deenergize the system. For obvious safety reasons, federal regulations[11] mandate that electrical repairs be made only on deenergized components. This requires that either a redundant path is provided across the region of failure or that electrical service is interrupted. The initial layout decisions thus depend on an economic trade-off between the (yearly depreciated) cost of supplying equipment needed for a parallel bypass, contrasted with the (lost revenue and other) costs due to service interruption of these processes. These costs are compared in Section 10.2.1 for a hypothetical "typical" mine.

Modifications. Electrical system modifications include advancement, retreat, and functional expansion of the system. Advancement and retreat are most obvious within the face regions and are thereby linked to the rate of face advancement or retreat and the flexibility of the electrical equipment. The rate of change of the location of the face depends on the type of extraction under way and mining conditions. For a 46-in. seam, three-entry, continuous development section, a "typical" rate varies from 100 to 150 ft per shift; for a five-entry conventional development section, the rate varies from 50 to 90 ft; for a 500-ft retreating longwall panel, the rate varies from 15 to 35 ft per shift; and for a 300-ft shortwall panel, the rate varies from 20 to 50 ft per shift. The second major variable determining the rate of electrical system advancement is the length of the trailing power cables associated with the face equipment. Cable length varies from 250 to 750 ft in length. Thus as the face advances, the power center must advance in discrete 250- to 750-ft increments. (This increment can be extended if mobile power distribution centers are connected to the power center.)

The last category of system modifications includes adding functions to the power system. An example of such an addition might be new safety monitoring equipment. Such modifications are always more easy accomplished in simple series systems than in complex redundant circuits.

Inspection and maintenance. The inspection and maintenance procedures for the electrical system are in part regulated by state and federal laws and regulations.[11] These requirements are only minimal requirements which generally describe the preventive-routine maintenance strategy (Section 6.1.4). Each mine or operating division must plan a scheduled maintenance program concentrating on unreliable portions of the electrical system. Face electrical equipment due to movement and greater environment stress is particularly troublesome. A higher frequency and degree of inspection is warranted for those components historically having a high failure rate per unit per year. Although each mine will vary in the severity and type of failures, the following can be used as an approximate guide:

Equipment	Percent of downtime failures*	Average rate (failures/unit/yr)
Trailing cables (all)	40	39
Control circuitry (all)	38	37
Methane monitor (all)	16	16
Motors	8.5	8
Other	Remainder	—

*Initial data.[10]

10.1.5 Optimize the System Value

The optimum benefit-to-cost ratio is achieved when all four functions are achieved at the lowest possible cost. For the electrical system, some decisions related to the economics of the power system include:

1. Selection of power delivery points and corresponding voltages (normally, higher voltages represent lower distribution and line costs but higher switchgear and transformer costs).
2. Level of reliability (both underground and in terms of requested standby power).
3. Level of maintainability.

These decisions are discussed in more detail following an analysis of the different types of power distribution layouts commonly used.

10.2 METHODS OF ELECTRICAL LAYOUT

The possible methods of electrical layout underground do not differ substantially from those available for industrial plant design aboveground. One such distribution system, the simple radial system (Fig. 10-2), was described in Section 10.1.1. Additional types of distribution arrangements include the secondary-selective system, the primary-selective system, the looped-primary system and the secondary-network system.[12]

With the exception of the secondary-network system, which is most commonly associated with utility company distribution networks, all of the arrangements noted above may be used underground. Figure 10-3 (a)–(d) shows the radial, secondary-selective, primary-selective, and looped-primary systems, respectively.

The *radial* layout is characterized by a single primary substation and feeder which is distributed underground as a common distribution bus. Individual load centers step down the voltage to utilization levels and secondary feeders and trailing cables distribute the power to the individual loads. The advantages of the radial layout are (1) simplicity of control and distribution, (2) low cost, (3) ease of expansion or modification, and (4) good voltage regulation. The major limitations of the radial arrangement are: (1) failure to the primary feeder causes failure of the entire system, and (2) maintenance and inspection procedures of the primary feeder require deenergization of the system, which is equivalent to system failure.

The *secondary-selective* system is characterized by two primary substations and feeders, each feeding a bus, which in turn supplies two different load concentrations. The two buses are linked by a normally open tie breaker. Under normal conditions the two legs operate as two distinct radial systems; however, if failure occurs in one primary feeder or if the feeder is deenergized for maintenance reasons, the tie breaker can be closed, allowing uninterrupted operation. The major advantage of the secondary selective layout is that redundancy of the primary feeder increases the reliability and maintainability of the entire system. The major disadvantage is that the initial cost of the system is twice that of the radial arrangement since the

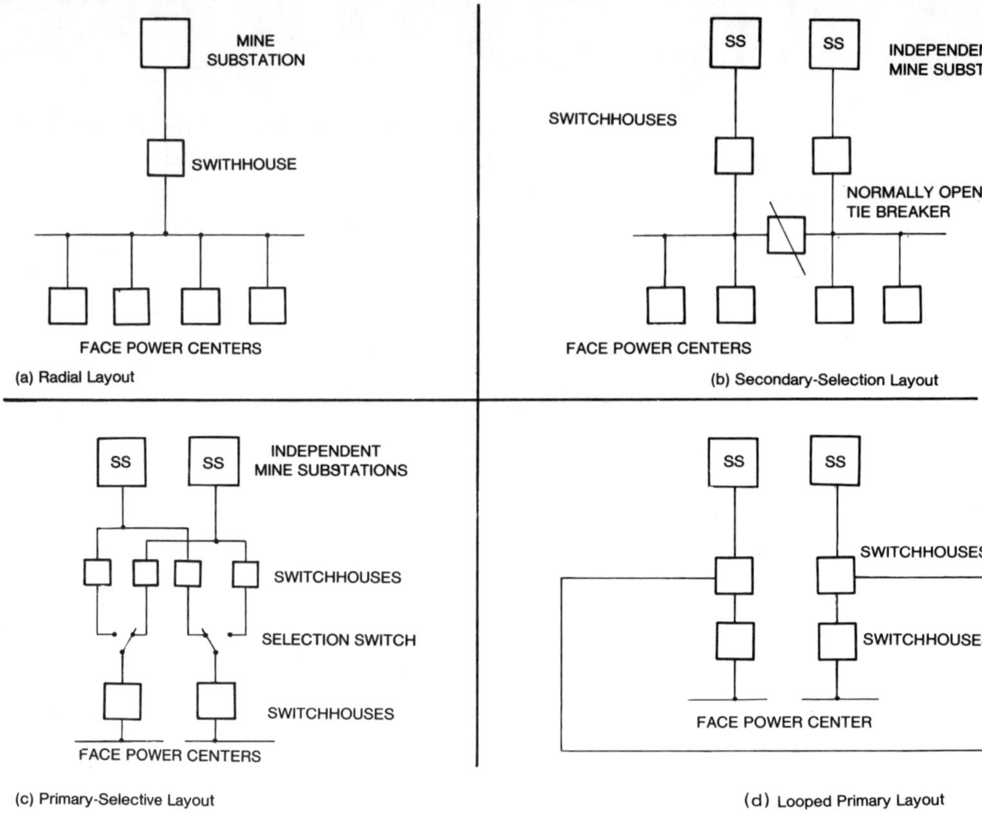

Figure 10-3 Major circuit layouts for underground power distribution. (a) Radial layout. (b) Secondary-selection layout. (c) Primary-selective layout. (d) Looped primary layout.

substations and feeder of each leg must be capable of carrying the same load as those of the radial system. However, the long-term operation and maintenance cost may offset this initial capital outlay since the primary feeder will operate at 50% derated condition, providing lower system downtime and reduced power losses.

The *primary-selective* system is similar to the secondary selective system in that two primary substations and feeders supply two load concentrations underground. The primary-selective has two single-throw load-interrupter switches which permit greater selectivity and load balancing for the system than is possible with the secondary selective layout. Roughly the same reliability is assured by both systems; however, the control circuitry and corresponding cost is slightly higher for this arrangement.

The *looped-primary* system provides two substations linked by a single primary feeder and a series of load-interrupter switches. The load demand upon either substation can be balanced somewhat by opening or closing the interrupter switches. Like the two selective systems, a degree of redundancy is possible. However, two major disadvantages to the looped primary layout inhibit its adoption: (1) a fault in the primary feeder causes all loads to fail until the fault is located and isolated by appropriate switching, and

(2) the possibility of power being energized from either side of a fault necessitates extreme safety measures which may negate any advantages the system has.

10.2.1 Layout Comparison

Table 10-1 summarizes the relative advantages and disadvantages of each layout scheme according to the system characteristics of (1) providing power, (2) ensuring safety, (3) maximizing reliability, (4) improving maintainability, and (5) optimizing value. It must be noted that each mine is different and must be evaluated individually. This ranking is presented for purposes of illustrating the need to consider each functional aspect of the system.

TABLE 10-1 COMPARISON OF ELECTRICAL LAYOUT STRATEGIES

	Characteristic*				
Strategy	Provide power	Ensure safety	Maximize reliability	Improve maintainability	Optimize value
Simple radial	2.5	3.5	2	2.5	2.5
Secondary-selective	3.5	3	3.5	3	2.5
Primary-selective	3	2.5	3.5	3	2.5
Looped-primary	2	1	3	1	2
Secondary network	Not employed underground				

*Scale: 0, function not provided; 1, function poorly attained; 2, function provision acceptable; 3, function provision exceeds minimum requirement; 4, function provision highly desirable.

Providing power. Each layout scheme is capable of providing power when and where needed. The differences in the relative rankings shown in Table 10-1 depend on the ability to selectively regulate the distribution, to simplify the controls required, and to regulate the voltage. The radial system is the simplest to control and provides good voltage regulation capabilities, but is not selective. Both the primary-selective and secondary-selective systems provide selectivity (with the former being the most selective) and good voltage regulation; however, both arrangements are somewhat more complex to control. The looped-primary system is not particularly selective, and suffers from a more complex control and regulation system than the radial system. Based on these factors, the secondary-selective layout is ranked first, followed by the primary-selective, the radial, then the looped-primary systems.

Ensuring safety. Each of the system arrangements must be equipped with roughly the same protective schemes discussed in Section 10.1.2. Thus the relative comparison of insurance of safety will be based on (1) an analysis of "dangerousness," assuming that failure occurs within the primary feeder network, and (2) an estimate of the complexity of the safety monitoring systems. If failure occurred within the distribution of the radial system, all sections would be without power. This is an inconvenient situation but not a potentially dangerous one in terms of electrical hazard.

If distribution failure occurs to either the primary-selective or secondary-selective system, the power to the sections can be selectively restored. If the failure resulted from an overload from the failure and if this were not corrected prior to restoration of power, a potentially dangerous situation might result.

An interesting result can be achieved with either the secondary-selective or the primary-selective system through layout. If the load in any given region were electrically split so that half would be attached to one primary feeder and half to the other, failure of one primary feeder would result in partial failure within any given region. Thus some power to any working section would always be assured, except in the most unusual circumstances.

If failure occurred to the primary feeder of the looped primary arrangement, the entire system would be down until isolation of the fault occurred. Although the downtime would be less than for the same type of failure in the radial system, the condition of power being supplied to either side of a fault would be unacceptable in most underground situations.

The monitoring complexity of the various layout schemes follows the same rationale of the operation simplicity analysis found above under "provide power." The radial system is the easiest to monitor, followed by the secondary-selective, the primary-selective, and the looped-primary systems, respectively.

Ensuring reliability. System reliability is evaluated relative to system simplicity, the use of redundant strategies, and the ability to use equipment derating during operation. The simplest arrangement is, of course, the radial system, followed by the secondary-selective, primary-selective, and the looped-primary networks.

Any of the systems can use redundant strategies by adding a bus to supply parallel operations; however, only the secondary-selective, primary-selective, and the looped-primary networks are designed for redundant operations. The relative rating of these three depends on the ability to switch or change the network arrangement at will. The "value" of redundancy as a layout strategy depends on a comparison between the different system layouts relative to (1) the rate of frequency of failure, (2) the mean time to repair (MTTR) failures, and (3) the cost of the failure. These are compared later in this section.

Like redundancy, system derating depends in part on the ability to divide the load between parallel systems. In reality, though, even the radial system can be significantly derated by merely overdesigning or purchasing larger-capacity components. Thus the relative ranking of the alternatives with regard to derating is roughly the same.

Ensuring maintainability. Maintainability for the various layout schemes is evaluated by comparing (1) the ability to repair, inspect, or maintain the system at no power, (2) the ease with which the system can be advanced or otherwise modified, and (3) the complexity of the system (requiring greater complexity of inspection equipment, more skilled personnel, and so forth). The radial system is the easiest to advance since this can be accomplished by simply extending the main feeder. The ranking of the other systems depends on their relative simplicity; thus the radial system is followed by the secondary-selective, primary-selective, and the looped-primary systems.

The ability to maintain the system at no power is related to the use of selective redundancy. Thus the system can be ranked as follows from first to last: the primary-selective, secondary-selective, radial, and looped-primary (ranked last since switching between the primary feeders is not provided).

Optimizing value. The value of the system (i.e., the benefit-to-cost ratio) depends on total "cost" of the system components, including the initial equipment costs, operation costs, and maintenance costs, whereas the relative "benefit" of each layout is reflected in continuity of service. To illustrate the role of value in decisions about selecting the optimum system, consider the following.

Example

Assume that the value of two different power layouts, a radial layout and a secondary-selective layout, are being compared for providing service to a mine. The estimated production capacity of the mine is 1500 tons per shift. Figure 10-4 is a block diagram contrasting the layout of the radial and secondary selective arrangements, showing the major components, their corresponding reliability indexes (figures are modified to better approximate mine failure rates, estimated to be 10 times the failure rate of industrial plants), and the mean time (hours) to repair failure estimates.[8-10]

Solution. The reliability of the primary feeder system radial layout is calculated based on Eq. 5-1.

$$R = \Sigma_1 R_1 R_2 \ldots R_n$$

$$R_A = (0.9618)(0.9763)(0.8110)(0.9400)$$

$$= 0.7158$$

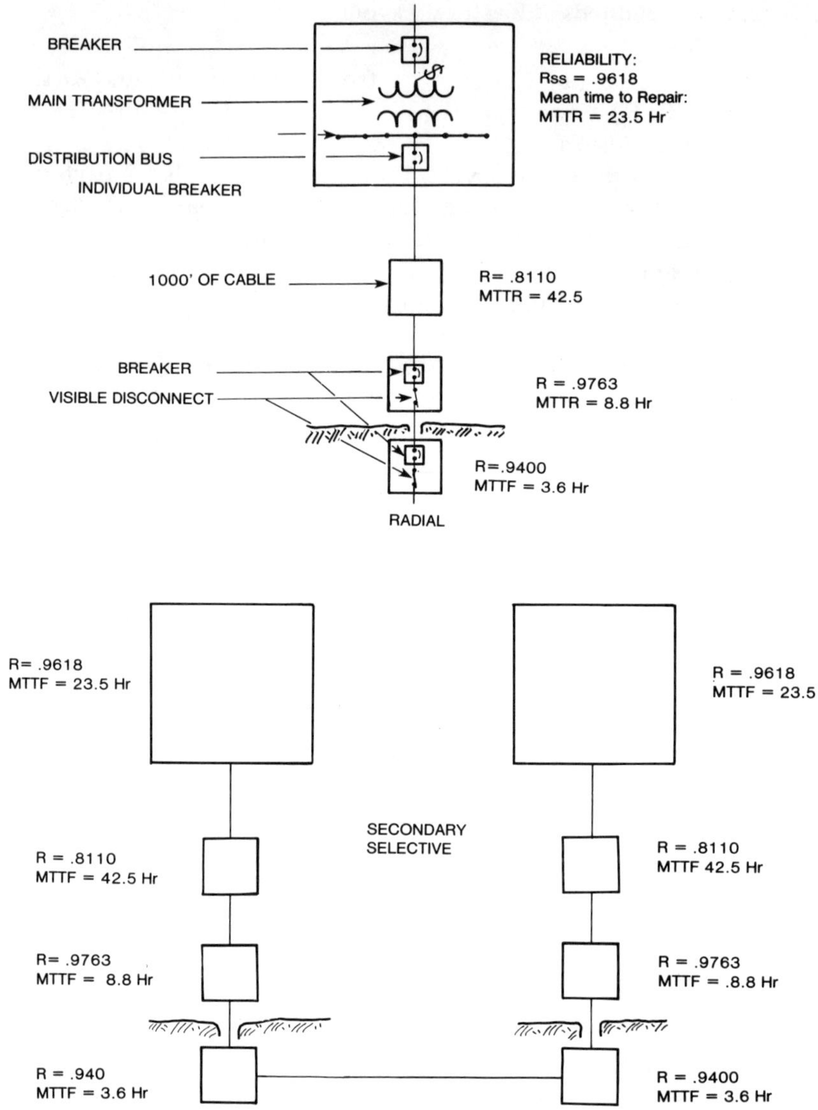

Figure 10-4 Comparison of radial and secondary selective arrangements.

The reliability of the primary feeder system in the secondary selective system is calculated according to Eq. 5-2.

$$R_B = R_1 + R_2 - R_1 R_2$$
$$= 0.9193$$

The estimated yearly failure rate for either system is calculated according to Eq. 10-5.

$$\text{FR} = 1 - \text{RI} \tag{10-5}$$

where FR = failure rate
RI = reliability index

Thus

$$FR_A = 0.2842$$

$$FR_B = 0.0807$$

The expected downtime (EDT) per year is calculated as the sum of the products of individual failure rates times the mean time to repair (MTTR) according to Eq. 10-6.

$$EDT = \Sigma\,(FR_i)(MTTR)_i \qquad (10\text{-}6)$$

where EDT = estimated downtime per year
FR_i = individual failure rate
$MTTR_i$ = individual meantime to repair following failure

For the radial system the EDT is calculated as follows:

$$EDT_A = 0.0382(23.5) + (0.0237)(8.8) + (0.06)(3.6) + 0.189(42.5)$$
$$= 9.167$$

EDT_B can be quickly approximated by the product of EDT_A and the ratio of reliability indices R_A/R_B such that

$$EDT_B = \frac{R_A}{R_B}(EDT_A)$$
$$= \frac{0.7158}{0.9193}(9.167) = 7.138$$

The relative difference between the two arrangements in terms of downtime is 2.029 hr/year. The significance of this difference must now be evaluated by calculating the relative effect that 2.029 hr/year has on company profitability. Although many costs are involved in downtime (see Section 6.1), we concentrate on the lost revenue that results. Assume for purposes of example that the coal being mined sells for \$25.00/ton. The effect of the downtime can be calculated as

$$\text{lost revenue/yr} = (2.029\ \text{hr/year})(1500\ \text{tons/shift})$$
$$= (0.125\ \text{shift/hr})(\$25/\text{ton})$$
$$= \$9511/\text{year}$$

The initial cost of layout *B* over and above the initial cost of layout *A* is calculated as the sum of the redundant components:

Breaker-feeder	\$31,000
Two high-voltage disconnect switches	16,000
Line cost	5,500
	\$52,500

Assuming that the price per ton of coal remains constant, an economical analysis of the two systems is now possible, taking into account depreciation, interest costs, expected life, recovery value, and so forth. Such an analysis is beyond the scope and intent of this book; however, assuming an annual cost of investment after taxes of 10%, the return of \$9511, or 18.1%, would be justified.

10.3 PROCEDURES OF ELECTRICAL SYSTEM LAYOUT

The procedures needed to complete the layout of the electrical system follow the same general format described for all layouts in Section 6.2. As with most layouts, the minimum information content is mandated by federal and state laws.[11] Each step of this procedure as it is applied to the electrical system is expanded below.

10.3.1 Functional Analysis

Functional analysis is a combined phase of obtaining as much information about the proposed or existing system as possible and assessing the electrical power needs of that system. The most convenient method of gathering and storing the information is to begin with a general mine layout or layout projections and depict the electrical data on the layout. The data are shown relevant to the functions of (1) providing the power, (2) ensuring safety, (3) maximizing reliability, (4) extending maintainability, and (5) optimizing value of the electrical system. The specific data needs depicted include, but are not limited to, the following:

Provide power (see Table 10-2)

TABLE 10-2 UNDERGROUND LOAD DEMANDS

Load description	Connected load	Demand factor[1-3]
Belt drive (48 in.)	245 kVA	1.0
Belt drive (36 in.)	148 kVA	1.0
Belt feeder	148 kVA-40(3)	0.9-1
Cutting machine	276 kVA	0.7
Charger (battery)	35 kVA	0.8
Continuous miner	445 kVA	0.6
Compressor (air)	39 kVA	0.3
Drilling machine	42 kVA	0.7
Lighting	190 kVA	1.0
Loading machine	190 kVA	0.7 (est.)
Longwall shearing machine	206-686 kVA	0.9
Power pack	206-343 kVA	1.0
Rock duster	28 kVA	0.2
Roof bolter	42 kVA	0.3
Welder		0.3
Resistance arc		0.2
Shuttle car	32 kW*	0.6
Pumps	hp (0.7457) kVA	0.3-1.0
Track	(depends on size)	(depends on duty cycle)
General		
Semicontinuous processes		0.6
Continuous process		0.9

*1 kW = 1 kVA.

1. The types and quantities of underground loads
   (a) Section loads (ac and dc)
   (b) Other clustered loads (ac and dc)
   (c) Major motors (e.g., belt drives)
   (d) Track haulage (duty cycle)
   (e) Lighting requirements
   (f) Future expansion sites
2. The types and quantities of surface loads
   (a) Ventilation fan(s)
   (b) Existing or proposed substations
   (c) Preparation plants
   (d) Offices and bath house
   (e) Maintenance facilities and shops
3. Existing power availability (line size, voltages, number of cables)
   (a) Utility distribution network
   (b) Mine site distribution system
4. Unusual starting or surge loads (e.g., welders, large motors)
5. Critical installation problems (e.g., poor drainage, high fault zones)

Insure safety

1. Evaluate high-risk areas (e.g., high load concentrations).
2. Perform a safety system analysis to determine the regions of system agent interfacing and evaluate the components contributing to possible dangerous conditions (part of this evaluation may be represented as a separate document for instance, a relational diagram).
3. Incorporate required safety monitors and equipment within the layout.

Maximize reliability

1. Assess the effects (see Section 6.1.3) of failure for each separate section of the mine. Identify those critical areas where continuity must be maintained: for instance, the mine fan. Reliability estimates for various components are given in Table 10-3.
2. Assess the existing reliability of the system to determine questionable practices as they influence reliability: high stress (current density) areas, complex controls, poor environment, and so forth.

Extend maintainability

1. Identify those environments in which system maintainability is adversely affected.
2. Evaluate present maintenance problems within the existing system.
3. Identify those areas in which the system is to be either advanced or retreated in the near future.

TABLE 10-3 ELECTRICAL COMPONENT RELIABILITY ESTIMATES

Component	Estimated reliability*	Estimated* MTTR (hr)
Substation (primary)	0.962	
Transformer (750–2500 kVA)	0.970	220
Circuit breaker (all)	0.944	11
Motor Starters		
0–600 V	0.970	70
600–15000 V	0.965	300
Motors		
ac—induction (0–600 V)	0.974	120
dc—all	0.900	50
High-voltage disconnect switch	0.994	4
with circuit breaker	0.976	15
Rectifier	0.942	400
Feeder (per 1000 circuit feet)	0.980	45

*Data approximated for illustrative purposes only; actual values depend on individual mine and adherence with Chapter 5 concepts.

Optimize value. Prepare detailed economic analysis, including trade-off assessment among initial, operational, and maintenance costs.

10.3.2 Performance Specification Objectives

From the data gathered and calculated, the next phase of preparing the electrical layout entails the preparation of performance specification objectives (PSOs) for the system. The PSO items are *measurable* statements of cost, input and output requirements, reliability and maintainability requirements, listing applicable codes and requirement installation time limits for the electrical system, and the major components of the system. It should be noted that the PSO items should for safety and completeness be prepared by a qualified (preferrably a registered) electrical engineer.

10.3.3 Generation of Alternative Solutions

The total possible system arrangements that can be proposed for delivering the power when and where needed is limited only by the imagination of the designer. However, the four types of systems discussed previously are the basis for most alternative solutions. A schematic line or block diagram for each alternative is now drawn depicting the total number of sections and loads needed to meet the input/output PSO. Following this the components, power centers, switch gear, rectifiers, and so forth, are depicted. The input/output ratings, impedance, and other component characteristics are shown. The size and length of power feeder and cables should also be shown.

Next, the reliability and maintainability index (at the appropriate load level) are depicted for each component in the system. Finally, the estimated cost for each component is shown. This set of procedures is followed for each alternative arrangement proposed.

10.3.4 Evaluation of Alternative Solutions

The proposals are evaluated by assessing the degree to which each alternative will meet the PSO items, comply with the regulations, codes, or laws, and conform with the "standard practices" of the mining industry (somewhat a measure of the ease of installation, level of technology, simplicity, and so forth).

10.3.5 Preparation of the Preliminary Design Layout

After one alternative arrangement has been selected, the design layout must be prepared. Using a general mine layout, show the locations of the system components relative to the underground projections and the other interfacing systems: ventilation, haulage, and drainage. Where applicable, scheduled installation dates should be shown on the layout to facilitate the coordination of the rate of mining and the advancement of the electrical systems.

10.3.6 Preparation of the Installation Layout

Following the completion of the general electrical system layout, preparation of individual layouts specifying the installation of permanent and semimobile electrical equipment is required. Detailed layouts are required for each belt drive, track rectifier, extraction section (continuous, longwall, or shortwall system), and so forth. These layouts are discussed in Section, 10.4.

10.3.7 Initiation of the Detailed System Design

A registered electrical engineer experienced in mine power system design is responsible for performing the detailed design of the electrical system, including providing short-circuit analysis and protective equipment coordination, calculating and specifying controller, monitoring, and sensory equipment, calculating and specifying input and output of components, preparing installation and maintenance schedules and specifications, and preparing (or assisting the purchasing department in the preparation of) equipment bids. The specific design calculations required are beyond the introductory nature of this text and will be reserved for subsequent courses in electrical distribution theory.

10.3.8 Implementation of the Design

Once the calculations and layouts are completed, the needed equipment received, and the installation sites (above and below ground) prepared, the electrical system design can be implemented.

10.4 SECTION LAYOUT

The layout of the electrical network within each working area depends on (1) state and federal regulations, (2) the nature of the equipment being used, and (3) the physical constraints of the working section being served. Section 10.4.1 presents an analysis of major electrical equipment and subsystems which constitute the electrical system. Section 10.4.2 presents the basic arrangements and installation guidelines for this equipment within differing sections of the mine.

10.4.1 Equipment Analysis

The electrical system, regardless of the type of circuit layout, consists of (1) main substation, (2) power feeders and cables, (3) underground couplers and switch gear, (4) underground transformation devices, and (5) utilization equipment. Each has its own set of functions and installation requirements, which will greatly influence the eventual layout.

Main substation. The primary purpose of the substation is to step down the utility's distribution voltage, ranging from 23 to 138 kV, to a level more appropriate for local and underground distribution, usually between 4160 V and 13.8 kV. Secondary functions of the main substation are to provide a distribution bus for feeding other loads, such as the mine fan, preparation plant, bath house, and so forth, and to ensure system safety, both load-side safety and line-side (the utility company's network) safety.

To provide these functions, the main substation is composed of a main transformer, feeder bus and various switch gear devices. One possible arrangement is detailed in the radial layout shown in Fig. 10-4. Circuit breakers are installed on incoming lines to protect the upstream distribution network from load-side faults. The power is then fed into the primary coil of the transformer. The output power is taken off the secondary coil and fed through an internal bus. Each circuit off-take from the bus is protected with a circuit breaker, visible disconnect switch, and lightning arrestors.

Installation. The size of the main substation depends on the size of the mine and the load being served. Smaller mines, for instance small drift mines requiring as little as 500 kVA, may transform the incoming power to

480 V or other utilization levels. For these smaller sizes the main substation is commonly pole-mounted. Larger power conversions, at times requiring as high as 50,000 kVA, require a fenced concrete pad or preferably a main substation building, which houses the transformer, breaker-feeder bus, switch gear, and protective devices (e.g., lightning and surge arrestors). Additional individual load substations may be needed with larger installations. These additional substations would require the same protective devices and switch gear as with the main substation, although they may not have the breaker-feeder bus arrangement.

Legal Constraints. The main substation or the immediate substation supplying the mine power must conform with CFR Title 30, Part 75.800, which carries the following constraints:[11]

1. The substation must be fenced or contained within a building with grounded access.
2. Common ground must be established at this point.
3. The substation must have a lightning arrestor and suitable protection against under voltage grounded phase, short circuit, and overcurrent.
4. The surface circuit breaker and visible disconnect must be located at least 100 ft from underground entry.

Distribution cables. The electrical conductors that comprise the distribution network are perhaps the single most costly component of the electrical system. Functions provided by the feeder cables include:

1. Provide a low resistance/impedance electrical path.
2. Prevent direct or indirect (magnetic effects) harmful contact between the energized electrical conductor and the surrounding environment or interfacing personnel (ensure safety).
3. Ensure long service life and continuity (ensure reliability).
4. Conform to standard installation and maintenance practices (ensure maintainability).
5. Provide a cost-effective way to supply power underground (ensure value).

For the electrical power cables the foregoing functions are accomplished through appropriate material selection and design. These two factors, materials and design, can be combined in thousands of alternative construction permutations; however, three principal designs are shown in Fig. 10-5: (1) the nonshielded single conductor, (2) the shielded single conductor, and (3) the shielded multiconductor cables.

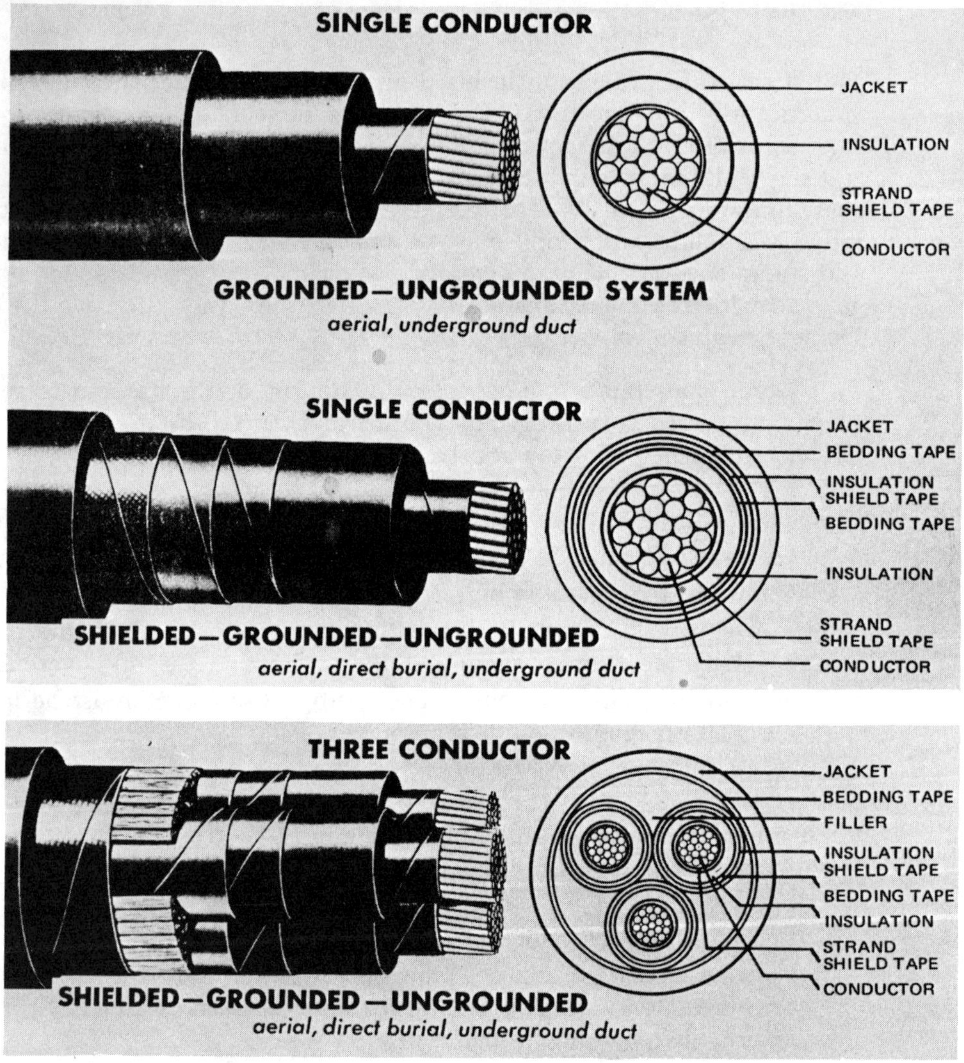

Figure 10-5 (a) Nonshielded single-conductor cable. (b) Shielded single conductor. (c) Shielded multiple-conductor cable. (Courtesy of PLM Products.)

Construction Elements.

Conductor. The *conductor* is the current-carrying low-resistance element of cables. The most common conductor materials are copper and aluminum. The selection of the most appropriate material for a given situation depends on the needed characteristics. Aluminum is lighter in weight and generally less costly, but copper has a higher conductivity, does not readily form insulating surface oxides, provides better mechanical properties over time (e.g., creep and relaxation), and is usually easier to install and maintain.

Another characteristic of the conductor is the degree of flexibility, which usually depends on the type and shape of the stranding used. The various shapes provide differing flexibility, weight to size (diameter) ratios, reactances and corresponding voltage drops, distribution of current density, and improved maintainability.

The last and perhaps most important conductor characteristic is conductor size, given in AWG (American Wire Gage) and Mcm (thousand circular mils, where 1 circular mil equals the area of a circle 1/1000 of an inch in diameter). The size of the conductor is directly related to the current-carrying capacity of the conductor or the kVA rating. Table 10-4 shows the commonly available conductors by size and equivalent diameter in inches.

TABLE 10-4 NOMINAL CONDUCTOR DIAMETERS COPPER OR ALUMINUM

Conductor Size AWG or kcmil	Class B Concentric Stranded Conductor Inches	Number of Wires	Compressed Round Concentric Stranded Conductor Inches	Compact Round Concentric Stranded Conductor Inches	Class H Concentric Rope Lay Stranded Inches	Number of Wires
14	0.073	7	0.071	—	—	—
12	0.092	7	0.089	0.085	—	—
10	0.116	7	0.113	0.107	—	—
8	0.146	7	0.142	0.134	0.167	133
6	0.184	7	0.179	0.169	0.210	133
4	0.232	7	0.225	0.213	0.266	133
2	0.292	7	0.283	0.268	0.355	133
1	0.332	19	0.322	0.299	0.378	259
1/0	0.373	19	0.362	0.336	0.442	259
2/0	0.418	19	0.406	0.376	0.477	259
3/0	0.470	19	0.456	0.423	0.536	259
4/0	0.528	19	0.512	0.475	0.601	259
250	0.575	37	0.558	0.520	0.653	427
300	0.630	37	0.611	0.570	0.716	427
350	0.681	37	0.661	0.616	0.772	427
400	0.728	37	0.706	0.659	0.826	427
450	0.772	37	0.749	0.700	0.878	427
500	0.813	37	0.789	0.736	0.923	427
550	0.855	61	0.829	0.775	0.980	703
600	0.893	61	0.866	0.813	1.022	703
650	0.929	61	0.901	0.845	1.064	703
700	0.964	61	0.935	0.877	1.106	703
750	0.998	61	0.968	0.908	1.145	703
800	1.031	61	1.000	0.938	1.180	703
900	1.094	61	1.061	0.999	1.253	703
1000	1.152	61	1.117	1.060	1.320	703
1100	1.209	91	1.173	—	1.386	703
1200	1.263	91	1.225	—	1.446	703
1250	1.289	91	1.250	—	1.477	703
1300	1.315	91	1.276	—	1.505	703
1400	1.364	91	1.323	—	1.561	703
1500	1.412	91	1.370	—	1.617	703
1600	1.459	127	1.415	—	1.674	1159
1700	1.504	127	1.459	—	1.724	1159
1750	1.526	127	1.480	—	1.751	1159
1800	1.548	127	1.502	—	1.773	1159
1900	1.590	127	1.542	—	1.823	1159
2000	1.632	127	1.583	—	1.868	1159

Strand Shield Tape. Strand shielding is a layering of conductive material between the metallic conductor and the insulator. The purpose of this material insert is to reduce the potential or voltage concentrations in those air spaces between the strands that were unfilled by the insulator. If the voltage concentrations were not reduced, the air within the gaps would ionize, causing the insulation material to age and fail prematurely.

Insulation. The insulation is the second most important element of the cable. The purpose of the insulator is to present a barrier of high electrical resistance and high magnetic reluctance between the conductor and its surroundings. The material selected depends on the following characteristics:

1. Initial physical, mechanical, and electrical properties (including but not limited to)
   (a) Stiffness properties
   (b) Strength properties
   (c) Toughness properties
   (d) Shore hardness
   (e) Crack and tear resistances
   (f) Insulation resistance
   (g) Track and (electrical) erosion resistance
   (h) (Mechanical) wear and abrasion resistance
   (i) Dielectric strength
2. Environment and aging reactions (including but not limited to)
   (a) Moisture effects
   (b) Acidic attack
   (c) Ozone and oxidation resistance
   (d) Ultraviolet and other radiation degradation
   (e) Corona and ionization resistance
   (f) Temperature effects

Jackets. The external covering or *jacket* is placed over the insulation to provide mechanical and weathering protection for the insulation as well as additional electrical insulation.

Conductor/Insulation Shields. Underground high- and medium-voltage power cables must carry a grounded *conductor shield,* which provides for system deenergizing and a fault-to-ground return path in the event that the cable is severed. Two types of conductor shields are available. The SHD type (which must be used on high voltages greater than 1000-V circuits) surrounds each conductor with a protective shield, while the second, the SHC type (used on medium voltage, 660–1000 V), surrounds the three clustered power conductors with a single, grounded conductor shield.

In addition to providing a fault-return path and deenergizing safety link, the SHD conductor/insulation shield provides a magnetic barrier that limits electromagnetic interference with communication systems, distributes the dielectric field across the conductor (which reduces voltage stresses and surge potentials), and confines the dielectric field to the insulation surrounding the cable.

Bedding Tapes. Bedding tapes are materials inserted in the construction of the cable to increase the mechanical properties (e.g., tensile strength) of the cable.

Filler Tapes. Filler materials are used to complete the circular cross section of the power cable and, more important, to exclude air, which may be ionized, from the voids and interstices between conductors, thereby reducing corona attack.

Installation. Because of the low reliability of electrical power cables, most installations are required to be in regularly inspected intake air courses. To further protect the cable, trenches are often recessed in the top, floor or rib to prevent equipment and personnel contact. In cases where dangerous contact is imminent (e.g., roof fall, haulage equipment, water), the power cables should be conduited. Cables that must support their own weight, such as bore hole cables, should be rated for vertical suspension even if only a portion of the installation is truly vertical.

Legal Constraints. To conform with CFR Title 30, Part 75:

1. Power cables must be located in a regularly inspected air course.
2. All connections must be located in intake air courses.
3. High-voltage cables must be of the SHD type of shielding.
4. Power cables must be covered, buried, or placed to afford protection.
5. Power cables must be protected by at least 4 in. of concrete when passed under a haulageway.
6. Bore hole cables must be surrounded by a 6-ft-high fence.
7. There can be no more than one permanent splice on any trailing cable.

Switch gear. Switch gear or switch houses include all devices used to energize or deenergize electrical power circuits, such as high-voltage disconnects or vacuum circuit breakers. The primary function is to provide a means of selectively removing power from various locations of the mine. Secondary functions include providing a visual method, usually a site window, for confirming the on–off status of the outlet circuit, providing a means for dividing the circuit, and protecting the load and line side of the device from undervoltage overcurrent, short circuit, or grounded phase (as all equipment must).

Installation. Switch houses are located at all junctions along the distribution circuit or at any location where section isolation may be valuable. Some of the more obvious locations include separately pillaring sections from development sections, longwall sections from other sections, isolation of the track rectification system, isolation of main belt from remaining loads, and so forth.

Legal Constraints. To conform with CFR Title 30, Part 75, visible disconnect switchhouses require:

1. Installation within 500 ft of the bottom of the point of electrical power entry.
2. Location within intake.

Load centers. Load centers, including face power centers, belt power centers, transformers, and rectifiers, usually constitute the last major manipulation of the parameters of the electrical power system before utilization. Except for the rectifier, the primary function performed by these devices is to transform the high distribution voltage to the lower utilization voltage required by face equipment. For the rectifier, although this voltage step down is usually required, the primary function is to rectify or transform the alternating current and voltage to direct (all positive) current and voltage. Secondary functions of load centers can include (1) visual methods of confirming the on-off status of the power circuit; (2) internal protection against short circuit, overload, undervoltage, grounded phase or voltage surge; and (3) an internal bus for supplying several circuits (1 or more than 20) with the appropriate power.

To accomplish these functions, each load center consists of a transformer, distribution bus protective devices, and trailing cable disconnects, all enclosed as a unit within a heavy-gage steel enclosure. The physical size of the power center varies with the load demands and the physical size limitation of the rooms. Power centers typically range in service load capability from 25 to 1000 kVA. Since the size of the room housing the equipment is variable, most manufacturers carry a wide range of cabinet sizes.

Installation. Power centers, battery-charging stations, and rectifiers generate large quantities of heat during the conversion process. To help dissipate this heat, these centers should be elevated, (leveling on blocks), thereby permitting freer circulation of air around the unit.

Since power centers, rectifiers, and battery-charging centers are classified as nonpermissible, they must be located at least 15 ft from the last open crosscut.

Legal Constraints. To conform with CFR Title 30, Part 75:

1. Permanently mounted power centers (e.g., belt power centers), rectifiers, certain oil-filled transformers, and some battery-charging stations must be established on a separate split of air vented to the return.

2. Units must be located away from haulageways and have safe roofs.
3. Units must be at least 15 ft from the last open crosscut.

Utilization equipment. The primary purpose of the utilization equipment is to convert the electrical energy into mechanical, radiant, or other form needed to provide some operation. Depending on the type of utilization equipment being analyzed, secondary functions include (1) providing internal control or regulation of circuit parameters as might be needed for motor starting and control; (2) providing internal distribution as might be needed for lighting, motors, relays, or other functioning; and (3) in some cases, providing internal rectification to supply dc motor needs.

Power requirements of utilization equipment varies from a few milliwatts for safety monitoring circuits to 1000 kVA demands of large longwall sections. Table 10-3 summarizes some of the more common equipment demands that might be encountered underground.

Legal Constraints. The major legal requirement for utilization equipment is the permissibility rating mandated by the U.S. Bureau of Mines for all electrical equipment used in the last open crosscut.

10.4.2 Section Component Arrangements

Major electrical load concentrations underground are conveyor belt drives, track haulage networks, room and pillar mining sections, longwall mining sections, and shortwall mining sections.

Conveyor belt. The motors that drive the conveyor belts underground are either squirrel cage induction motors or wound rotor induction motors, which although more expensive, provide greater starting and speed control. Since starting torque of induction motors is a function of the voltage squared, control of the voltage drop from the belt power center and control panel to the drive motor is important. To facilitate this control, the *IR* drop from the power center to the drive motor must be as low as possible; thus the power center should be located as close to the belt drive motors as possible. To avoid installation difficulties and possible safety problems associated with crossing the belt with cables, the belt power center and belt motor controller should be located on the same side of the belt as the drive motor.

During operation the power center will generate heat resulting from the power losses associated with tranforming the higher distribution voltages to the lower utilization voltage. To dissipate this heat, the belt power center should be elevated to permit free circulation of air around the power center enclosure. To ensure air circulation around the power center and motor controller and since the installation is considered permanent, the installation should be on a separate air split.

As with all power system components, the belt power center and motor controller need be leveled to ensure proper operation of control relays and switches, located in a dry crosscut, assured accessibility for operation inspection and maintenance, and protected from ground movement.

Track networks. These major components constitute the track haulage power network: (1) the power rectifiers, which transform and rectify the high-voltage ac power to the lower voltage dc power (less than 550 V); (2) the trolley wire, which distributes the dc power along the track entries, and (3) the track, which provides the ground return circuit.

The track rectifiers are placed in crosscuts adjacent to the track entry and are established on a separate split of air vented directly to the return air course. Like other major electrical components described earlier, the rectifiers are permanently installed within dry areas, free of obstructions and protected from ground movement.

The location of the rectifiers depends on the *IR* voltage drop along the system. Mainline haulage will require a greater number of rectifiers per mile than secondary routes. Two rectifiers located per mile on the mainline route and one rectifier per mile on secondary routes would be fairly commonplace.

The trolley line is an uninsulated copper "track" suspended above the rail and offset at least 6 in. outside the track gage line (as required by state and federal law). The dc power is tapped from this line through a conductive pickup slip or roller, which completes the electrical connection between the stationary power wire and the moving locomotive motor. The motor is electrically grounded by the locomotive wheels. The contact between the wheels and the return circuit of the rails completes the electrical circuit for the track.

In addition to the trolley wire, a common industry practice is to install a parallel trolley feeder wire, which functions to reduce the total current carried by the trolley wire and the total voltage drop of the system, and significantly improves the reliability of the system. The feeder wire is usually insulated and is attached to the trolley line at 20- to 40-ft intervals.

The trolley line is suspended from the top by insulated hangers, which rigidly anchor the line to prevent excessive sag caused by gravity or lateral movement caused by the tracking forces of the trolley slip. Obviously, the more uniformly the trolley line corresponds to the rail, the less these tracking forces will, and the less wear and strain will be experienced by the trolley wire. The 6-in. offset of the wire relative to the gage is usually kept to the "tight" side of the track, and it is usually the rail on this side of the track that is assumed to act as the principal return circuit. By so locating the power wire and the return, safety will be improved because more "space" exists between the electrical system and most trafficking of equipment and people.

The trolley line and trolley feeder wires must, in accordance with state and federal regulations, be sectionalized using "cutout" switches every 2000 ft. and at every turnout or branch circuit. These dc circuit breakers provide overcurrent protection and a means of removing power from a section of track as might be necessary in an emergency. Finally, all trolley lines and trolley feeder wire, like all nonpermissible components of the electrical distribution systems, must be kept at least 15 ft from the last open crosscut of working section and 150 ft from all pillaring or longwall faces.

The last major component of the track electrical systems is the rail, which functions electrically as the ground return circuit. The rail circuit is a particularily vulnerable segment of the system, since it is subject to high mechanical loading and wear, ground movement (which varies with the presence of water and loading), and high electrical resistance at rail junctions or breaks. The effect of a poor return circuit resulting from these factors is excessive motor commutator-brush sparking and wear, which are the major dc motor maintenance problems. To improve the return circuit continuity, designers often install a separate ground feeder parallel to the rail and attached to the "return" rail at intervals varying from 250 to 500 ft. Like the trolley feeder line, this redundant ground reduces the *IR* voltage drop, reduces the current carried by the rail, and improves the reliability of the system. In addition to the parallel grounding, a short copper wire is connected across each rail joint on the return side and another wire is connected between the two parallel rails at intervals up to 300 ft. These procedures are called *track bonding* and are necessary steps for improving the ground circuit continuity.

Room and pillar mining. Room and pillar mining includes both conventional and continuous mining operations. Although the mining operations and procedures differ between the two, the components and layout of the two electrical systems are equivalent. The major components shown in Fig. 10-6 are the face power center, distribution center, and utilization equipment.

The power center must be located at least 15 ft from the last open crosscut. The maximum distance from the face depends on the length of trailing cable and whether or not a distribution center is used to extend the service range. Although the maximum trailing cable length is 750 ft, most equipment used today requires the face power center to be within 500 ft of the face. Because of the mobile nature of most face power, they need only be located within the intake air course, as opposed to being established on separate split. This interpretation does vary between states, as do many of the electrical laws and regulations, so you need to be cognizant of the applicable statute in your region. The power center is often located close to the belt feeder conveyor such that shuttle cars linking the face and conveyor belt rewind their trailing cables as they approach the feeder. As

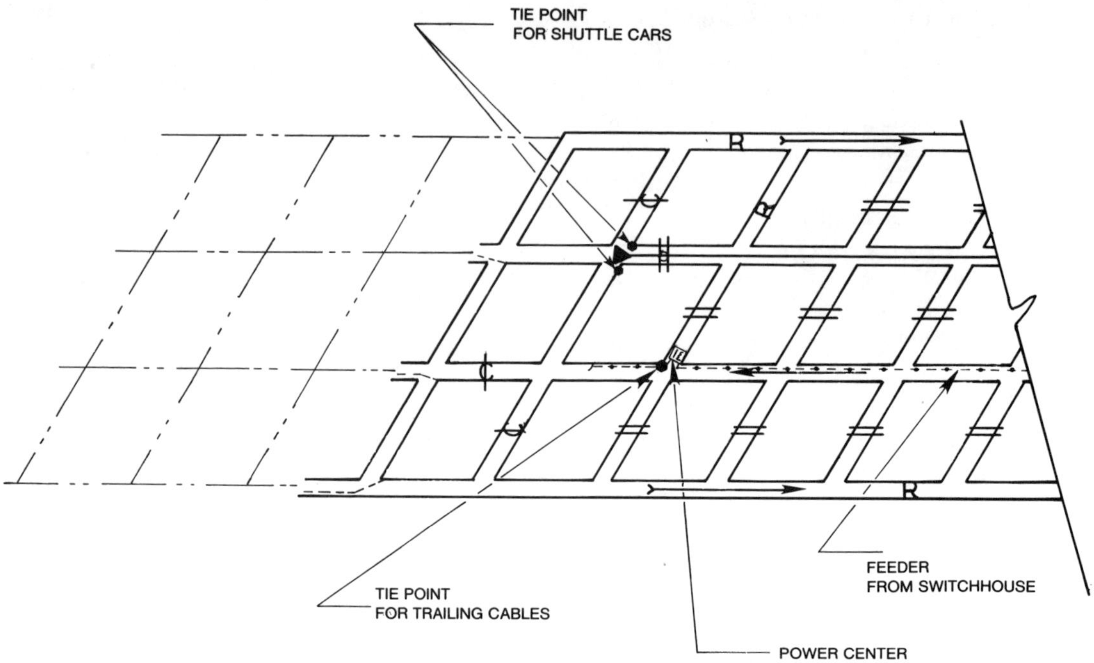

Figure 10-6 Possible face layout for room and pillar mines.

they approach the feeder, the angular velocity of the rotating spool decreases, so that if they must pass the power center along their service route, the inertia loading associated with the reversal of the direction of spool travel will be minimal. Ideally, the route of the shuttle car should be such that the takeup spool does not need to reverse, since too rapid a reversal is a major cause of spool failure.

The distribution center is a mobile junction box located near the last open crosscut. The distribution center is used to provide a convenient multiple power connect for equipment used at the face. It extends the effective range of the power center and decreases the frequency of power center relocation. Both the power center and the distribution center need to be located away from haulage routes and personnel concentrations but within accessible, dry, and well-ventilated locations.

The utilization equipment within each section in part determines the eventual size of the electrical power system. The power ratings and corresponding demand factors for utilization equipment underground are shown in Table 10-3. It will be left as an exercise for you to determine the needs of various sections.

Longwall layout. The power demands of a typical longwall section are perhaps the greatest of any single region underground. Figure 10-7 depicts the layout of major electrical supply and utilization equipment needed for the longwall operation. Power from the primary distribution system is

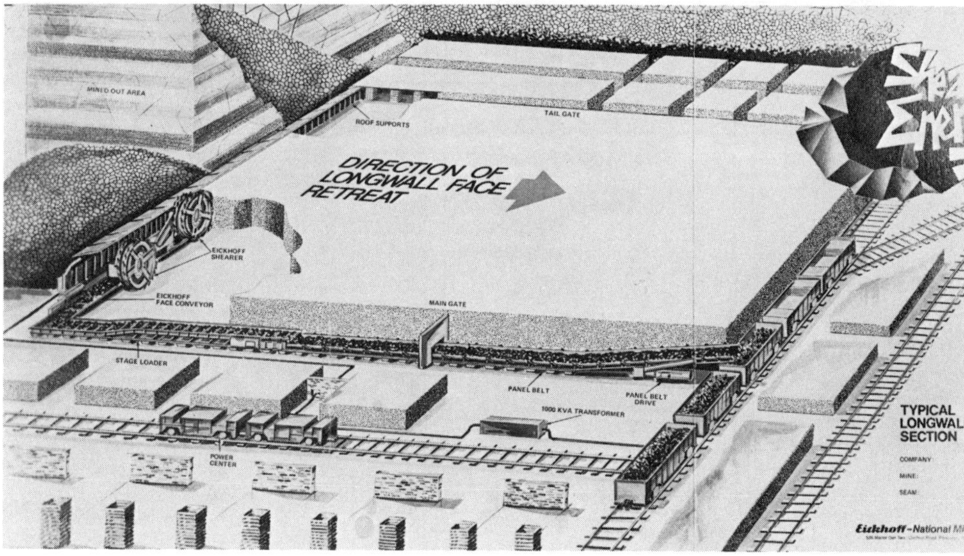

Figure 10-7 Layout of electrical support equipment for a longwall section (Courtesy of Eichoff America Corp.)

routed to the 1000-kVA (typical) section transformer, where the distribution voltage is stepped down to the maximum utilization voltage, usually 960 V. The power is then routed to a power center, where additional transformation of the voltage is applied as needed for auxiliary pumps and equipment. The power at the appropriate voltage levels is then routed to the bank of major motor starter-controllers, including those for the shearer, the face conveyor, the stage conveyor, the pump power pack, and various independent pumps.

The power center, motor controller bank, and power pack can be track-mounted, as shown, for ease in retreat, but more often than not these components are skid-mounted and periodic retreat is effected using a dc roof bolter to drag the units 100 ft to 150 ft from the established face.

Shortwall layout. The shortwall system requires a power layout similar to that of the longwall system; however, the initial 1000-kVA transformer located at the section mouth is usually not required, since the continuous miner, the loader, the shuttle cars, and roof support power pack operate at roughly the same voltage levels. CFR Title 30 requires that power centers, controllers, and other nonpermissible equipment be maintained at least 15 ft from the face. Continuous miners for shortwall sections usually operate with the maximum 750-ft trailing cable. With a 200-ft face and equipment in the second entry, the power center and controllers can be positioned at a distance up to 400 ft from the lateral face, thereby limiting the frequency of power system retreat. Figure 10-8 shows one possible layout for a shortwall face region.

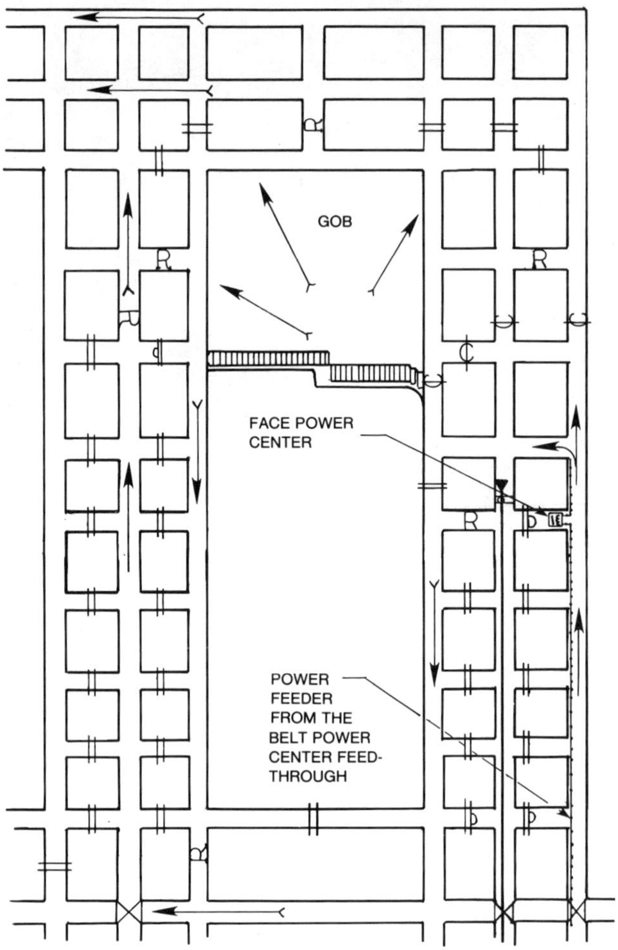

Figure 10-8 Shortwall electrical plan.

REFERENCES

1. Walter C. Johnson, *Transmission Lines and Networks*, McGraw-Hill Book Company, New York, 1950.
2. Harold H. West, "AC Distribution Systems," in Arthur B. Cummins and Ivan A. Given, eds., *SME Mining Engineering Handbook*, Vol. 2, Society of Mining Engineers, AIME, New York, 1973.
3. James E. Elswick and Frank R. Schwamberger, *Electrical Hazards in Mining*, National Mine Health and Safety Academy, U.S. Dept. of the Interior, U.S. Government Printing Office, Washington, D.C., 1978.
4. Gary L. Powell, *The Safety Grounded System*, Mine Extension Service, West Virginia University, Morgantown, W. Va., 1972.
5. *Distribution System Protection Manual, Bulletin No. 71022*, Power Systems Division, McGraw-Edison Co., Canonsburg, Pa.

6. *Proceedings of the Third W. V. U. Conference on Coal Mine Electrotechnology*, West Virginia University, Morgantown, W. Va., 1976.
7. Mac A. Carnes, *Permissibility—Electric Face Equipment*, Safety Manual No. 16, MSHA, U.S. Department of the Interior, U.S. Government Printing Office, Washington, D.C., 1979.
8. W. H. Dickinson (Chairman) et al., Report on Reliability Survey of Industrial Plants, Part 1, Reliability of Electrical Equipment, Industrial and Commercial Power Systems Committee, IEEE, May 1973.
9. Personal communication, Pemco Corp., Bluefield, W. Va., March 1980.
10. Personal communication, E. K. Stanek, Dept. of Electrical Engineering, West Virginia University, Morgantown, W. Va., May 1980.
11. *Federal Mine Safety and Health Act of 1977*, Public Law 91-173 as amended by Public Law 95-164, U.S. Government Printing Office, Washington, D.C., 1977.
12. John A. Hart (Chairman) et al., *Electric Power Distribution for Industrial Plants*, 4th ed., IEEE Standards Publication No. 141-69, Industrial Plants Power Systems Subcommittee, Institute of Electrical and Electronics Engineers, Inc., New York, 1969.

PROBLEMS

1. Needed is 15,000 W of dc power supplied at 250 V. What is the current requirement?
2. A 90-lb track has a resistance of 0.0057 ohm/1000 ft. What is the voltage drop if 220 A is delivered 2 miles? What is the power loss?
3. A 90-lb track and trolley wire have a combined resistance of 0.0134 ohm/1000 ft. If 500 A is drawn at an average voltage of 250 V, determine the length of track producing a 5% voltage drop.
4. How much power must be generated to provide 22.5 kW of power to a utilization point 10.2 miles underground if the resistance of the feeder is 0.009 ohm/1000 ft and the following voltages are possible: 550 V, 4160 V, or 7200 V? What is the efficiency of each voltage level? What is the voltage drop for each?
5. Calculate the total reliability of system shown in Fig. 10-8. If a secondary-selective system were used, what would the reliability be?

Appendix: Mine Mapping Symbols

A.0 INTRODUCTION

The symbols used to represent features on mine maps have yet to be standardized, unlike the more conventional drafting, topographic, and most geologic mapping symbolization. Often, each individual mining concern develops its own set of "mapping standards" which they believe best depicts the underground features. The symbols selected by one company may be slightly different from or in some cases exactly opposite those selected by a neighboring mine. For instance, the intake (fresh air) air courses are most commonly shown with a simple arrow (see Fig. A-1); however, some mines are just as likely to use the same symbol to represent the return (stale air) air course. Such a situation becomes intolerable as the status of mine drafting departs from that of an "art" form and approaches that of technologic communication. This appendix attempts to portray the most commonly used symbols as "standard" in an effort to compile a consistent index of mining symbolization. Guidelines governing the selection of appropriate mine map symbols were described in Section 1.2.3. To review, standard symbols must be (1) simple and clear in composition, (2) obvious in their descriptiveness, and (3) easily drawn.

Figure A-1 Standard intake and return symbols.

This appendix presents purpose, placement, physical construction, and symbol as it would appear on a mine map. Major classes of symbolization are (1) ventilation symbols, (2) haulage symbols (3) drainage symbols, and (4) electrical power and communication symbols.

A.1 VENTILATION FEATURES AND SYMBOLS

Within an underground mine, proper ventilation is one of the most time-consuming but critical functional activities. The ventilation system:

1. Supplies fresh air (O_2) to those working underground.
2. Dilutes and carries away harmful gases and dusts.
3. Regulates (to greater or lesser extent) the humidity and temperature underground.

To perform these functions adequately, the ventilation system relies upon a number of components, devices, and structures. These features combine to direct, restrict, and regulate the flow of air through the various entries and workings.

A.1.1 Air Circuit Symbols

The *air circuit* is a continuous stream of air that changes in gaseous and particulate composition as it passes through working sections. Coal dust, methane, and other gases are the major sources of airstream contamination. As a result of this contamination, the air circuit is conventionally divided into two segments, an *intake*, which is fresh air flowing toward the working sections, and the *return*, which is contaminated air flowing away from working sections. Various symbols have been used for these two circuits; however, the two symbols considered standard are shown in Fig. A-1.

Companies often use color-coded arrows or color-coded headings to differentiate between the intake and return air courses. Although effective particularly for those individuals not familiar with reading maps, these methods cannot be simply reproduced, and are therefore not considered standard drafting practice. In all cases symbols should be used to represent features. Colors are used only to highlight and enhance the symbols. Other symbols are also used, and you should be somewhat familiar with some of the nonstandard variations shown in Fig. A-2.

A third trait of the ventilation circuit represented by a symbol is the division of the air current into two or more separate airstreams called *splits*. Splitting of ventilation circuits creates parallel paths. Splits should be shown at all locations where they occur. The symbol and schematic diagram for ventilation circuit splits are shown in Fig. A-3.

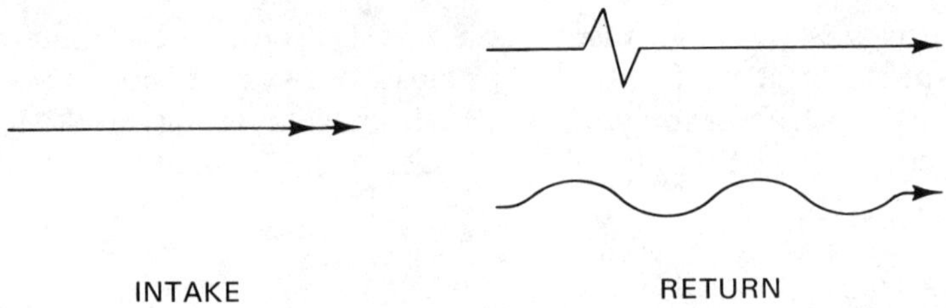

Figure A-2 Nonstandard intake and return symbols.

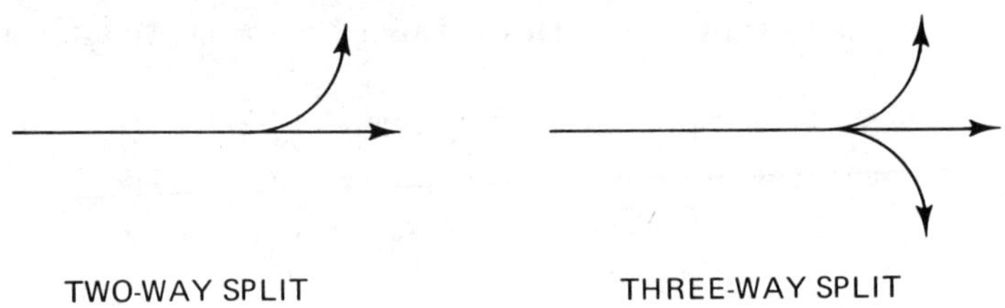

Figure A-3 Standard air split symbols.

A.1.2 Mine Fan

The mine fan causes air to flow through the mine. The amount of airflow (Q) caused by the fan depends on the characteristics of the fan and the resistance of the mine. Various types of fans have been used. Today, most mines use the "axial flow" fan design, with variable pitch blades. A schematic diagram of an axial flow fan is shown in Fig. A-4.

Figure A-4 Schematic diagram and component nomenclature.

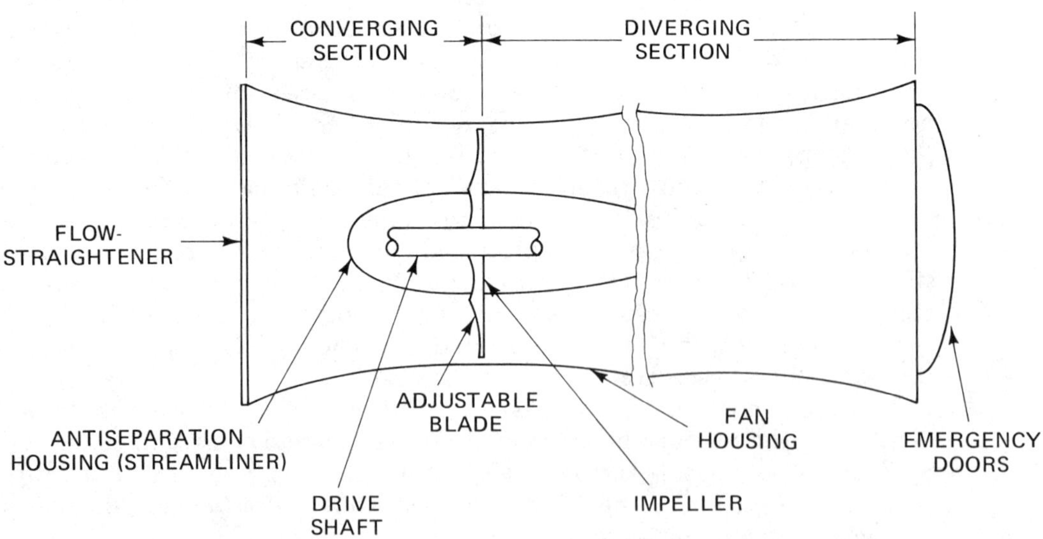

The installation of the mine fan is important for proper operations and drafting purposes. Installation is also controlled by both federal and state regulations. The mine fan must be:

1. Installed on the surface.
2. Installed in a fireproof housing and connected to the main mine opening by fireproof air ducts.
3. Enclosed in an "explosion-walled" area.
4. Offset more than 15 ft from the main opening.
5. Equipped with explosion doors (or weak housing walls) in line with the main mine opening (or explosion paths) and having a cross-sectional area greater than that of the total explosion paths.

A photograph of a typical installation, together with the symbol for the main mine fan, is shown in Fig. A-5.

A.1.3 Diverters and Separators

The greatest number of ventilation structures either divert the airstream or separate one air course from another.

Figure A-5 Mine fan.

Diverters. Devices used to divert the airflow include (1) line brattice curtains, (2) check curtains, and (3) ventilation tubing.

Line brattice curtains are used in the rooms, entries, or headings leading to the face. Their function is to confine the airstream to a narrow passage adjacent to one of the ribs. By confining the intake airstream in this fashion, the air may be directed into the corners and across the face, providing proper sweeping action of air across the face. Line brattice is constructed of a noncombustible clothlike material (e.g., jute, plastic-treated canvas), which is suspended from the top and secured at the bottom. For effective ventilation, the end of the curtain must be within 10 ft of the face. Figure A-6 shows the standard drafting symbol, the way in which the symbol would appear on a map (for a blowing system), and a photograph of the line curtain.

Check curtains are used in areas close to the face operations. In addition to diverting or stopping the airflow, they serve to permit the passage of workers, equipment, and materials (e.g., haulage of product). Check curtains are easily installed and dismantled. Being made of the same nonflammable clothlike materials as line curtains, check curtains are suspended from the roof and secured or weighted at the bottom. They differ from line curtains in that they are normally split in one or more overlapping sections to allow for the movement of workers and equipment through them. The standard symbol, the way in which the symbol appears on a map, and a photograph of the check curtain in the mine are shown in Fig. A-7. (Note: For map consistency and as a drafting guideline, any symbol that has a letterlike orientation, such as ₵, should be placed so that it can be read from either the bottom or the right-hand side of the map.)

Ventilation tubing is used in those rooms leading from the last open crosscut to the face. The tubing is used in conjunction with auxiliary blower

Figure A-6 Line brattice curtain. (Courtesy of MSHA.)

Figure A-7 Check curtain. (Courtesy of MSHA.)

fans instead of line brattice curtains to ventilate the face. The advantages of this method over line curtains are that (1) the tubing is mounted from the top and does not extend to the bottom of the heading, thus more equipment and personnel maneuverability can result; (2) the end of the tubing can be directed into hard-to-ventilate areas; and (3) the fans can be mounted with simple scrubbers that remove airborne face dust from the airstream. Like the line curtain, the end of the ventilation tubing should stay within 10 ft of the face. Figure A-8 shows the standard drafting symbol, the way in which the symbol would appear on a mine map, and a perspective schematic of the feature as it might appear in the mine.

Separators. Ventilation devices and structures that function to stop the flow of air or to prevent the intake air from "short circuiting" into the return air circuit are called *separators.* Separators include (1) temporary stoppings, (2) permanent stoppings, (3) overcasts, (4) undercasts, and (5) mine doors.

Stoppings, both permanent and temporary, function to stop the flow of air and prevent short circuiting. *Temporary stoppings* are constructed of noncombustible material which can be simply and quickly installed. They prevent "substantial" movement of air in those areas where frequent changes of the air circuit are likely. In practice, temporary stoppings span crosscuts

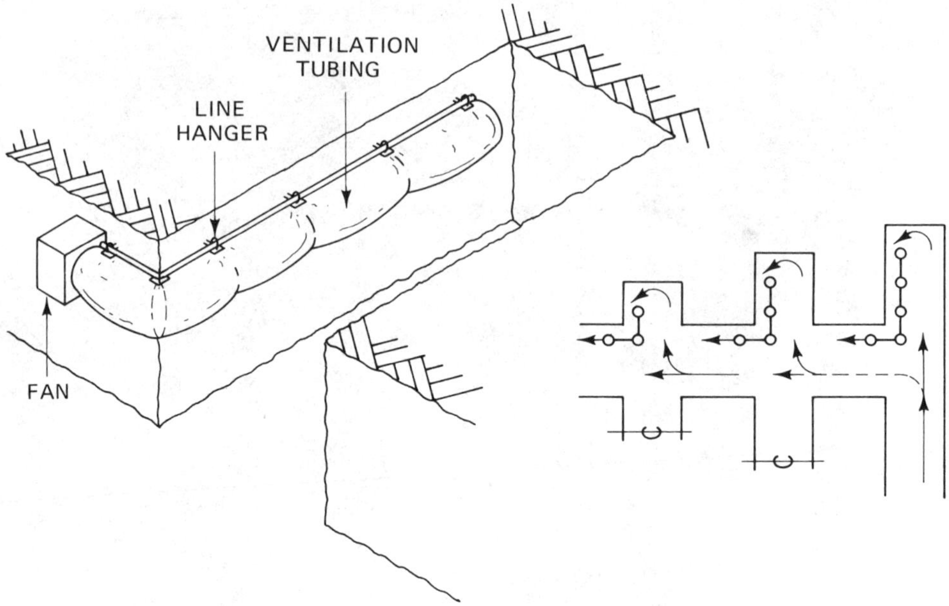

Figure A-8 Schematic diagram ventilation tubing and symbol.

adjacent to and out from the last open crosscut. Figure A-9 shows the standard drafting symbol, the way in which the symbol would appear on a mine map, and a photograph of the stopping in the mine.

Permanent stoppings are the most numerous human-made structures encountered within the mine. Permanent stoppings, as the name implies, are permanent walls built from top to bottom and rib to rib in any crosscut

Figure A-9 Temporary stopping. (Courtesy of MSHA.)

or heading that needs to be blocked. Permanent stoppings, constructed of concrete, block, and brick, provide (1) the stopping of air (or water) across a passage, (2) the segregation of intake air from return air, and (3) the sealing of one portion of the mine from another. The sequence of procedures needed to construct a typical permanent stopping is as follows:

1. Prepare the footer for the structure.
2. Build the stopping; blocks are usually laid up dry, but for more air-tight construction, block should be wetted first.
3. If an access is required, start the opening after the third course of block and close the opening after the seventh course of block.
4. Wedge out a top to prevent leakage and tighten the structure.
5. Plaster the stopping on the higher-pressure intake side.
6. Hang an access door, if required.

The average cost of a permanent stopping in the spring of 1978 was \$60 for materials and \$112 for labor. Figure A-10 shows the standard symbol, the way in which that symbol would appear on a drawing, and a photograph of a stopping in a mine.

Overcasts are used at intersections where the intake and return air

Figure A-10 Permanent stopping. (Courtesy of MSHA.)

courses must necessarily cross one another. The principal function of an overcast is to allow the crossing of return air over intake air while maintaining their segregation. Overcasts allow the mine engineers to make frequent splits in the intake air, which is necessary in modern mining operations. Several types of overcasts exist; however, the two most prevalent are the high-capacity overcast and the pipe overcast.

The high-capacity overcast requires that a portion of the top be removed and that a permanent structure, much like a bridge, be placed within the intersection. This "air bridge" allows the return air to flow above and across the intake air without mixing. High-capacity overcasts, built of block, concrete, brick, and metal decking, are constructed according to the following sequential procedures:

1. Predrill the intersection top.
2. Cut down the top (overcast region).
3. Bolt and wire mesh the open top.
4. Load out fallen waste rock.
5. Prepare the footer for the structure.
6. Build walls (much like permanent stoppings, between ribs on the return air circuit) by laying block flat and wet.
7. Lay H-beams between the walls and place Q-decking across these beams.
8. Lay one course of block on the Q-decking.
9. Close off large leaks with wooden wedges.
10. Seal the structure with cement, plaster, and rock dust.
11. Construct ladders within return air circuits.

The average cost of a high-capacity air overcast as of the spring of 1978 was $855 for materials and $3900 for labor (including supervisory costs). Figure A-11 shows the standard drafting symbol, the plan, and profile views of the feature as it would appear in a mine setting.

Pipe overcasts provide the same functions as do the high-capacity overcast except that pipe overcasts lack the ability to inspect the return air circuit freely. The pipe overcast, as the name implies, is constructed with one or more pipes (used for the return) extended between two permanent stoppings which span the return circuit. Pipe overcasts have the following advantages compared with high-capacity overcasts: (1) less expensive materials and significantly lower labor costs, (2) quicker and easier to construct or change if the ventilation plan were to change, and (3) no changes (and resulting discontinuities) are required in the top. Disadvantages in addition to those previously noted are: (1) restricts the volume of return (an thereby increases the pressure drop); (2) reduces the head clearance on the intake haulage. Figure A-12 shows the standard drafting symbol, the plan, and profile views of a pipe overcast.

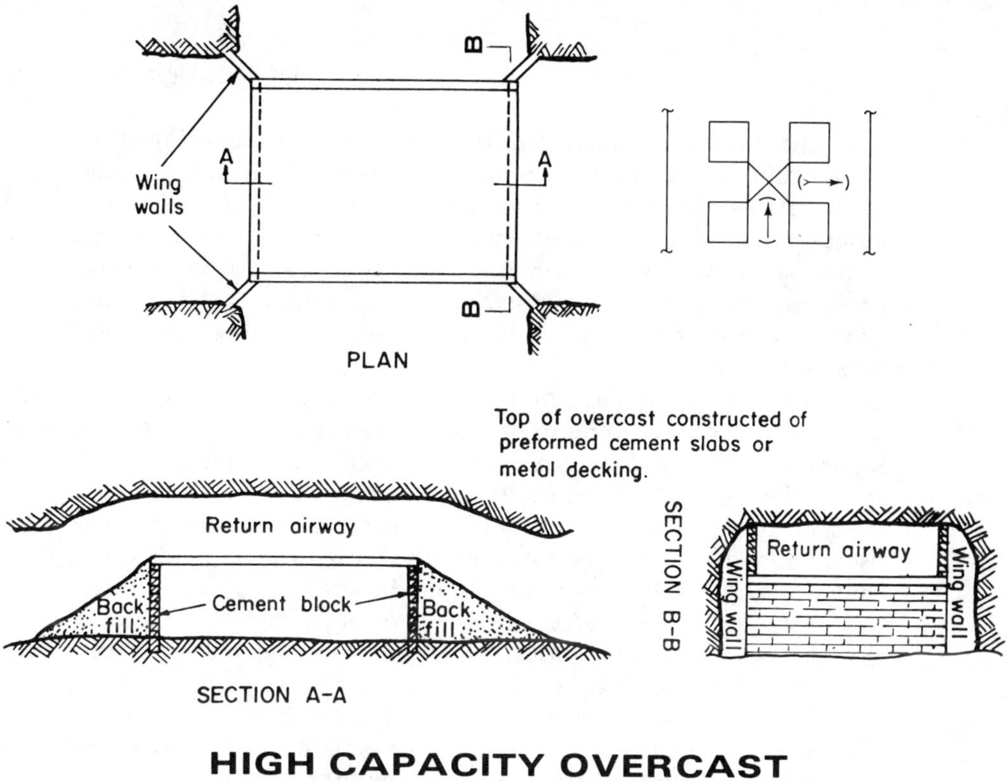

Figure A-11 High-capacity overcast. (Courtesy of MSHA.)

Figure A-12 Pipe overcast. (Courtesy of MSHA.)

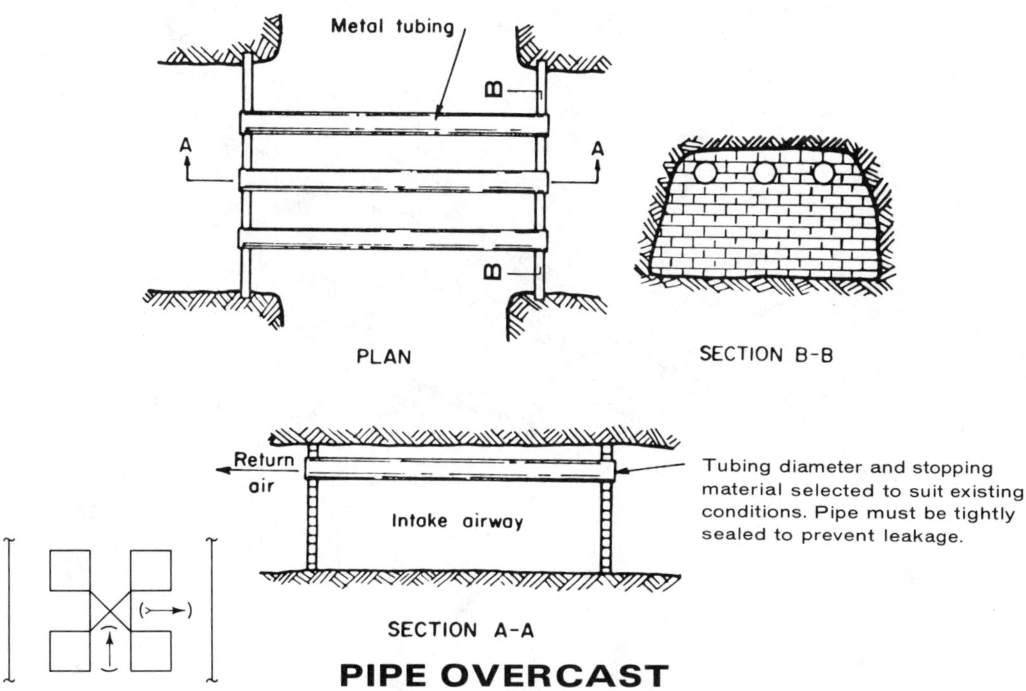

The *undercast* provides the same general functions as the overcast except that the bottom is removed instead of the top. Return air is then taken below the intake airway. Undercasts are most often used in conjunction with conveyor belts, which must be operated on a separate isolated intake. However, since the low spot that is created presents a drainage problem, the undercast is used sparingly. Figure A-13 shows the standard symbol, the way the symbol would appear on a mine map, and a perspective view of the structure.

Mine doors are large hinged "doors" which span the width of the entry. The purpose of the mine door is to prevent the flow of air in the entry or to separate intake and return airways while permitting frequent haulage between the two entries. Mine doors are metal or treated wooden panels hung so that the higher-pressure intake air always acts to close and therefore "seal" the doors. Doors are normally hung in pairs, called an *airlock*, such that one door will always be closed before the other is opened. By using an airlock, no mixing of the intake and return can occur. Figure A-14 shows the standard drafting symbol, the way it would appear on a mine map, and a perspective schematic of a mine door.

The *box check* is a structure used in conjunction with a conveyor belt. The functions of the box check are to isolate the conveyor belt from working sections while permitting the movement of the belt and coal through the device. The construction of the box check at the head section is permanent, much like a permanent stopping with a hole in the middle, while at the tail

Figure A-13 Schematic sketch of the undercast. (Courtesy of WVU Mine Extension Service.)

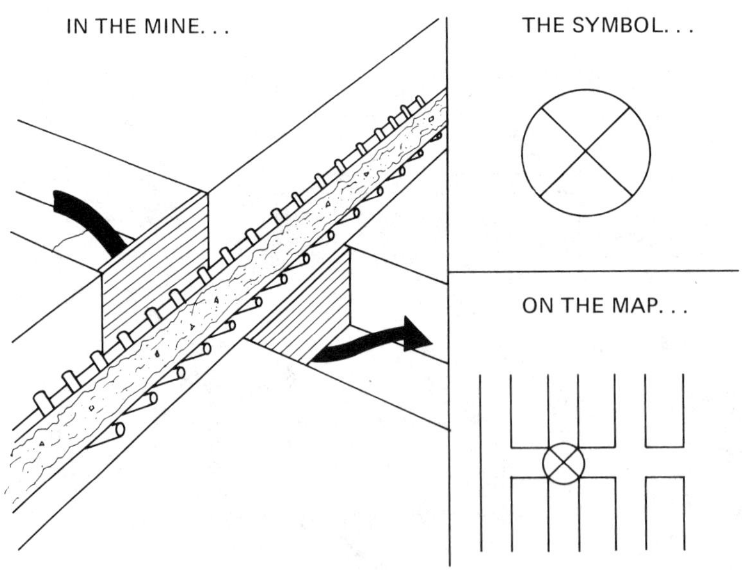

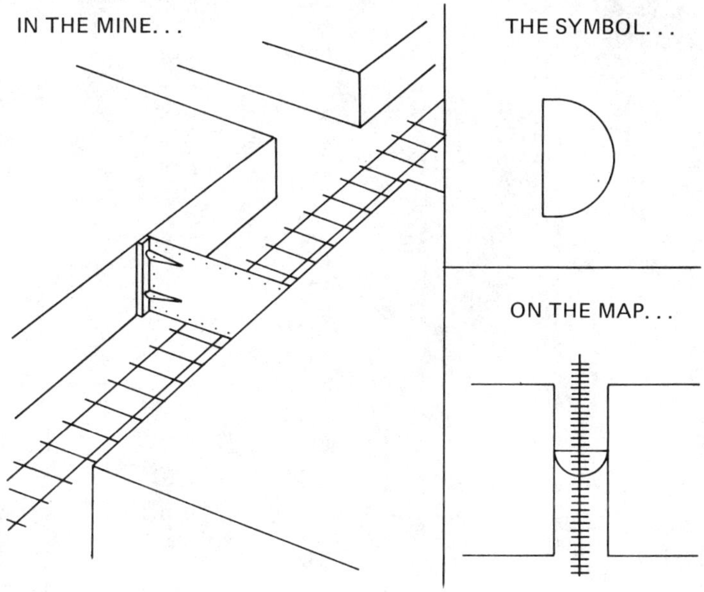

Figure A-14 Schematic sketches of a mine door. (Courtesy of WVU Mine Extension Service.)

section the construction is more temporary, since this end is periodically advanced along with the workings. Box checks are always installed in pairs to form a type of airlock along the length of the belt. Figure A-15 shows the standard drafting symbol, the way the symbol would appear on a mine map, and a perspective schematic diagram of a box check.

Regulation devices. The last section of major ventilation devices covers devices that regulate the quantity of air that flows through a given area. Regulation devices include regulators and partial stoppings.

Regulators are needed at or near the end of all air splits to provide a method of adjusting the quantity of air flowing into that section. All regulation devices operate on the principle that the volume of airflow in any given region depends on the resistance of that airway and the resulting pressure drop from the inlet to the outlet. The resistance of any passageway can only be practically changed by *decreasing* the flow area within that passage. Regulators therefore allow the mine foreman to vary the flow area and thus the resistance for each separate split. Regulators are placed in the returns and far enough away from the face so as not to interrupt the normal haulage and transportation. Figure A-16 shows the standard drafting symbol, the way in which it would appear on a mine map, and a photograph of the regulator in an entry.

Figure A-15 Regulator. (Courtesy of MSHA.)

Figure A-16 Schematic sketches of a box check. (Courtesy of WVU Mine Extension Service.)

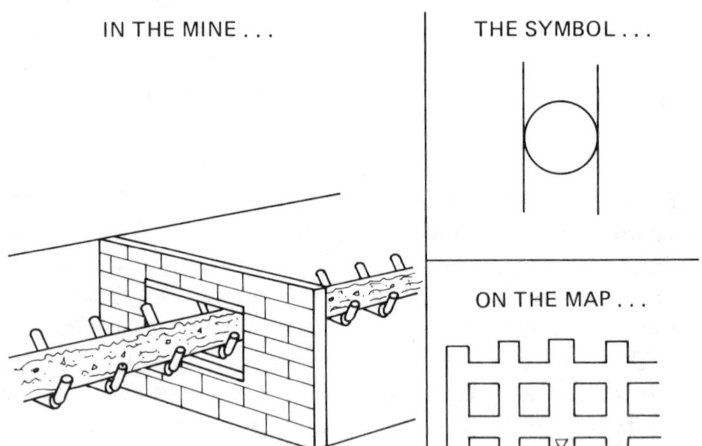

The *partial stopping* is a permanent stopping, which has had a specific number of blocks removed to allow a regulated amount of air to flow through. The symbol and location for this type of regulator is the same as for a standard regulator.

A.2 TRANSPORTATION FEATURES AND SYMBOLS

Another critical subactivity of the mining operation is transportation. Proper transportation design and layout are of upmost importance in achieving and maintaining acceptable levels of productivity.

The transportation system can best be analyzed by subdividing it into three interrelated segments: (1) face haulage and transportation, (2) intermediate haulage and transportation, and (3) mainline haulage. Face haulage is that which occurs between the face and the intermediate loading station. Face haulage is normally handled by shuttle or scoop car in room and pillar mining, armored face and extensible belt conveyors in longwall mining, and shuttle cars or extensible belt conveyors in shortwall mining. Intermediate haulage is haulage that spans the distance between the active workings and the mainline haulageways. Normally, these systems are either conveyor, tracked mine cars, or trackless tractor-trailer systems. The mainline haulage is the system that extends from underground to the surface. This system commonly uses either tracked locomotive and mine cars or conveyor systems (shafts require special systems, such as skip hoists or elevator systems interfaced with either mine car or conveyor systems).

A.2.1 Face Haulage

Rarely do mine maps depict mobile face haulage equipment (e.g., scoop car, shuttle car, or tractor trailer). When needed, the haulage path is shown using a dashed line similar in construction to a section line. Longwall and shortwall conveyor haulage is shown on the map, since it interfaces continuously with the intermediate haulage system. Figure A-17 shows a typical armored-extensible belt system used for face haulage for a typical longwall system.

A.2.2 Intermediate and Mainline Haulage: Track Systems

Coal transport by mine car is still the most popular form of underground haulage. The symbols needed to represent the extent of underground track layout are (1) track symbols, (2) turnout symbols, (3) transfer (loading and dump) stations, and (4) speciality symbols.

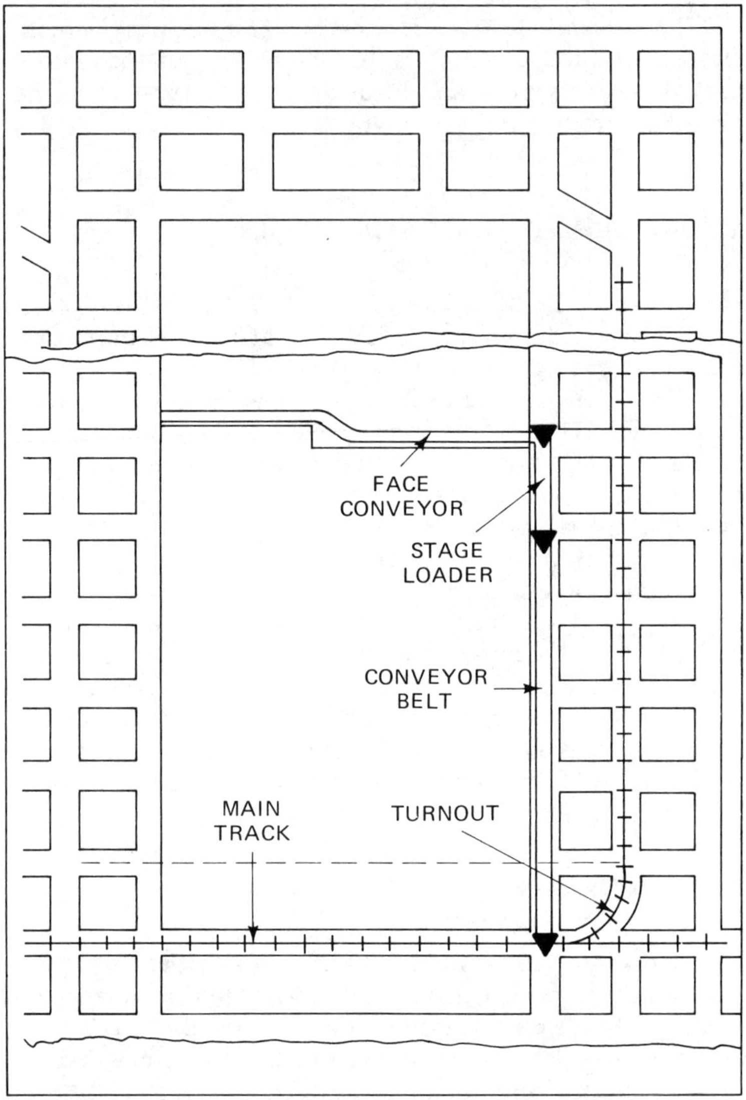

Figure A-17 Longwall haulage system (ventilation not shown).

Various track arrangements can be used. Depending on the nature of roof and haulage conditions, both single- and double-track systems can be used within an entry. Double track has the advantage of increasing productivity and simplifying scheduling. Figure A-18 depicts some of the more prevalent arrangements.

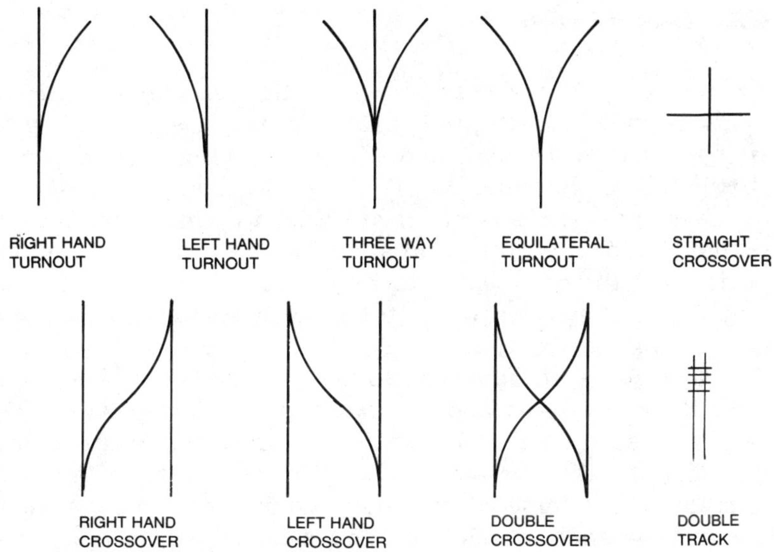

Figure A-18 Major trackwork arrangements.

A.2.3 Turnout Symbols

Turnouts are locations where one track interfaces with another track, permitting:

1. The feeding of a branch (haulage) track (e.g., mainline to intermediate).
2. The feeding from a haulage track onto a storage, holding, maintenance, or transfer station branch.
3. The feeding from one track to a parallel track.

Several types of turnouts are used underground, including single, double, and triple turnouts. Figure A-19 shows the layout for a typical turnout and the symbol.

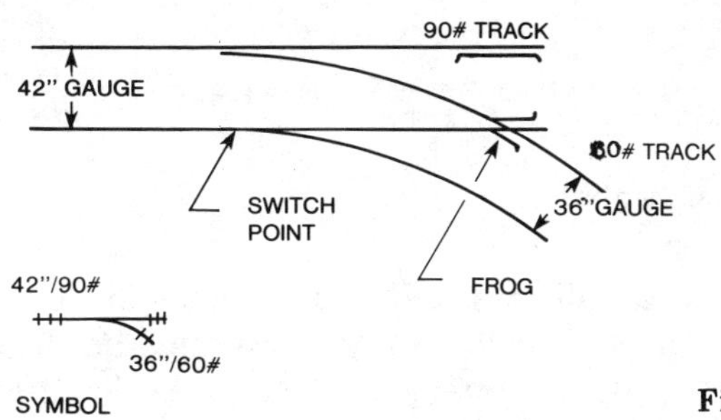

Figure A-19 Turnout.

A.2.4 Transfer Stations

Transfer stations are locations where mine cars are either loaded or unloaded underground. Underground loading usually begins with a feeder conveyor, which transfers the coal from either the intermediate belt or an underground storage pit to the mine car. The track auxiliary equipment used for loading includes car retarders, which are used to control the movement of the car within the loading zone, and a car index spotter or stop, used to locate the position of the car relative to the loading chute. A car spotter automatically positions successive empties for loading so that the locomotive is free for performing other duties.

Unloading stations are somewhat similar to loading stations in that both the car retarder and car spotter are used to position the cars within the region of an unloading chute. Several methods of unloading are employed, including rotary dumper, which physically rotates the entire car; bottom dumping car, a mine car in which the bottom can be opened, thereby allowing the material to fall out; and self-dumping cages, in which the car is laterally transported to the dump site, tilted, and then returned. The symbols for the various types of unloading stations are basically the same. Figure A-20 shows the mine map symbol for various features used in conjunction with transfer stations.

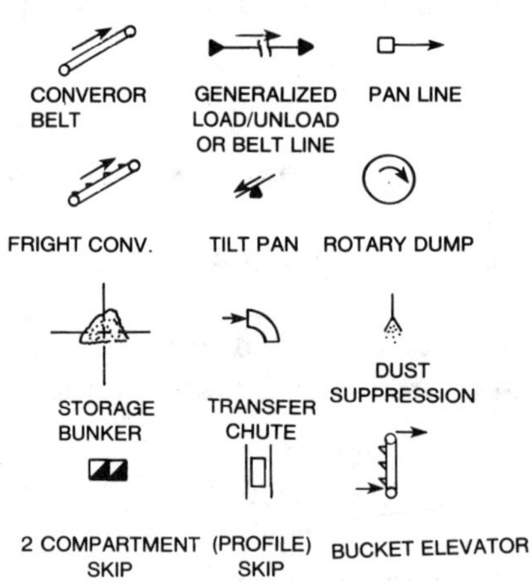

Figure A-20 Transfer station features and the generalized symbol.

A.2.5 Special Symbols

A variety of auxiliary track equipment can be represented on a mine map. Depending on the scale of the map and the degree of detail desired, these components may be required. Figure A-21 shows some of the more common auxiliary track symbols.

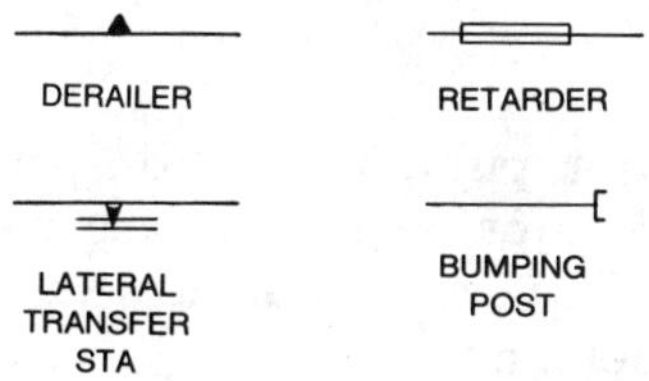

Figure A-21 Ancillary track feature.

A.2.6 Intermediate and Mainline Haulage: Conveyor Belts

Conveyor belts provide certain installation and operational advantages over the mine track system of haulage. Belts may be suspended from the entry roof by cables, thereby eliminating concern over soft or uneven bottom. Additionally, belts can be operated at grades exceeding 30%, whereas mine cars are limited to grades less than 18% (even for very short distances). Thus as the "quality" of an entry declines, the application of conveyor belt systems increases.

Although the specific belt layout is complex, the representation of the belt system on the mine map is not. Only two components are customarily depicted, loading and unloading stations and the belt. The loading and unloading station symbols for belt haulage are identical to the generalized transfer symbols shown in Fig. A-20. The belt usually extends in a straight line between these points.

A.2.7 Mainline Haulage: Shaft Systems

Vertical shafts represent the fastest, simplest, and usually the most economical (in terms of installation, operation, and maintenance) method of providing transport between surface facilities and underground workings when those workings are under substantial cover.

Shafts, like other entries, vary in geometric configuration, ranging from rectangular, to oval, to fully circular. Rectangular configurations allow the most efficient utilization of space; however, circular openings provide the best ground stability. Most haulage shafts, regardless of geometry, are compartmentalized, which permits multiple uses for the shaft, such as personnel elevators placed adjacent to coal haulage skip hoists.

Figure A-20 shows the most common symbol for a haulage/supply shaft, together with a schematic profile sketch of a shaft installation and an illustration of how the symbol would appear on a map.

A.3 DRAINAGE FEATURES AND SYMBOLS

Drainage is the functional operation concerned with preventing mine water from adversely affecting mining operations. Underground the only features that are expressly represented are (1) major sumps, which are used for collection; and (2) major pumping stations, which are used to distribute water collected in the sumps to the surface and the shaft used to contain seam to surface piping. Figure A-22 depicts major drainage features as they are represented on a general mine map.

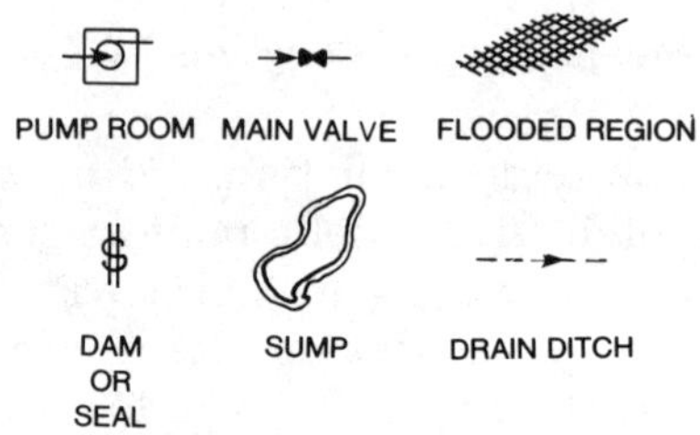

Figure A-22 Drainage symbols.

A.4 ELECTRICAL POWER DISTRIBUTION FEATURES AND SYMBOLS

Electrical power is the principal energy source used in underground coal mining today. Almost universally, electrical power is supplied by regional utility companies. The supply voltage can range from 480 V to over 23 kVA. The quantity supplied depends on the size, location, and capacity of the utility's generation network and on the size, location, and needs of the mine power system.

Most mine maps depict only "major" components of the electrical system, leaving the details of the distribution network for schematic representation. The major components symbolically represented include (1) electrical drops (bore holes), (2) substations (both surface and underground), (3) major power centers, and (4) rectifiers. The layout and specific functioning of each component are detailed in Chapter 10. Figure A-23 shows the mine map symbol of each of the electrical components listed above.

Figure A-23 Electrical distribution symbols.

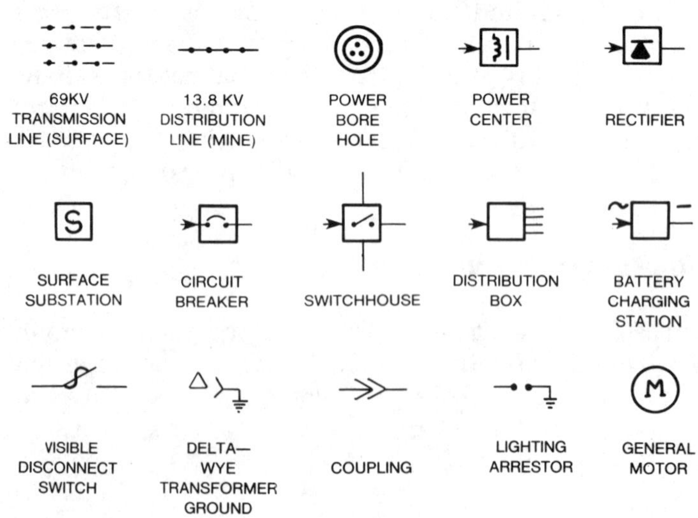

Index

A

B

C

Coplaning, 43-44

D

E